Sustainable Forestry

Philosophy, Science, and Economics

Chris Maser

$S{_L^t}$

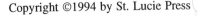

Printed and bound in the U.S.A. Printed on acid-free paper.

Library of Congress Cataloging-in-Publication Data

Maser, Chris.
    Sustainable forestry : philosophy, science, and economics / Chris
Maser.
      p. cm.
    Includes bibliographical references.
    ISBN 1-884015-16-6 (acid-free paper)
    1. Sustainable forestry. I. Title.
SD387.S87M375 1994
634.9—dc20   94-5192
                                       CIP

Phone: (407) 274-9906
Fax: (407) 274-9927

S^{t}L

Published by
St. Lucie Press
100 E. Linton Blvd., Suite 403B
Delray Beach, FL 33483

DEDICATION

To Zane,

my wife and learning partner,
I dedicate this book with love, respect, and gratitude.
You are my constant challenge to grow, to change, to become...

CONTENTS

Foreword   xi

Preface   xv

Author   xix

Acknowledgments   xxi

PART I
WHAT IS A FOREST? 1

1   Can the Notion of a Forest Be Contained in a Definition? 3
   Unpredictability   4
   Long-Term Trends   12
   Diversity   20
      For Want of a Squirrel, A Forest Is Lost   26
      What Happens to the Forest When It Loses the Grandparent Trees?   29
   Interrelated Processes   36
      A Forest Is More than the Sum of Its Ingredients   37
   All Elements Are Neutral   39
      The Squirrel   40
      The Fungus   41
      Squirrel–Forest Relations   45
         The Fungal Connection   46
         The Sporocarp Connection   47
         The Squirrel Connection   50
         The Pellet Connection   51
   A Flexible, Timeless Continuum of Species   57
   Landscape Pattern   58
      A Lesson from Black Bear   61
      Why Are Patterns Across the Landscape Important?   66
   The Longest Living Being on Earth   68
   A Unique Entity   71
   Self-Sustaining, Self-Repairing   72

PART II
AS WE THINK, SO WE MANAGE 75

2   Forestry Past, Present, and Future 77
Some Causes of Trouble in the Forestry Profession   79
    The Economic Myth of Forestry   79
    Dogmatization of Forestry   79
    Limitations of Science   80
        The Lure of Grants   81
        Attachment to a Single Hypothesis   81
        Science Can Only Disprove Information   81
        Method-Oriented vs. Problem-Oriented Questions   82
        Objectifying Nature   82
    Informed Denial   82
    University Training   83
Foresters of Yesterday, Today, and Tomorrow   87

3   The Cycle of an Agency 91
The Inception and Function of an Agency   92
We Are the Agency   93
Stages in the Cycle of an Agency   93
Dysfunction   96
Homeostasis   99
Boundaries   103
Coping Mechanisms   103
    Anger and Aggression   104
    Appraisal   104
    Defensiveness   105
    Denial   105
    Filters   106
    Rationalization   107
    The Upshot   107
Breaking the Dysfunctional Cycle   108

4   Conflict Is A Choice 113
The Enemy in the Courtroom   115

5   Change: The Universal Constant 121
The Dynamics of Change   123
Changes We Can Control   124
Changes We Cannot Control   124
Change in Human Terms   125
    Change by Exception   125

Incremental Change   126
Pendulum Swing   126
Paradigm Shift   126
Coercion   127
Inner Dictate   127
Can We Stop Change?   127
I Cannot Change Circumstances; I Can Only Change Myself   128

PART III
PLANNING: THE BRIDGE FROM CONFLICT TO VISION 133

6   Vision: The Frontier Beyond Conflict 135
Collective Vision: A Lesson from Insects   138
Rational Thought: A Requirement for Rational Planning   142
Rational Planning Requires Our Total Presence   145
Modifying Our Belief Systems   145
decisions, Decisions, DECISIONS   147

PART IV
FUNDAMENTAL ISSUES: THE SILENT DILEMMA 151

7   Technology, Science, and Uncertainty 155
Lessons from History   157
   Anasazi   157
   Easter Islanders   158
   Mayans   159
   What Will History Record About Us?   160

8   Short-Term Economic Expediency 165
Greece   167
Mediterranean   167
Middle East   168
Forest Decline   168
   Europe   175
   Asia   182
   United States   185
Recent Experiences   188
   Slovakia   188
      Definition of Native Forest   189
      Why the Native Forest Is Important   189
      Areas of Concentration in Management   189
      Plan of Action: The Forest Plan   190
      In Conclusion   193
   Japan   194

9   Sustainable Forests = Sustainable Harvest 199
Forest-Dependent Industries   200

10   Why Old Growth? 205
The Value of Old Growth   205
Old Growth as a Living Laboratory   210

11   Genetics, Adaptability, and Climate Change 213
Genetic "Improvement"   213
Climate   217
The Untested Product   220
The Value of Native Forests   222
A Question of Affordability   224

12   A Forest Is a Living Organism, Not a Machine 227
A Forest Is Cyclic, Not Linear   230
Cyclic Forests and Linear Models Do Not Match   233
Biological Sustainability Is Cyclic   235
Faster Is Not Necessarily Better   237
The *Invisible* Makes the Difference   239
Defile Not the Land   243
    Soil   243
    Special Cases and Common Denominators   244
Air: The Global Commons   247

13   Water: An Inescapable Necessity 249
How We Think About Watersheds   255
Roads and Water   257
Balancing the Stresses   259

14   Planning with Half-Used Data 265
Use *All* Available Data   266
There Is No Magic Hinge   269
Forestry Practices Affect the Ocean   271
    By Chris Maser and James R. Sedell
The Forest as a Living Trust   274

PART V
SUSTAINABLE FORESTRY 277

15   Where Are We Headed? 279
What Is Natural?   281
When Is a Native a Native?   282
A Reassessment of Our Control of Nature   284

Sustainable Forestry as Conscious Evolution   289
The Questions We Ask   289
The Hope We Plant   293

**16   Sustainable Forestry Through Adaptive Ecosystem
Management Is an Open-Ended Experiment** 303
*By Chris Maser, Bernard T. Bormann, Martha H. Brookes,
A. Ross Kiester, and James F. Weigand*
Rationale Behind the Experiment   303
   "Management": A Combination of Ecological Principles
     and Societal Values   305
   Sustainable Forestry Is More than a Local Issue   306
   A Clarification of Terms   308
Framing the Experiment   311
   Scientific Premises in Ecosystem Management   316
      1.   Both science and society are influenced by individual and
           societal values that are sometimes recognized but often
           forgotten   316
      2.   Natural sciences recognize people as part of the eco-
           system; social sciences recognize biological and physical
           constraints   318
      3.   Science has roles as both forecaster and conscience   319
      4.   The structure and use of science limits its application for
           ecosystem management   320
      5.   Ecosystems are artificial constructs and therefore must be
           selected and classified with caution   320
      6.   Ecosystems are fundamentally complex beyond our under-
           standing and at best difficult to predict   321
      7.   Diversity is essential to adaptability   321
      8.   Ecosystem patterns and processes appear, and must be
           studied, at different geographic and time scales; reconciling
           these differences is difficult   322
      9.   Local conditions may override or obscure general patterns
           and processes; the general may not contain the specific   322
     10.   Ecosystem science at large scales relies on ecosystem
           management for empirical evidence   323
     11.   The entire system must be managed within its context   323
     12.   Good can come from management   323
   Societal Premises in Ecosystem Management   324
      1.   People choose goals for ecosystem goods, services, and
           conditions based on their strong desires, which they perceive
           as needs; goals are set and evolve by cultural and political
           processes   324

2. People group themselves into communities of common or complementary values to further their personal goals   325

3. Diversity, complexity, and the changing nature of human communities create uncertainty about societal priorities and demands for ecosystem products, both now and in the future   326

4. In a society of diverse and often discordant communities, conflicts are likely to develop when too many goals for ecosystem management evolve in competition with one another, resulting in some people not getting what they want   327

5. Ecosystem management decisions benefit from societal vision and goals that are clear, concise, informed, and integrated and that recognize and accept that ecosystems are complex beyond our comprehension   328

6. Policymakers translate scientific information about ecosystems and societal goals into treaties, laws, government bureaucracies, planning, and budgets   330

7. No matter how people attempt to anticipate and control outcomes and effects of decisions, policy for ecosystem management will always be determined under conditions of uncertainty and ignorance   331

8. Unexpected ecosystem events and shifts in societal demands for ecosystem products require changes in societal institutions, which will in turn bring changes to society   332

Conducting the Experiment   333

What Is Adaptive Ecosystem Management?   334

Why Is the "Adaptive" in Ecosystem Management Necessary?   335

Decision and Planning Processes   337

Design and Analysis of the Management Experiment   339

17   Today's Decisions, Tomorrow's Consequences 341

Appendix: Common and Scientific Names 349

Endnotes 353

References 359

FOREWORD

The obstacles to discovery—the illusions of knowledge—are also part of our story. Only against the forgotten backdrop of the received common sense and myths of their time can we begin to sense the courage, the rashness, the heroic and imaginative thrusts of the great discoverers. They had to battle against the current "facts" and dogmas of the learned.

Daniel J. Boorstin
The Discoverers

Amid the cascade of helpful new books and papers on the science of sustainable forest management and the art of environmental conflict management, this heartfelt call for reform by Chris Maser stands out as unique. Maser is a discoverer charting a new course, and his thesis is compelling. Here, an experienced forest ecologist asks all the battle-weary forestry policy combatants without exception to adopt a radical new mindset or attitude, for the sake of our forests, ourselves, and future generations.

Specifically, Maser recommends that parties to the forest management debate:

- Shelve their legislative drafts and legal briefs based on a view of those with other primary interests as the enemy
- Be gentle with one another, "do[ing] whatever we do with love"
- Be gentle with nature, viewing it as a "Thou" to be revered rather than an "it" to be exploited
- Adopt an attitude of humility, because of our limited knowledge of ecological processes—which is definitely *not* your typical, hard-edged, oh-so-familiar "save the last stand" or "owls versus jobs" bugle call to battle, but something new, different, and well worth considering.

His appeal for gentler, kinder human relations ("heal the social rupture") and for the application of good science ("biologically sustainable forestry") is indeed timely. In my view, it is the right prescription. Unfortunately, important but radical new ideas like this often take a long time to become accepted, and their prophets often go long without honor.

Recall, for example, President Woodrow Wilson's July 1918 vision of a League of Nations, which was spurned by the U.S. Senate:* "the Past and Present are in deadly grapple and the peoples of the world are being done to death between them. An organization of peace must be established to make it certain that the combined power of free nations will check every invasion of right." Similarly, Chris Maser sees *forest management* "Past and Present" as being "in deadly grapple," and he calls on the combatants—the forest products industry, the environmental protection community, and the forestry profession—to adopt a new *modus operandi* to take the place of their fruitless, seemingly never-ending political/legal/media relations trench warfare.

"What we need, " he says, "is a collective dream large enough to encompass and transcend all our small, individual dreams in a way that gives them meaning and unity." To transform that dream into reality, he recommends the use of "sustainable forestry" to produce both wood fiber and wilderness for society in perpetuity. The iron fist within Maser's velvet glove—the pill that will be hard for many to swallow—is his call for forestry as a profession to be "broken out of its encrustation of economic dogma" (which he says erroneously assumes all ecological variables to be economic constants) and "catapulted into ecological truth...if the 20th century exploitation of forests is to become the 21st century healing of forests."

Saving the planet's productivity and thereby improving the chances of accommodating the needs of a growing human population over time will require, he contends, (1) the recognition of the primacy of ecological principles and (2) their integration into our political decision-making process. This is certainly easier said than done, given the current primacy of economic principles, but at last the ecologists' goal has been clarified.

It will take time for Maser's proposal for forest management reform to win broad recognition and support, just as Wilson's proposal for a new forum for the peaceful resolution of disputes among sovereign nations took years to bear fruit, and for much the same reasons: the loss of total independence of action on the part of the various players and the elusiveness of compromise, exacerbated by the lack of rewards for its practitioners.

From personal experience, I know that environmental group and industry trade association board members encourage their staff members to pick fights

---

*Hoover, Herbert. 1958. *The Ordeal of Woodrow Wilson,* McGraw-Hill, New York, pp. 24–25, 293.

and win them, on Capitol Hill and in court, and rarely applaud their staff members for participating in negotiated (compromise) settlements, such as those facilitated by the Keystone Center.* Yet such off-the-record head-knocking efforts to attempt to reach consensus on sticky environmental policy issues are important and necessary. In my view, the participants should be encouraged and rewarded, because this process can be used to break the policy gridlock and help chart paths through mine fields of opposing views to reach agreements on improved forest practices and therefore a better future for all.

I for one support reform of forest practice in the direction advocated by Maser, for to vote (by inaction) to continue with business-as-usual forestry clearly is a vote to steal from our grandchildren any chance of experiencing the quality of life generations now living enjoy. The continuation of forest management business as usual (characterized by Luna Leopold in 1990 as "stressed by a plague of special interests and a disdain for equity") could well result in widespread forest depletion. A vote for no change in forest practice therefore represents a questionable act by any caretaker or trustee of a living trust, such as a forest, because at the current rate of exploitation we will rapidly run out of old, high-volume trees. To liquidate the old-growth forests, upon which both our existing forest products industries and thousands of species of plants and animals depend, is to liquidate much of our inherited biological diversity.

"Unfavorable" questions must be asked and "unfavorable" inquiries must be pursued, Maser argues, to provide the checks and balances necessary to steer our technological culture away from catastrophic failure. More specifically, he observes, "we did not design the [original] forest; we do not have a blueprint, parts catalog, or maintenance manual with which to understand and repair it; [therefore] how can we afford to liquidate the old growth that acts as a blueprint, parts catalog, maintenance manual, and service station—our only hope of understanding the sustainability of the redesigned plantation forest?"

We have been "mining the old-growth forests" while we have "exceedingly little understanding of young-growth forests, especially their sustainability over time," Maser observes. Yet, "we are marching ahead as though we know what we are doing—marching from complex old-growth forests designed by Nature toward simple, uniform Christmas tree-like plantations designed by humans— jeopardizing our forests for lack of data and lack of patience with Nature's design." Maser's prescribed alternative is "sustainable forestry, with the focus on landscape management and forest health rather than on the level of forest product harvest…on the requirement for an ecologically sustainable forest in which the biological divestments, investments, and reinvestments are balanced in such a way that the forest is self-maintaining in perpetuity."

---

*See, for example, Final Consensus Report of the Keystone Policy Dialogue on Biological Diversity on Federal Lands, The Keystone Center, Keystone, Colo., 1991, 96 pp.

I have heard Chris Maser roundly criticized for being impractical and impatient. Economists ask who will pay the bills to implement his go-slow approach—the cost of finding the needed ecological facts and the cost of applying the research results on private as well as public lands. Ecologists respond that they can see the forest management future on the current trajectory, and it does not work.

What price will our descendants pay for our failure to make a correction in the course of forest management now and, instead, are presented with an impoverished landscape?

Or don't we care?

M. Rupert Cutler
The River Foundation
Roanoke, Virginia

PREFACE

The seemingly limitless quantity and quality of resources favored the staunch capitalistic individualism with which North America was forged. Although we still want the boundless supply of high-quality products of our historical abundance, times have changed. We are now a society that is being forced to learn ways of living within bounds of limited and diminishing resources. No longer can we simultaneously maximize both quantity and quality. Past abundance has become present limitations, and if we are not careful, present limitations will become future scarcities. Nevertheless, we continue to fight over who will get the last of the limited resources rather than cooperatively striving to perpetuate them.

With the passing of each decade our society becomes increasingly complex and confused. Our attention is riveted on manipulating the world for products as we lose sight of human dignity. Tragically, it is human dignity—the barometer of our well-being—that is being neglected.

In a world of rapidly growing population and finite resources, competition becomes severe for whatever is perceived to be limited in abundance. The pace of life quickens and yet never seems to be fast enough. All of this leads to frustration, which often manifests itself in cynicism. Inherent in the choice of whether or not to be cynical is the question of having the courage to accept responsibility for one's own actions

Another tendency of human beings faced with frustration is to defend a point of view, which is synonymous with individual survival. There are, however, as many points of view as there are people, and everyone is indeed right from his or her vantage point. Therefore, no resolution is possible when each person is committed only to winning agreement with his or her position. The alternative is to recognize that "right" vs. "wrong" is a judgment about human values and is not winnable. It is best, therefore, to define the fundamental issue and focus on it. An issue, usually perceived as a crisis, becomes a question to be answered, and in struggling toward the answer, both positive and negative options not only become apparent but also become a choice.

In field and forest, in desert and ocean, Nature makes the choices. In human society, the choice is ours. That we tend to avoid change does not diminish the positive option; it only means that we have chosen to argue for our limitations instead of for our potential. To opt for potential always involves an element of risk. Both individually and collectively, we tend to focus on and cling to a view of pending doom because of the emotional discomfort in the unpredictability of risk.

Although individualism is good, even necessary, in the embryonic stages of an endeavor, it must blend into teamwork in times of environmental crisis. Setting aside egos and accepting points of view as negotiable differences while striving for the common good over the long term is necessary for teamwork. Unyielding individualism represents a narrowness of thinking that prevents cooperation, possibility thinking, and the resolution of issues. Teamwork demands the utmost personal discipline of a true democracy, which is the common denominator for lasting success in any social endeavor.

Even if we exercise personal discipline in dealing with current environmental problems, most of us have become so far removed from the land sustaining us that we no longer appreciate it as the embodiment of continuous processes. Attention is focused instead on a chosen product and the outcome of management efforts. Anything diverted to a different product is considered a failure.

It is now time to re-evaluate our philosophical foundation of forestry and society and how we integrate the two. It is also time to re-emphasize human dignity in our decisions, so that we can broaden the philosophical underpinnings of management to include the ecological sustainability of forests and grasslands, oceans and societies, rather than only the exploitation of a few selected commodities they produce. Emphasis on human dignity will help foster understanding and teamwork, which in turn nurtures mutual trust and respect.

What we need is a collective dream large enough to encompass and transcend all our small, individual dreams in a way that gives them meaning and unity. If we dare to dream boldly enough, our special interests will both create and nurture the whole—a healed, healthy, sustainable forest.

Now, more than ever, we must recognize that we are part of the human family and must trust and respect one other as if human dignity truly were the philosophical cornerstone of society. We must also recognize and accept that ultimately we have one ecosystem that simultaneously produces a multitude of products. And we, as individuals and generations, as societies and nations, are both inseparable products of and tenants of that system, as well as custodians for those who follow.

Much has changed in recent years. Global warming and ozone destruction have become major issues. The timber industry in the United States, which has all but liquidated the old-growth forests, has lost much of its political clout. The forestry profession is undergoing painful changes as a result of challenges to its traditional philosophical foundation. To the north, the Canadian public is increasingly challenging both the old paradigm of their professional foresters and how the timber industry operates. And I have changed.

Today, I see even more clearly that the past, the present, and the future are all contained in this instant and that the choice of what we do and where we go is ours. If perchance we make a mistake, we can always choose to choose again. As long as there is choice, there is hope. As long as there is hope, the human spirit will ever aspire to its highest achievement. It seems appropriate, therefore, that I consider the forest in light of these changes.

I have written this book for three reasons. The first is to examine, as best I am able, what we are doing today in the name of "forestry," why we are doing it, and how and why we must change if we really want sustainable forests in the future.

There comes a point in the history and evolution of every individual, profession, agency, and society when change is necessary if that individual, profession, agency, or society is to continue to evolve. And it all begins or ends with the willingness of the individuals—who collectively *are* the profession, the agency, and society—to change.

As historian Arnold Toynbee concluded, great civilizations of the past failed because people would not, or believed they could not, change their way of thinking to meet the changing conditions of their world. Thus, with a rigidity of thinking and overpopulation, they stripped the productive capacity of their land and forests, conforming to the currently acceptable knowledge of the day. The profession of forestry today stands at such a juncture. It can move forward only to the extent that individuals within the profession accept the new philosophies and methods of management demanded by a rapidly changing culture.

The second reason I have written this book is simply to point out that we must have a biologically sustainable forest before we can have a physically sustainable yield of forest products. I see no sustainable forests as I look around the world because human society is summarily cutting them down and, where possible, replacing forests with fast-growing tree plantations.

Although liquidation of the world's forests in favor of fast-growing tree plantations to maximize quick profits seems rational enough in the short term, it may not be so rational in the long term. Throughout our habitation on earth we have developed our technologies based on the *assumption* that the autonomous biosphere, which produces the environment needed for life, is beyond our ability to destroy. The day has arrived, however, when the life system itself is within reach of our technology and our power to disarrange and destroy. This is a danger no age has ever faced before.

The third reason I have written this book is to propose a new paradigm for forestry, one based on the special part of us we seldom see when we look into society's mirror. That view, so carefully hidden from ourselves, is our ability to love, trust, respect, and nurture both one another and the land. To have sustainable forests, for example, we must change our thinking, and to change our thinking, we must transcend our own special interests and our own local, regional, and national boundaries and encompass all interests in the forest as a global community—present and future.

To better discuss sustainable forestry, I have divided this book into five parts. Part I is a brief look at a few of the ecological characteristics of Nature's forest and our disregard for Nature's design. Part II is an examination of why we insist that our short-term economic design is better than Nature's long-term ecological design. Part III focuses on land-use planning as the vehicle that can lead from conflict to vision, the window into the future. Land-use planning is crucial because where there is no vision, the people perish in the end—witness history. Part IV examines some of the fundamental issues, the causes of the uncomfortable societal crises with which we are having to deal but which we are currently ignoring in favor of arguing over the symptoms. Part V is a suggested new paradigm of forestry, one that integrates Nature's design of a forest with society's cultural necessities in an attempt to achieve biologically sustainable forests and sustainable forestry.

There is throughout the text a necessary mixture of science and philosophy, because scientific knowledge without a philosophical underpinning is at best useless. New practices are nonsense in terms of the old way of thinking. I have therefore done my best to marry science and philosophy in a way that will give birth to sustainable forestry in the near future for all of the future.

Although I have focused on the Pacific Northwest for many of the examples because it is the area of much of my experience, the principles and concepts discussed in the text are generally applicable to all forests. It is thus critical that we both understand and accept the effects we cause by redesigning global forests, because we are simultaneously redesigning the structural and functional processes of the world. At risk is human survival on earth and in the universe. If we disarm the world of all weapons of war and continue to pollute and kill the global ecosystem that sustains us, our extinction becomes only a matter of time.

If the 21st century is to be the time of healing the forests of the world, we must be willing to work together as people who care about our own future and that of our children and their children. To heal the forests, we must be willing to share openly and freely any and all knowledge necessary to achieve that end. In addition, we must be willing to cooperate with one another in a coordinated way, for cooperation without coordination is empty.

We must broaden our view of forests and forestry, not only because we need healthy forests for an abundance of quality water and to replenish the world's oxygen but also because we need healthy soils to produce wood fiber for fuel, building materials, and other products. With this in mind, let's begin our journey into the forest and forestry.

AUTHOR

Chris Maser spent over 20 years as a research scientist in natural history and ecology in forest, shrub steppe, subarctic, desert, and coastal settings. Trained primarily as a vertebrate zoologist, he was a research mammalogist in Nubia, Egypt (1963–64) with the Yale University Peabody Museum Prehistoric Expedition and was a research mammalogist in Nepal (1966–67) for the U.S. Naval Medical Research Unit #3 based in Cairo, Egypt, where he participated in a study of tick-borne diseases. He conducted a three-year (1970–73) ecological survey of the Oregon Coast for the University of Puget Sound, Tacoma, Washington. He was a research ecologist with the U.S. Department of the Interior Bureau of Land Management (1975–87), during which he studied old-growth forests in western Oregon, and a landscape ecologist with the Environmental Protection Agency (1990–91).

Today he is an independent author as well as an international lecturer and facilitator in land-use disputes, vision statements, and sustainable community development. He is also an international consultant in forest ecology and sustainable forestry practices. He has written over 230 publications, including *The Redesigned Forest* (1988), *Forest Primeval: The Natural History of an Ancient Forest* (1989), *Global Imperative: Harmonizing Culture and Nature* (1992), and *From the Forest to the Sea: The Ecology of Wood in Streams, Rivers, Estuaries, and Oceans* (1994, with James R. Sedell). Although he has worked in Canada, Egypt, France, Germany, Japan, Nepal, Slovakia, and Switzerland, he calls Corvallis, Oregon, home.

ACKNOWLEDGMENTS

I consider a book to be a progress report on the personal growth and world view of the author; I thus find no such thing as a definitive treatise. I also find that truth generally lies somewhere in the middle of extreme points of view, because, unlike the proponents of polarized positions, "truth" has no agenda—it simply is.

I have in the past been chastised for saying one thing one year and something different the next. I change my mind because my perception of the truth changes as I grow older and my world view expands. I now know and accept that while I seek "The Truth," what I find is but a constantly changing illusion of it, an illusion that is at best relative. Yet each time I peel a layer off the onion of illusion, I come a little closer to the Mystery of Truth, and therein lies the purpose of both science and society. Science is a process for seeking The Truth and society is a process for living The Truth. Neither is perfect as we practice them.

I say this because in my struggle to find "The Truth," I find only my truth (that which to me at this moment is true), and I freely acknowledge that it is at best only an illusion of that which is "true." Nevertheless, my truth is all I have to offer, and I offer it with humility, because I do not have the answers. In fact, only now am I beginning to understand some of the questions.

Many people over many years have helped me to arrive at the ideas embodied within these pages, ideas that are but a point of departure for that which lies beyond. The following people have been kind enough to specifically read, comment on, and improve various parts of the text in one of its many iterations toward completion: William H. Moir (Research Ecologist with the USDA Forest Service Rocky Mountain Experiment Station, Fort Collins, Colo.), Robert F. Tarrant (Former Director of the USDA Forest Service Pacific Northwest Experiment Station and Professor of Forest Science, Oregon State University, Corvallis), Jean Holt (Science Editor for the USDI National Park Service, Corvallis, Ore.), and Susan Hoffman (School Counselor, Corvallis, Ore.).

James R. Sedell (Stream and Fisheries Ecologist, USDA Forest Service, Pacific Northwest Research Station, Corvallis, Ore.), Bernard T. Bormann (Research Plant Physiologist, USDA Forest Service, Pacific Northwest Research Station, Corvallis, Ore.), Martha H. Brookes (Editor, USDA Forest Service, Pacific Northwest Research Station, Corvallis, Ore.), and A. Ross Kiester (Supervisory Math Statistician, USDA Forest Service, Pacific Northwest Research Station, Corvallis, Ore.) each graciously consented to co-author portions of chapters of this book.

It has been a rare privilege to work with Sandy Pearlman, one of the very best editors I have ever worked with. I have seldom had anyone take a manuscript and make it flow as clearly and easily as she has.

Zane, my wife, deserves special acknowledgment, for she alone has suffered not only my long hours at the computer but also months of preoccupation with the ideas secreted between the covers of this book.

To all of you, I offer my sincere gratitude for your unfailing generosity.

Part I

WHAT IS A FOREST?

You can't condemn that which you understand.

Goethe

1

CAN THE NOTION OF A FOREST BE CONTAINED IN A DEFINITION?

On November 14, 1986, my hometown paper, *The Gazette-Times,* carried a story entitled "Changing a Forest's Character." The story opened: "Managers of the Siuslaw National Forest are introducing a plan today to gradually change the complexion [character] of the Coast Range forest...." What does it mean to change the character of a forest? Are character and appearance synonymous? Are character and forest products synonymous? Are character and forest processes synonymous?

Character is defined as a conventionalized graphic device placed on an object as an indication of ownership, origin, or relationship. In the case of forests, this means logging roads, single-species monocultures, genetically altered trees, etc. Thus, by definition, we are redesigning our forests from Nature's blueprint of a forest to humanity's cultural blueprint, most often to that of a plantation. What does it mean to redesign the forest? In order to answer this question, the word *forest* must first be defined.

Several attempts have been made to define *forest*. The Society of American Foresters described a *forest* as "a plant community predominantly of trees and other woody vegetation."[1] *Webster's Third International Dictionary* is somewhat more specific: "A dense growth of trees and underbrush covering a large tract of land (specif.) an extensive plant community of shrubs and trees in all stages of growth and decay with a closed canopy having the quality of self perpetuation or of development into an ecological climax."[2] The International Union of Forestry Research Organizations offers three definitions: "1. (ecology) Generally, an ecosystem characterized by a more or less dense and extensive tree cover. 2. (ecology) more particularly, a plant community predominantly of trees and other woody vegetation, growing more or less closely together. 3. (silviculture/management) An area managed for the production of timber and other forest produce, or maintained under woody vegetation for such indirect benefits as protection of [water] catchment areas or recreation."[3]

These so-called definitions have been offered by foresters and ecologists, yet each is not a definition, but rather a simplistic characterization of a forest. I even wrote a whole book, *Forest Primeval,* some years ago in an effort to define for myself the concept of an ancient forest, but I could not. I found that I could not define a forest because I could not define a tree, or life, or soil, or death.

In my endeavor to define the notion of a forest, I discovered that words or collectively language are but metaphors for that which we cannot reach with our minds because the intellect sees but a fraction of the world and in itself is incomplete. What is missing is the language of the heart, for which there can be no definition. A forest, therefore, while it can be variously characterized, can never be defined because a forest—as does everything else—touches the totality of the Universe, most of which is far beyond our meager comprehension.

With the preceding in mind, and with a great deal of humility, I shall, as best I can, suggest a few ecological characteristics of a forest. This does not mean, however, that I am in any real way describing what a forest is from any perspective other than my own. That is, I am describing the way in which I see a forest intellectually after having spent many years as a scientist trying to figure out just what I mean when I use the word *forest.* In the end, I must conclude that each reader will decide for herself or himself what a forest is. Even with the same data, each individual will reach a different conclusion. And that, it seems to me, is as it needs to be in an ever-evolving world.

UNPREDICTABILITY

In the summer of 1486, a few years before the epic voyage of Columbus, a fire swept through a watershed in the northern Coast Range of Oregon and killed most of the existing forest on the east side of a large stream. The fire was stopped by the stream. The old-growth Douglas-fir* forest survived on the west side of the stream, and seeds from these trees were blown across the stream into the burnt area during an early winter storm. Most seeds were eaten by small animals, but some remained to germinate. Of those that germinated, only a few survived the centuries.

In 1836, when the stand was 350 years old, a severe windstorm blew down several of the trees, some of which hit other trees as they fell. One tree was hit hard enough that its protective bark was removed near its base; the wound was only 6 inches wide and 12 inches long (Figure 1). Although small compared to

---

*Scientific names are provided in the Appendix.

FIGURE 1   An ancient Douglas-fir wounded when the tree in the foreground struck it while falling.

the size of the tree, the wound attracted wood-boring beetles, which chewed their way into the wounded area. In so doing, the beetles exposed the wood to spores of fungi, which began attacking the tree's tissues. The fungus spread, decomposing the tissues and weakening the tree over the next century and a half.

By 1911, when the tree was 425 years old, it had a large weakened area ten feet above the ground. That year, a newly formed queen carpenter ant set up housekeeping in the area of the old wound. As the ant colony grew, new tunnels were continuously chewed in the sapwood by worker ants intent on expanding the colony. A pileated woodpecker (Figure 2) found the ant colony in 1970 and pecked a large, squarish hole into the mainstream of ant life, where it returned periodically over the years to dine on the ants.

In 1986, a windstorm with gusts reaching 70 miles per hour blew over a small group of Douglas-fir trees that had been weakened by a root rot fungus. One fell against the tree inhabited by the ants, which was now 500 years old. It broke off in the area of the old wound and fell diagonally upstream into the water. There it rests until 1996, creating habitat for fish and other organisms.

The tree moves gently up and down in place as winter storms and high water come and go. Then, in the year 2000, a warm chinook wind coincides with a 15-foot accumulation of wet snow above 4000 feet elevation. The warm rains that accompany the sudden melting of snow swell the stream and flush the fallen tree into the river. It comes to rest on a gravel bar, where it lies for the next 15 years, until 2015. During those years, the tree traps silt carried by winter floods, which in turn allows a thicket of alder trees to become established on the protected downstream side of the fallen tree.

The winter of 2015 is heralded by a series of violent rainstorms, causing severe flooding that sweeps the tree down river and out to sea. Heavy seas and strong winds move the tree northward several miles and cast it onto a beach. Barkless and battered, the tree lies bleaching in the sun; still visible, however, are the holes made by the pileated woodpecker 45 years earlier (Figure 3).

A winter storm in 2100 again washes the tree out to sea, where it floats for a year and a half before it is seen by a 25-year-old tuna fisherman. Setting his nets, he makes a bumper catch under the floating tree, which is the only shade-producing structure for hundreds of miles in the open ocean.

In the winter of 2103, the tree is again carried toward land and this time is deposited in a mudflat along an estuary. A storm in 2110 moves the now water-logged tree toward the mouth of the estuary, where it finally sinks. Marine, wood-boring invertebrates, such as gribbles, are attracted to the tree and penetrate its wood. The sunken tree is fragmented over the years as the marine invertebrates tunnel their way throughout it. In addition, much of the wood fiber has been excreted in the feces of gribbles and other organisms that live in the wood. During this time it has served as cover and habitat for gribbles, shrimp, crabs, and fish, to name but a few.

FIGURE 2   The pileated woodpecker, the largest woodpecker in the Pacific Northwest, feeds largely on carpenter ants that chew their colonial tunnels in dead wood. (USDA Forest Service photograph by Evelyn L. Bull.)

FIGURE 3   A drifted tree from a distant forest and time awash on a beach. The large holes where a pileated woodpecker excavated carpenter ants while the tree was still in the forest remain evident.

A series of violent rainstorms in 2190 cause severe flooding, and the sunken tree begins breaking into chunks that wash back and forth with the tides. By 2250, all that remains of the tree is the substance of its cells, called lignin, which is now part of the organic material that enriches the floor of the continental shelf off the Pacific Northwest coast.

If one event in this story were changed, the entire history of events would be different. For example, if a mouse had chosen to eat that particular seed, the fisherman would not have made his bumper catch of tuna 764 years later. If, on the other hand, the tree had not been wounded in 1836, it might have been logged in 1950. The tuna fisherman would therefore not have had his bumper catch, but instead could have been born in a house made from the lumber of the tree cut 225 years earlier. Let's change just one more event. If the tree that fell in 1986 had hit the tree inhabited by the ant colony at a different angle, it would not have fallen into the stream. It would therefore have decomposed on land and been recycled through bacteria, fungi, beetles, and myriad other pathways into the changing forest over the next 200 to 300 years (2186 to 2286). Again, because the tree would not have fallen into the stream, the tuna fisherman would have missed an excellent catch on that particular day in 2102. These are only three of the countless possibilities that might have changed the entire course of events.

To predict means to foretell, and in order to foretell we must be able to foresee. Nature's design, however, is a continual flow of cause-and-effect relationships that precisely fit into one other at differing scales of space and time and are constantly changing within and among those scales (Figure 4).

When dealing with scale, scientists have traditionally analyzed large, interactive systems in the same way that they have studied small, orderly systems, mainly because their methods of study have proven so successful. The prevailing wisdom has been that the behavior of a large, complicated system could be predicted by studying its elements separately and by analyzing its microscopic mechanisms individually—the traditional, linear thinking of Western civilization that views the world and all it contains through a lens of intellectual isolation. During the last few decades, however, it has become increasingly clear that many chaotic and complicated systems do not yield to such traditional analysis.

Instead, large, complicated, interactive systems seem to evolve naturally to a critical state in which even a minor event starts a chain reaction that can affect any number of elements in the system and can lead to a catastrophe. Although such systems produce more minor events than catastrophic ones, chain reactions

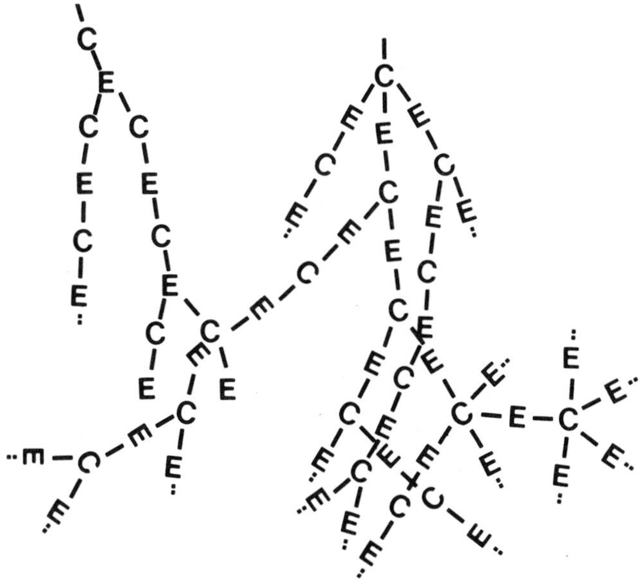

FIGURE 4 Each cause has an effect, and each effect is the cause of another effect, *ad infinitum.*

of all sizes are an integral part of the dynamics of a system. According to the theory call self-organized criticality,[4] the mechanism that leads to minor events is the same mechanism that leads to major events. Further, such systems never reach a state of equilibrium, but rather evolve from one semi-stable state to another.

Not understanding this, however, analysts have typically blamed some rare set of circumstances (some exception to the rule) or some powerful combination of mechanisms when catastrophe strikes, again often viewed as an exception to the rule. Thus, when a tremendous earthquake shook San Francisco, geologists traced the cataclysm to an immense instability along the San Andreas fault. When the fossil record revealed the demise of the dinosaurs, paleontologists attributed their extinction to the impact of a meteorite or the eruption of a volcano.

Although these theories may well be correct, systems as large, complicated, and dynamic as the Earth's crust and the ecosystem can break down under the force of a mighty blow as well as at the drop of a pin. Large, interactive systems perpetually organize themselves to a critical state in which a minor event can start a chain reaction that leads to a catastrophe, after which the system will begin organizing toward the next critical state.

Another way of viewing this is to ask a question: "If change is a universal constant in which nothing is static, what is a natural state?" In answering this question, it becomes apparent that the balance of Nature in the classical sense (when disturbed, Nature will return to its former state after the disturbance is removed) does not hold. For example, although the pattern of vegetation on the Earth's surface is usually perceived to be stable, particularly over the short interval of a lifetime, the landscape and its vegetation in reality exist in a perpetual state of dynamic balance—disequilibrium—with the forces that sculpted them. When these forces create novel events that are sufficiently rapid and large in scale, we perceive them as *disturbances.*

Perhaps the most outstanding evidence that an ecosystem is subject to constant change and disruption rather than being in a static balance comes from studies of naturally occurring external factors that dislocate ecosystems. For a long time, says Dr. J.L. Meyer of the University of Georgia, we failed to consider influences outside ecosystems. Our emphasis, she said, was "on processes going on within the ecosystem" even though "what's happening [inside] is driven by what's happened outside." Ecologists, she points out, "had blinders on in thinking about external, controlling factors,"[5] such as the short- and long-term ecological factors that limit cycles.

Climate appears to be foremost among these factors. By studying the record laid down in the sediments of oceans and lakes, scientists know that climate, in the words of Dr. Margaret B. Davis of the University of Minnesota, has been "wildly fluctuating" over the last two million years, and the shape of ecosystems with it. The fluctuations take place not only from eon to eon, but also from year to year and at every scale in between. "So you can't visualize a time in equilibrium," asserts Davis. In fact, says Dr. George L. Jacobson, Jr. of the University of Maine, there is virtually no time when the overall environment stays constant for very long. "That means that the configuration of the ecosystems is always changing," creating different landscapes in a particular area through geological time.

In connection to change, Professor John Magnuson made a wonderful observation about foreseeing cause and effect. He said that all of us can sense change: the growing light at sunrise, the gathering wind before a thunderstorm, or the changing seasons. Some of us can see longer-term events and remember that there was more or less snow last winter compared to other winters or that spring seemed to come early this year. It is an unusual person, however, who can sense, with any degree of precision, the changes that occur over the decades of his or her life. At this scale of time, we tend to think of the world as static and typically underestimate the degree to which change has occurred. We are unable to sense slow changes directly and are even more limited in our abilities to interpret their relationships of cause and effect. This being the case, the subtle processes that act quietly and unobtrusively over decades are hidden and reside in what Magnuson calls the "invisible present."[6]

It is the invisible present, writes Magnuson, that is the scale of time within which our responsibilities for our planet are most evident. "Within this time scale, ecosystems change during our lifetimes and the lifetimes of our children and our grandchildren."

It must be noted here that while it is possible to envision such serious accidents of human misjudgment as the meltdown at the nuclear plant in Chernobyl, Ukraine, or Iraq's invasion of Kuwait, the ultimate potential destruction of the planet with respect to human life will not be as apparent. Instead, it will occur slowly and silently, like the pollution of our air, our soil, and our water—in the secrecy of the invisible present.

The story of the floating tree in this chapter is about Nature's forest, but what about our social forest—the one we are trying to regulate in the invisible present? To regulate is to govern or direct according to rule; to bring under control of law or constituted authority; to bring order, method, or uniformity. What does it mean to design a regulated forest? Who has the authority to dictate how the forest will function? When decisions are made, how can unforeseen events (such as an ice storm, a volcanic eruption, a lightning strike, or metal fatigue in logging machinery) be regulated? How can we regulate what will happen in 1, 10, 100, 500, or 1000 years?

We have seen what can happen by changing a single event in our story. What makes us think we know enough to dare to alter the entire dynamics of Nature's design over whole landscapes for all time when we cannot even predict what will happen from one year to the next?

The world of Nature is a constantly changing dynamic set of relationships; the older I get and the more I study, the clearer this becomes. There are lags between the time decisions are made and the time their consequences become evident. Such lags in time are the rule in ecological systems; they separate cause and effect, which confuses our interpretation of Nature's world and makes it seem fickle and unsettled. Unless we adopt an attitude of far greater humility than we profess today, our lack of understanding of both events and processes in the invisible present will continue to be exceedingly costly to human society.

It is thus essential that we become aware of our individual and social tendency to suppress gradual trends. This will require a concerted, sustained effort at all levels of society to keep attention focused on the subtle changes in our environment that threaten our collective future.

LONG-TERM TRENDS

A trend is a line of general direction or movement. In Nature's forest, a trend is defined by a multitude of interacting factors, which include:

1. **Location of event:**   on land, in water, in air, in the tropics, at the North or South Pole, in a valley…

2. **Size of event:**   inside the tip of a tree root, on an acre, over a landscape, over a continent…

3. **Duration of event:**   ten seconds, an hour, a year, a century, a geological epoch…

4. **Time of event:**   day, night, season, year…

5. **Frequency of occurrence:**   hourly, daily, seasonally, annually…

6. **Distance between events:**   an inch, a foot, a yard, a mile, 1000 miles, 10,000 miles…

7. **Uniformity of event:**   uniform, roughly connected, disjunct…

8. **Type of event:**   physical, biological, combination…

The infinite variety of interactions among these factors creates an infinite variety of short-term trends that fit into a longer-term trend (Figure 5A), that fit into a still longer-term trend, *ad infinitum* (Figure 2B). Studying short-term trends (those that can be detected and perhaps understood) and projecting them over time may allow some degree of predictability of Nature's actions.

There are two cautions, however. First, we must accept that all of these trends are ultimately cyclic and that their governing principles are neutral and impartial; the shorter the trend, therefore, the more imperative is our acceptance of Nature's

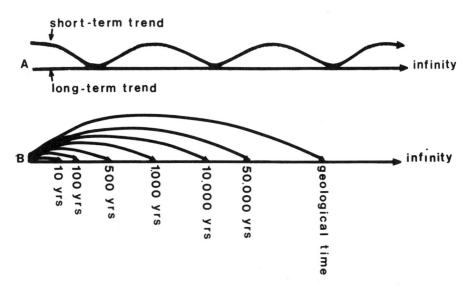

FIGURE 5   Longer-term trends are composed of interrelated, dynamic, shorter-term trends.

neutrality and impartiality. Governing principles, whether biological or physical, are always neutral and impartial. On the other hand, when we assign values to Nature's actions that are based on our perceptions of "good" and "bad," we interject into Nature's design the artificial variable of partiality that clouds our vision. We have robbed ourselves of our ability to predict the future by rejecting Nature's impartial neutrality.

Second, short-term trends must be viewed in relation to long-term trends and long-term trends in relation to even longer-term trends. The more we project the present into the future, the better we understand the present. The more we trace the present into the past, the better we understand the present. A knowledge of the past tells us what the present is built on and what the future may be projected on, but this is true only if we accept past and present as a cumulative collection of our understanding of a few finite points along an infinite continuum—the trend of the future.

Here I suggest you fasten your seat belt for what only at first appears to be a "side trip." On this journey, however, there are no side trips, any more than in worldly terms there are side effects; there are only direct causes and effects. Because wholeness demands a vast grasp, we embark on a quick backward sweep of the events that led us to the present Western, dominant view of the world.

Although Nature provides a degree of predictability over time (through the ability to recognize and read a trend), we remain stubbornly committed to the concept of an absolute. The concept of an absolute probably arose in response to human fear of unknown but observable natural forces. This fear may have given rise to religious ideas conceived in "the necessity of defending oneself against the crushingly superior force of nature."[7]

In the sixth century B.C., the roots of Western science arose in the first period of Greek philosophy, a culture in which science, philosophy, and religion were united. A split in this unity began with the assumption of a Divine Principle that stands above all gods and people. Thus began a trend of thought that ultimately led to the separation of spirit and matter and to the dualism that characterizes Western philosophy. Because the Christian church supported Aristotle's view that questions concerning the human soul and God's perfection took precedence over investigations of the material world, Western science did not develop further until the Renaissance, when Sir Francis Bacon in the late 1500s gave humanity the ability to experiment and defined what is now called the scientific method. "Look," he told the world, "There is tomorrow. Take it with charity lest it destroy you."

Separation of knowledge and intuition began with the ancient Greeks. Western philosophy embraced what is considered to be rational knowledge: that which is derived from experience with objects and events in one's immediate environment. Such knowledge belongs to that realm of the intellect that discriminates, divides, compares, measures, and categorizes. It creates a world of apparent distinctions whose opposites can only exist in relation to one other.

Abstraction, according to physicist Fritjof Capra, is a crucial feature of knowledge, because in order to compare and classify the immense variety of sizes, shapes, structures, and phenomena that surround us, we must select a few significant features to represent the incomprehensible milieu. We thus construct an intellectual map of reality in which things are reduced to their general outlines. This results in knowledge being a system of abstract concepts and symbols characterized by the linear, sequential structure that is typical of our thinking and speaking.

The natural world, on the other hand, is one of infinite variables and complexities, a multidimensional world without straight lines or completely regular shapes, where things do not happen in sequences, but all at once, a world (as modern physics shows us) where even empty space is curved. It is clear that our abstract system of conceptual thinking is incapable of completely describing or understanding this reality and therefore is necessarily limited.[8]

Thinking and knowledge in Western culture have become so linear that we have forgotten that everything is defined by its relationship to everything else (Figure 6). Nothing exists by itself; everything exists in relation to something else. For example, in harvesting trees, a stand is divided and isolated once a sale boundary is delineated. From then on, it is "managed" as a linear sequence of

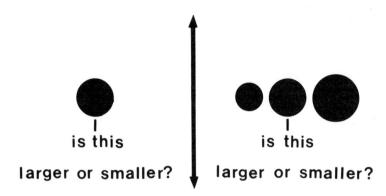

is this

is this

larger or smaller?   larger or smaller?

FIGURE 6   Everything we visualize is defined by its relationship to everything else.

special cases—stands of timber that are isolated in time and space in our minds but nowhere in the landscape.

To illustrate this point, consider the example of two old-growth timber sales disputed in federal court. The assertion was repeatedly made that these were "just two sales" and would not have an impact on the remaining old-growth forest, which was "ample." Yet no one knew how much old-growth forest existed, where it was, or how it was distributed over the landscape. No one knew how the delineated stands related to one another, let alone to the rest of the old-growth forest, or how the rest of the old-growth forest related to any particular stand.

We can begin to understand the significance of the remaining old-growth forest only when we relate the delineated units to one other in a local area as well as across a landscape. The problem is that we see a timber sale only as a numerically defined, commercial absolute. In so doing, we fail to see its relation to the forest. In fact, we fail to even see the forest.

Failing to see the forest for the trees is dangerous, because changes in the spatial patterns of land use, which grossly alter habitats through time, may well be crucial to understanding the dynamics of landscapes and will have implications for many ecological processes. Changes in the patterns of landscapes may also be related to the flows of materials and energy across landscapes, such as the processes of erosion and the movement of sediments. Characterization of the relationships between changing patterns on the landscape and how those changes affect ecological processes is particularly important if we are to develop a more complete understanding of landscape dynamics and our effects on them.

While we have progressed much in our understanding of Nature through the intellectual pursuit of science, we have lost much of our connection with Nature through materialism. We see Nature as something to conquer and control. Our

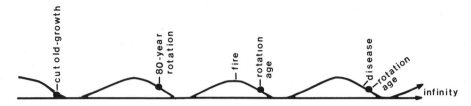

FIGURE 7 Dynamic, short-term trend superimposed on a dynamic, long-term trend, with a false absolute (rotation age) fixed in time and space. As shown, there is no guarantee that rotation age will be reached, despite planning.

fear of Nature causes us to select a single isolated point (such as a stand of old-growth trees) on the continuum of change, draw an illusionary line around it, quantify it (the volume of wood fiber), and define it as an unchanging, absolute entity in time and space.

An unchanging absolute and a dynamic relationship are diametrically opposed and mutually exclusive; in addition, it is both arrogant and unwise to demand from Nature what Nature is not designed to give. A forest may therefore be summed up as a continuum of causes and effects that appear random in the short term and patterned in the long term. We thus forfeit our understanding and any predictability of the trends Nature has provided as long as we blindly commit ourselves to dealing in terms of artificial absolutes (Figure 7).

To understand this concept, let's look at a brief history of forest fires in the American southwest, particularly in Arizona, from the year 1700 to the present. Fire is a physical process through which Nature originally designed forests in the western United States.

That is not, however, the way Gifford Pinchot (first chief of the U.S. Forest Service) saw it as he rode through park-like stands of ponderosa pine along the Mogollon Rim of central Arizona in the year 1900. It was a warm day in June as Pinchot rode his horse to the edge of a bluff overlooking the largest continuous ponderosa pine forest in North America. The pine-scented forest, without a logging road to scar the ground or a chain saw to tear the silence, was a sight to behold.

"We looked down and across the forest to the plain," he wrote years later. "And as we looked there rose a line of smokes. An Apache was getting ready to hunt deer. And he was setting the woods on fire because a hunter has a better chance under cover of smoke. It was primeval but not according to the rules."[9]

The forest over which Pinchot gazed on that June day in 1900 was 300 to 400 or more years old, composed of trees that had germinated and grown throughout their lives in a regime characterized by low-intensity surface fires sweeping repeatedly through their understory. These fires, which occurred every few years or so, consumed dead branches, stems, and needles on the ground and simultaneously thinned clumps of seedlings growing in openings left by vanquished trees. Although fire had been a major architect of the park-like forest of stately pines that Pinchot admired, he did not understand the significance of fire in designing the forest.

Every forest ecosystem evolves inevitably toward a critical state in which a minor event sooner or later leads to a catastrophic event that alters the ecosystem in some way. As a young forest grows old, it converts energy from the sun into living tissue, which ultimately dies and accumulates as organic debris on the forest floor. There, through decomposition, the organic debris releases the energy stored in its dead tissue. A forest is therefore a dissipative system in that energy acquired from the sun is dissipated gradually through decomposition or rapidly through fire.

Of course, rates of decomposition vary. A leaf rots quickly and releases its stored energy rapidly. Wood, on the other hand, rots much more slowly, often over centuries. As wood accumulates, so does energy stored in its fibers. Before suppression, fires burned frequently enough to generally control the amount of energy stored in accumulating dead wood by burning it up, thus protecting a forest for decades, even centuries, from a catastrophic killing fire.

Eventually, however, a forest builds up enough dead wood to fuel a catastrophic fire. Once available, the dead wood needs only one or two very dry, hot years with lightning storms to ignite such a fire, which kills the forest and sets it back to the earliest developmental stage of grasses and herbs. From this early stage, a new forest again evolves toward old age, accumulating stored energy in dead wood and organizing itself toward the next critical state—a catastrophic fire—which starts the cycle over.

Therefore, a 700-year-old forest that burned could be replaced by another, albeit different, 700-year-old forest on the same acreage. In this way, despite a series of catastrophic fires, a forest ecosystem could remain a forest ecosystem. In this sense, the old-growth forests of western North America have been evolving from one catastrophic fire to the next, from one critical state to the next.

Although Pinchot knew about fire, he was convinced that it had no place in a "managed forest." Fire was therefore to be vigorously extinguished, because conventional wisdom dictated that ground fires kept forests "understocked," and more trees could be grown and harvested without fire. In addition, surviving trees, like the ones Pinchot had seen in Arizona, were often scarred by the fires, and this kind of injury allowed decay-causing fungi to enter the stem, thus reducing the quantity and quality of harvestable wood. Finally, any wood not used for direct human benefit was an economic waste.

At this point, linear, commodity-oriented thinking entered into the profession of forestry in the United States, and Pinchot's utilitarian conviction that fire had no place in a managed forest became both the mission and the metaphor of the young agency that he built. "Managed" in this sense came to mean any forested acre wherein someone perceived an economic interest in the trees. It was this notion of linearity—of economic waste if a potential commodity was not used by humans for the demonstrable benefit of humans—that introduced the "invisible present" into the profession of forestry so long ago.

In Pinchot's time and place in history, he was correct and on the cutting edge, and the ecological problems caused by such thinking were unbeknownst to him. Nevertheless, incorporation of these ideas into forestry began to take its toll. Only now, decades after the instigation of fire suppression, has the significance of changes in forest structure, composition, and function become evident.

Recent evidence shows, for example, that some ponderosa pine forests in northern Arizona had only 23 trees per acre in presettlement times. This presettlement density is in stark contrast to the current density of approximately 850 trees per acre, with predominantly small diameters.

The *increase* in density of trees is estimated to have caused: (1) a 92 percent drop in the production of grasses and forbs, (2) a 31 percent reduction in stream flow, (3) a 730 percent increase in accumulated fuels on the forest floor, (4) a 1700 percent increase in volume of saw timber, (5) a decrease from 115 to –8 in the index of scenic beauty, and (6) a habitat shift from open, savannah-like conditions in presettlement times to dense forest.[10]

This increase in the density of trees may also result in the decreased vigor and increased mortality of all trees, especially those of the oldest age classes. Finally, the increased closure of the canopy, the vertical continuity of fuel in the form of trees in the understory (a fuel ladder), and the high loads of fuel on the ground result in a severe hazard of crown fires that will kill the forest. Such fires probably were exceedingly rare and localized in presettlement times, although ground fires were common before the settlers introduced grazing by domestic livestock and the Forest Service introduced the suppression of fire.

Only now, decades after Pinchot's instigation of the suppression of one of the most prevalent forces of Nature—fire—has the significance of changes in the structure and composition of forests in many areas become evident. During the last 80 to 100 years, since the advent of fire suppression, there has been a general increase in the number of trees and an increase in the amount of woody fuels on the forest floor. There also has been a decrease in the extent of quaking aspen (which often resprouts from roots following fire) and a corresponding increase in those species of trees that are more tolerant of the shaded conditions in closed-canopy forests. Some of these shade-tolerant trees have grown into the forest canopy and form a ladder up which a fire can burn from near the ground to the tops of large trees.

Although the role of fire in the physiological and ecological requirements of individual plant species may be relatively clear, there is greater difficulty in determining how fire regimes design whole forests. Most historical studies, termed chronosequences, are hampered by effects of unknown events that can result in erroneous interpretations. It is therefore particularly important to study major ecological processes in an integrated fashion because such mechanisms are interdependent, and the variability in fire regimes is more likely to be important to plant communities than are the mean values computed from some arbitrary period of fire history.

For example, unusually long periods without fire may lead to establishment of fire-susceptible species. The simultaneous occurrence of such fire-free periods and wetter climatic conditions may also be extremely important to such species as ponderosa pine that have episodic patterns of regeneration (occur as specific, discrete episodes) as opposed to plants whose regeneration patterns are continual. Therefore, while statistical summaries of fire histories are useful in a general comparison of fire regimes in different forests, the influence of fire on a particular ecosystem is strongly historical. Some forests are more a product of unusual periods of climate and fire frequencies in the short term than they are of average or cumulative periods of climate and fire frequencies in the long term.

Although it is possible that climatic change could account for the increased number of large "wildfires," change in forest structure and composition is the most likely cause. Intensive study of historical fires has failed to document any cases of fire killing a forest by burning through tree tops in the ponderosa pine forests of the American southwest prior to 1900. In contrast, however, since 1950 numerous fires exceeding 5000 acres have totally razed the forests down to mineral soil. The intensity of these fires is attributed to the amount of woody fuels on the forest floor and to dense stands of young trees within the forests— both of which have come about since 1900.

Because of the dynamic nature of evolving ecosystems and because each system is constantly organizing itself from one critical state to another, an ecosystem can only be "managed" for its possible evolution, not for an absolute value of anything, such as a given sustained yield of products. The only sustainability for which we can manage is whatever ensures the ability of an ecosystem to adapt to evolutionary change (such as global warming) in a way that may be favorable for us.

Today, a major task facing scientists and managers is to re-educate themselves and the public. Smokey Bear (not to mention Walt Disney's Bambi) has done a resounding job of convincing people that a forest fire is terrible and wasteful. The new message must be that fires set by careless people can be needlessly destructive. On the other hand, fires under specific conditions are both beneficial and necessary to the long-term ecological health and biological sustainability of western forests.

The only constant feature in a forest is change. Each forest evolves in response to short- and long-term variations in its environment, such as climate, over which humans have no control. Humans can neither arrest nor control changes wrought by Nature. If we could, the quivering balance through which Nature has produced the very forest we value and want to protect would be upset.

It now seems obvious that the effort to eliminate fire from forests is an economic rule made by humans. One major obstacle to changing this rule is our illusion of knowledge, which economic imagination draws with certainty and bold strokes while scientific knowledge advances slowly by uncertain increments and contradictions.

The challenge for today's forestry profession is to sit humbly in the forest to discover the rules by which forests have evolved and lived out the millennia. In fact, were the far-sighted Gifford Pinchot alive today, knowing what we now know about fire, he would undoubtedly say, "Fire in managed forests is primeval and is according to the rules. We must therefore learn all we can about its role in managing western forests for the benefit of our forests and our citizens."

DIVERSITY

Diversity is having or being capable of having a variety of forms or qualities, of being composed of unlike or distinct elements. *Stability* is defined as the strength to stand or endure, the property that causes an entity to develop processes that restore its approximate original condition prior to a disturbance. It follows that at some point diversity underlies stability; conversely, stability is an outgrowth of diversity (which was discussed thoroughly in *Global Imperative*[11] and will not be repeated here). Four questions lead to examples that demonstrate this point.

1. What happens when just one part is removed?   A helicopter crashed in Nepal some years ago and killed two people. A helicopter has a great variety of pieces with a wide range of sizes. The particular problem here was with the engine, which is held together by many nuts and bolts. Each has a small sideways hole through it so that a tiny "safety wire" can be inserted and the ends twisted together to prevent the tremendous vibration created by a running engine from loosening and working the nut off the bolt. The helicopter crashed because a mechanic forgot to replace one tiny safety wire that kept the lateral control assembly together. A nut vibrated off its bolt, the helicopter lost its stability, and the pilot lost control. All this was caused by one missing piece that altered the entire functional dynamics of the aircraft. The engine had been "simplified" by one piece—a small length of wire.

Which piece was the most important part in the helicopter? The point is that each part (structural diversity) has a corresponding relationship (functional diversity) with every other part, and they provide stability only by working together within the limits of their designed purpose.

2. What happens when a process is "simplified"?   A newly elected mayor of a city whose budget is overspent guarantees to balance the budget; all that is necessary, in a simplistic sense, is to remove some services whose total budgets add up to the overexpenditure. In a "simplistic sense" is used here because it is not quite that simple. What would happen, for example, if all police and fire services were eliminated? Would it make a difference, if the price were the same and the budget could still be balanced, if garbage collection were eliminated instead?

The trouble with such a simplistic view is in looking only at the cost of and not at the *function* performed by the service. The diversity of the city is being simplified by removing one or two pieces or services, without paying attention to the functions performed by those services. To remove a piece of the whole may be acceptable, provided we know which piece is being removed, what it does, and what effect the loss of its function will have on the stability of the system as a whole.

3. What happens when so-called renewable resources are "mined?"   Because mining is an entirely different function from farming, it is necessary to look back into history to see how this unlikely connection was made. The European farmers of early North America, with their small, diversified family farms, provided increased structural diversity and, therefore, increased habitat diversity through a good mix of food, cover, and water.

The many small, irregular fields with a variety of crops created an abundance of structurally diverse edges, and tillage offered a variety of soil textures for burrowing animals. Uncultivated fence rows and ditch banks provided strips of habitat that acted as homes for some species and provided travel lanes for others between fields, hedge rows, and farm woodlots. All this resulted in small crop yields because of competition from other plants and because insects, birds, and mammals all took a share of the crop.

In modern agricultural practice in North America, however, large fields often are planted with a single species, which results in a greatly simplified environment. Such monocultures are basically unstable because they lack the checks and balances of a natural, diverse ecosystem. Modern agricultural crops, therefore, require constant human care (cultivation) and control (insecticides, rodenticides, and/or herbicides) if production is to be as desired.

Replacement of small farms by large farms dependent on mechanization and specialized crop monocultures caused a drastic decline in habitats within and adjacent to croplands. Because of the decreased crop stability (increased crop vulnerability) that resulted from the greatly simplified "agricultural ecosystem," farmers are more and more inclined to view native plants and animals as actual or potential "pests" to their crops. In essence, intensive agriculture is the controlled encouragement of a foreign organism—a crop—to grow rapidly and purely at a high density and to suppress any other organism that might be seen as competing with it.

The erratic economics of agriculture is considered to be the primary impetus behind increasing crop specialization in North America. In addition, governmental influence on modern agriculture, in attempting to maintain low-cost food production, has largely made the small, diversified farm uneconomical. Consequently, most have disappeared. In turn, in order for modern agriculture to survive economically, modifications in farming strategies have been necessary, and drastic changes in land use have resulted: (1) increased specialization of farms (growing fewer crops in larger fields), which causes amalgamation of small, individual fields; (2) increased size of individual farms due to large, specialized corporate farms replacing small, diversified family farms; (3) increased use of modern machinery, which is more easily and more economically operated in large fields; (4) increased clearing of fence rows to gain more land for agriculture (one mile of fence row may occupy about half an acre); (5) increased use of large sprinkler irrigation systems, which eliminate uncultivated irrigation ditches and their banks; (6) replacement of uncultivated earthen irrigation ditch banks with concrete; and (7) federal aid to farmers through the Agricultural Stabilization and Conservation Service for various types of land "reclamation."

What are Nature's penalties for ecologically unsound agriculture? One obvious penalty is loss of fertile topsoil. W.C. Lowdermilk, a soil scientist with the Soil Conservation Service, wrote, "If the soil is destroyed, then our liberty of choice and action is gone, condemning this and future generations to needless privations and dangers." He also composed what has been called the "Eleventh Commandment":

> Thou shalt inherit the Holy Earth as a faithful steward, conserving its resources and productivity from generation to generation. Thou shalt safeguard thy fields from soil erosion, thy living waters from drying up, thy forests from desolation, and protect thy hills from overgrazing by thy herds, that thy descendants may have abundance forever. If any shall fail in this stewardship of the land thy fruitful fields shall become sterile stony ground and wasting gullies, and thy descendants shall decrease and live in poverty or perish from off the face of the earth.[12]

In a very real sense, the dire prediction in Lowdermilk's "Eleventh Commandment" has and is becoming a reality in many parts of the world, including the United States—perhaps especially the United States. As the last few years have shown, modern agriculture emphasizes high technology, such as expensive applications of fertilizers and toxic pest controls. In addition to continued, extensive soil erosion, such as takes place in the wheat-raising country of northeastern Oregon and eastern Washington, the single most widespread source of nonpoint pollution in streams, rivers, and groundwater is chemicals from intensive agriculture.

Contamination of our water supply is such a serious problem that the Office of Technology Assessment prepared a two-volume document on *Protecting the Nation's Groundwater from Contamination.* Of the 245 substances known to occur in groundwater, 68 compounds are listed as connected with herbicides or pesticides, and the list did not even consider fertilizers, rodenticides, or fungicides used in agriculture.[13]

When a system is so simplified and specialized that it becomes vulnerable to otherwise innocuous native biota, its stability has been lost through simplification—a loss of structural diversity and its interrelated processes that allowed the system to function according to its design. When this happens, humans try to correct the problem through artificial means, such as chemicals; this rarely works, however, and the chemicals end up contaminating the world's waters and other parts of the ecosystem as well.

Sustainable farming advocate Wes Jackson said that "Modern agriculture views the land as a quarry to be mined; it uses up the topsoil, groundwater, and petrochemicals in pursuit of edible products." He goes on to say that the result is "the least efficient and most devastating form of agriculture ever invented."[14] To Jackson's way of thinking, *farming is the number one environmental problem,* more serious than the disposal of toxic wastes or even the destruction of the ozone layer.

Statistics bear him out. Nine tons of topsoil per acre per year is lost in American agriculture. That is the equivalent of one inch of topsoil every 16 years, soil that may have taken anywhere from 300 to 1000 or more years for Nature to create. In addition, 80 to 90 percent of our water is consumed by agriculture, and very inefficiently at that. Further, people in some areas of the Great Plains are using water ten times faster than it can be replaced. And the groundwater that is replaced is being polluted by the runoff of some 40 million tons of commercial fertilizers, which not only are dangerous to human and nonhuman health but also are energy-intensive and use up nearly one fifth of the country's natural gas. Thus agriculture as an industry is the leading polluter of the streams, rivers, seas of the world, and groundwater and is heavily dependent on fossil fuels.[15]

Civilization after civilization has risen on the strength of its natural resources and then fallen as those resources were abused and exhausted. Have things really changed much from this scenario? Is it not the case that such natural resources as soil and old-growth forests are still being mined wherever they occur worldwide?

To mine is to dig under to gain access to or to cause the collapse of valuable material. The problem with mining natural resources is that mining, by definition, can only go one way: exploitive extraction without replacement. Mining "renewable" natural resources, by definition, sooner or later makes them "non-renewable," because mining reinvests none of the natural resource capital into that part of the ecosystem from which it came. No living system can function indefinitely without a balanced input and output of energy, and mining only extracts. Even President Franklin Delano Roosevelt observed that "We treat timber resources as if they were a mine." We are thus actually mining our own civilization when we undercut the resource base on which it grows.

4. What happens when a system is simplified through "management?" Let's take a look at a forest in the Pacific Northwest. We begin with an old-growth Douglas-fir forest because old growth is the commodity that has been liquidated over vast acreages in order to meet a "sustained" yield of wood fiber. An old-growth forest is one successional stage in the continuum of successional stages in a forest cycle; it is, however, the most stable stage, where stability is defined as the ability of a community to withstand catastrophe or to approximate its "original" state after severe alteration. Although there have been some attempts to derive an ecological definition of old-growth Douglas-fir forests in western Oregon and western Washington, the incalculable diversity of these old forests makes a uniform, general description or definition impossible. Inherent in these attempts to define old-growth is the continual recurring discovery that the apparent ecological stability of these old forests in time is somehow related to their diversity.

The attributes of forests are composition, structure, and function. Composition refers to the array of plant and animal species, which includes shifts in abundance as well as the presence or absence of species. Structure refers to their arrangement in space and time. Function refers to the ecological processes and the rates at which they occur, such as production of organic matter and cycling of nutrients through the various pathways. One example of a functional change with succession is the increased or decreased complexity of the pathways that route dead organic material throughout the changing plant community.

Old-growth coniferous forests in the Pacific Northwest developed over long periods that were essentially free from such catastrophic disturbances as wildfire. They occupied vast expanses of the presettlement landscape. The elimination of these forests began because they represented both a valuable resource (large volumes of virtually free, high-quality wood) and a hindrance to agricultural development. Consequently, their liquidation began early and has been virtually assured.

Today, little old-growth forest exists outside of public lands, and liquidation continues. For example, the first record of logging in what is now the Willamette National Forest of western Oregon occurred in 1875. Ninety percent of the timber cut during the first three decades of this century occurred below 4000 feet elevation. From 1935 through 1980, the volume cut doubled every

15 years. By the 1970s, 65 percent of the timber cut was above 4000 feet elevation.[16]

Old-growth coniferous forests differ significantly from young-growth forests in species composition, structure, and function. Most of the obvious differences can be related to four structural components of the old-growth forest: large live trees, large snags (standing dead trees), large fallen trees on land, and large fallen trees in streams.

On land, this large, dead, woody material is a critical carryover component from old-growth forests into young-growth forests. When snags are removed from short-rotation stands following liquidation of the preceding old growth, 10 percent of the wildlife species (excluding birds) will be eliminated; 29 percent of the wildlife species will be eliminated when both snags and fallen trees (logs) are removed from intensively "managed" young-growth forests.[17] As pieces are continually removed from the forest with the notion of the simplistic uniformity that is termed "intensive timber management," the ultimate simplistic view of modern forestry—the plantation or "Christmas tree farm"—comes closer to reality.

What do these four examples have to do with redesigning Nature's forest? First, recall that in the story of the helicopter, the tiny, missing safety wire responsible for the crash was neither obvious nor visible when the aircraft was flying. Where is the safety wire in the forest? The irony is that it will not be found until it is missing and something drastic happens. Second, like the city mayor, pieces are being removed because prevailing economic models say it is okay to do so even without understanding what the pieces do.

Third, the native forests are being mined, even with exceedingly little understanding of how they function, especially in terms of their sustainability over time. We are, however, proceeding as though we know what we are doing—moving from complex, diverse native forests designed by Nature toward simple, uniform Christmas-tree-like plantations designed by humans—solely to meet economic desires. The reasons given for these actions, such as jobs and community stability, do not alter the fact that forests may be in jeopardy for lack of data and lack of patience with Nature's design.

Fourth, a forest is thought of and "managed" similar to modern agriculture. We simplify (specialize) and intensify with no knowledge of the consequences, straining to have ever more of everything simultaneously, but at what cost—loss of soil fertility, reliance on chemicals that pollute the land and water, and many acres that may not again grow a forest in the foreseeable future!

As "forest management" intensifies, it comes closer and closer to merging with intensive agriculture, particularly in three respects: (1) the increasing attempt to purify and specialize the crop tree, (2) the increasing move toward monoculture, which decreases diversity and therefore stability, and (3) the increasing view of plants and animals that exert any perceived negative effect on crop trees as pests. This view *necessitates* an increasing outlay of capital, time, energy, and such materials as fertilizers, pesticides, plastic tubing to protect seedlings, and on and on. Intensified management ensures that many normal biological processes will be viewed as management-created competition, which conflicts with production goals and will call for continued artificial simplification of the forest. The most ubiquitous and irreversible environmental problem society already faces is the loss of biological diversity.

For Want of a Squirrel, A Forest Is Lost

If diversity of species is essential, how do species act in concert as a whole? "For want of a squirrel, a seed was lost; for want of a seed, a tree was lost; for want of a tree, a forest was lost; for want of a forest...." What exactly does this all mean? It means that the tremendous biodiversity of the tropical rain forests could be lost without cutting a single tree! How?

Let's consider the rain forest in Gabon, Africa, where to biologist Louise Emmons its fascination lies in "its stunning complexity." In this forest, says Emmons, "You can stand anywhere and be surrounded by hundreds of organisms that are all 'doing something,' going about their living in countless interactions—ants carrying leaves, birds dancing, bats singing, giant blue wasps wrestling with giant tarantulas, caterpillars pretending they are bird droppings, and so on."[18]

In Gabon, Emmons found that nine species of squirrels all live together in one forest. Each is a different size; three have specialized diets or habits, which leaves six that feed on nuts, fruits, and insects and could therefore be potential competitors for food. A closer look, however, reveals that three of the six species—one large, one medium, and one small—live exclusively in the canopy of the forest, where the largest one (a "giant" squirrel) feeds primarily on very large, hard nuts while the smaller ones eat proportionally smaller fruits and nuts. The other three species—again one large, one medium, and one small—live exclusively on the ground, where they eat the same species of fruits and nuts as do their neighbors in the canopy, except that they eat the fruits and nuts that have fallen to the ground.

The forest in Gabon is evergreen. Fruit can be found on the trees throughout the year, but any one species of tree produces fruit for only a short period each year. To support three species of squirrels, eight species of monkeys, and eight species of fruit-eating bats (and so on) in the canopy, this forest must have a wide variety of species of trees and lianas (high-climbing, usually woody vines), each of which produces fruits and nuts in its own rhythm. The varying sizes of the fruits and nuts can support different sizes of squirrels with different tastes, whereas these same fruits and nuts when they fall to the ground can feed a whole analogous array of species.

Just how rich in species is a tropical rain forest? Al Gentry of the Missouri

Botanical Garden has for many years been counting the species of trees and lianas in tropical rain forests. The richest site he has found thus far is a plot 2.5 acres in extent near Iquitos, Peru, where he counted an incredible 283 species of trees over 4 inches in diameter. He found 580 trees of this size in the 2.5-acre plot, which means that there was an average of only two individual trees per species and an astounding 58 species among the first 65 individual trees counted.

Worldwide, tropical rain forests seem to have from about 90 to 283 species of large trees within every 2.5 acres, and this does not include the other plants and the animals. Even the "poorest" tropical rain forest has an average of about five individual trees per species every 2.5 acres.

In contrast, a dry tropical forest, such as occurs in northern India, has about half as many species of trees as does a wet tropical forest. The richest forests in the United States have about 20 species of trees over 4 inches in diameter, with an average of about 30 individuals per species, in each 2.5 acres of ground. Most temperate forests, however, are much poorer than this.

It thus seems clear that tropical rain forests are amazingly rich in species of trees, but not just any trees: in particular, those trees whose fruits are eaten and dispersed by birds and mammals. Not surprisingly, therefore, tropical rain forests are also rich in species of mammals and birds, but not just any mammals and birds: in particular, those that eat fruits and disperse their seeds. There are, for example, 126 species of mammals within a single area of forest in Gabon and 550 species of birds within a single lowland site in the Amazon basin of Peru. Further, the life cycle of each species is interdependent on the life cycles of the other species. The enormous number of vertebrate animals appears to be supported by the large number of species of plants that act as sources of food year round.

If all this biodiversity is to be maintained, each individual tree must succeed in leaving offspring. Seeds and tender young seedlings are among the richest foods available to forest animals, and their succulence greatly increases their chance of being eaten by the large numbers of hungry animals searching for food around the bases of fruit- and nut-bearing trees. Similarly, such organisms as fungi, worms, and insects soon accumulate where the seeds and seedlings are concentrated and spread from one seed or seedling to another.

Under such circumstances, seeds carried away from such concentrations of hungry organisms are more likely to succeed in germinating. Another major benefit of seeds being carried away from the parent tree is the availability of a wide variety of places with different conditions into which a seed is likely to fall. A new condition might offer a pocket of better soil on a mound created by termites or in a spot where a dead tree has created a hole in the canopy that emits sunlight.

It is certainly no accident that about 80 to 95 percent of the species of trees in tropical rain forests produce fruits that are dispersed by birds and mammals. By dispersing those seeds, the birds and mammals also maintain the rich diversity of species of trees, which not only formed their habitat in the first place but also perpetuate it. This is an ideal example of a self-reinforcing feedback loop.

Many species of trees in the tropical rain forests, especially those which germinate in the dark understory, have large seeds that carry enough stored energy to grow leaves and roots without much help from the sun. Such fruits and seeds are often so large that only proportionately sized birds and mammals can swallow or carry them. In Gabon, for example, monkeys dispersed 67 percent of the fruits eaten by animals in the area studied by Emmons.

Seed-dispersing animals, such as large birds and large monkeys, are critical in replacing the large trees and lianas of the forest canopy and thus helping them survive. These animals are, however, the first species to disappear when humans hunt for food. These species, along with elephants, have already been hunted so heavily that they have either been drastically reduced in numbers or eliminated completely over vast areas of the African forest, and a similar situation exists in the tropical rain forests of Central and South America.

For the most part, foresters have overlooked how the interdependency of plants and animals affects the biodiversity of a plant community. Elephants in the Ivory Coast, for example, disperse the seeds of 37 species of trees. Of those, only seven species had alternate means of dispersal (by birds and monkeys). Of the 201 individual trees in one study area, 83 species were dispersed by elephants.

In one forest where humans had eliminated elephants a century earlier, few juvenile trees of the elephant-dispersed species were left, and the two major species had no offspring at all. One of these two species is the single most important species for the two largest squirrels that Louise Emmons studied in Gabon—one eats the large, hard nuts in the canopy and the other eats the same nuts once they have fallen to the ground.

Once the large species of birds and mammals are gone, the stunningly rich tropical rain forests will change and gradually lose species of trees, lianas, and other plants. Smaller seeds dispersed by wind will replace large seeds dispersed by large animals. Those species of plants whose seeds grow in the shaded understory will not survive, and the land will gradually be forested by fewer, more common species.

As the forests become poorer in species of plants, the number of species of birds, mammals, and other creatures will correspondingly decline. All the complex, interconnected, interdependent feedback loops among plants and animals will gradually simplify. The species that comprise the feedback loops will be lost forever—and the feedback loops along with them. This is how the evolutionary process works. Ecologically, it is neither good nor bad, right nor wrong, but those changes may make the forest less attractive and less usable by other species, including humans, that used to rely on its products.

The same types of self-reinforcing feedback loops that take place in tropical rain forests occur also in the temperate coniferous forests of the world, and they represent the same four basic elements of diversity: genetic, species, structural, and functional. Genetic diversity is the way by which species adapt to change; it is the hidden diversity that is so often subjected to "secret extinctions." The most important aspect of genetic diversity is that it can act as a buffer against the variability of environmental conditions, particularly in the long term. Healthy environments can act as "shock absorbers" in the face of catastrophic disturbance.

Here looms a critical concept: the past function of an ecosystem determines its present structure, and its present structure determines its future function. This means that structure is defined by function and function is defined by structure. As the composition of species in an area is altered, so too is its function in time altered. Over time, this new arrangement of species will respond to conditions differently than the original arrangement of species would have.

What Happens to the Forest When It Loses the Grandparent Trees?

When the old-growth forests in the Pacific Northwest have been liquidated, no more old trees will stand as living monarchs (Figure 8), to die and stand as

FIGURE 8   A large, live old-growth Douglas-fir. (USDA Forest Service photograph by J.M. Trappe.)

FIGURE 9   A large, standing, dead old-growth ponderosa pine, which has stood in northeastern Oregon for several years. Note the woodpecker holes in the snag in the lower foreground. (Oregon Department of Fish and Wildlife photograph.)

large dead trees (Figure 9), and to topple as large fallen trees (Figure 10) and lie for centuries decomposing, providing a kaleidoscope of habitats, and performing their myriad functions as they recycle and reinvest their biological capital into the soil from which they and their generation grew. The forest, the standing large dead tree and the large fallen tree, which are only altered states of the live old-growth tree, will go the way of the oldest living thing on Earth, the old-growth monarch of the forest, and become extinct.

With the extinction of the old-growth forest shall go such species as the

FIGURE 10   A large, fallen old-growth Douglas-fir. The man standing next to the fallen tree is about six feet seven inches tall.

northern spotted owl (Figure 11) and the marbled murrelet, which have evolved in concert with that particular habitat. In fact, the owl and the murrelet have adapted to particular features of that habitat.

The northern spotted owl nests in tall, broken-topped, old-growth Douglas-fir trees. The marbled murrelet, a seabird, nests on carefully selected large, moss-covered branches at least 100 feet up in old-growth trees, with other branches close overhead to protect the nest site. The murrelet's nest tree is located several miles inland from the coast, where it feeds. Because they are so specialized in the selection of their reproductive habitats, neither the owl nor the murrelet is capable of adapting to the rapid changes wrought by the liquidation of the old-growth forest.

An interesting twist to the story is that not only will species of plants and animals become extinct with the liquidation of the old-growth forests, so too will the "grandparent trees." As young trees replace liquidated old trees in crop after crop, the ecological functions performed by the old trees, such as creation of the "pit-and-mound" topography on the floor of the forest with its mixing of mineral soil and organic top soil, become extinct processes because there are no more grandparent trees to blow over.

The "pit" in pit-and-mound topography refers to the hole left when a tree's

FIGURE 11   The spotted owl is closely tied to old-growth forests in the Pacific Northwest. (Oregon Cooperative Wildlife Research Unit photograph by Gary Miller.)

roots are pulled from the soil (Figure 12), and "mound" refers to the soil-laden mass of roots, called a rootwad (Figure 13), suddenly projected into the air above the floor of the forest. The young trees that replace the grandparent trees are much smaller and are different in structure. They cannot perform the same functions in the same ways.

FIGURE 12 "Pit" of pit-and-mound topography caused when an old Sitka spruce tree blew over in a severe wind.

FIGURE 13 Rootwad of an old-growth Sitka spruce along the Oregon Coast.

FIGURE 14 Western hemlock growing rapidly out of a large, centuries-old, buried fallen tree in an opening in the forest caused by the fall of two old-growth giants. (USDA Forest Service photograph by J.M. Trappe and the author.)

Of all the factors that affect the soil of the forest, the roughness of the surface caused by falling grandparent trees, particularly the pit-and-mound topography, is the most striking. It creates and maintains the richness of species of plants in the herbaceous understory and affects the success of tree regeneration.

Uprooted trees enrich the forest's topography by creating new habitats for vegetation. Falling trees create opportunities for new plants to become established in the bare mineral soil of the root pit and the mound. In time, a fallen tree itself provides a habitat that can be readily colonized by tree seedlings and other plants. Falling trees also open the canopy, which allows more light to reach the floor of the forest (Figure 14). In addition, pit-and-mound topography is a major factor in mixing the soil of the forest floor as the forest evolves.

The extinction of the grandparent trees changes the entire complexion of the forest through time, just as the function of a chair is changed when the seat is removed. The "roughness" of the floor of the forest, which over the centuries resulted from the cumulative addition of pits and

FIGURE 15   Fallen high-elevation old-growth Douglas-fir trees. The Douglas-fir with the bark on (center) fell as a live tree in a recent windstorm. The whitened Douglas-fir was a standing dead tree for many years before it blew over and fell across the live tree, perhaps in the same windstorm. Large fallen trees add the diversity of "roughness" to the floor of the forest.

mounds and fallen grandparent trees (Figure 15), will become unprecedentedly "smooth."

Water moves differently over and through the soil of a smooth forest floor devoid of large fallen trees that act as reservoirs, storing water throughout the heat of the summer and holding soil in place on steep slopes (Figure 16). The huge snags and fallen trees that acted as habitats are gone, as are the stumps of the grandparent trees with their belowground "plumbing systems," which guided rain and melting snow deep into the soil.

This plumbing system of decomposing tree stumps and roots comes from the frequent formation of hollow, interconnected, surface-to-bedrock channels that drain water rapidly from heavy rains and melting snow. As roots rot completely away, the collapse and plugging of these channels force more water to drain through the soil matrix, which reduces soil cohesion and increases hydraulic pressure, which in turn causes mass soil movement. These plumbing systems cannot be replaced by the young trees of plantations.

FIGURE 16 Remains of fallen old-growth Douglas-fir lying along the contour of a gentle slope. Note the soil held in place by the presence of the fallen tree (lower right corner). This old tree is saturated with water and acts as a reservoir throughout the year. Pencil is used to indicate scale. (USDA Forest Service photograph by J.M. Trappe and the author.)

INTERRELATED PROCESSES

Focusing on the products of the forest (wood fiber) rather than processes that produce the forest brings to mind Buddha's parable of the wealthy but foolish man:

> Once there was a wealthy but foolish man. When he saw the beautiful three-storied house of another man, he envied it and made up his mind to have one built just like it, thinking he was himself just as wealthy. He called a carpenter and ordered him to build it. The carpenter consented and immediately began to construct the foundation, the first story, the second story, and then the third story. The wealthy man noticed this with irritation and said:—"I don't want a foundation or a first story or a second story; I just want the beautiful third story. Build it quickly."
>
> A foolish man always thinks only of the results, and is impatient without the effort that is necessary to get good results. No good can be

attained without proper effort, just as there can be no third story without the foundation and the first and the second stories.[19]

In terms of sustainable forestry, the third story (products) cannot be built without dealing with the foundation (soil fertility), the first story (patience with Nature's time table), or the second story (consequences of the invisible present in terms of the cumulative effects of improved technology for the harvest and use of wood fiber in the absence of an understanding of belowground forest processes).

First, it must be understood that every human endeavor (such as sustainable forestry) can only be attained by way of the effort and discipline necessary to achieve the desired result. It must also be understood that Nature has designed everything with an *order* that *must be respected.* Again, Buddha provided a lesson over 2000 years ago:

> At one time the tail and the head of a snake quarrelled as to which should be the front. The tail said to the head:—"You are always taking the lead; it is not fair, you ought to let me lead sometimes." The head answered:—"It is the law of our nature that I should be the head; I can not change places with you."
>
> But the quarrel went on and one day the tail fastened itself to a tree and thus prevented the head from proceeding. When the head became tired with the struggle the tail had its own way, with the result that the snake fell into a pit of fire and perished.
>
> In the world of nature there always exists an appropriate order and everything has its own function. If this order is disturbed, the functioning is interrupted and the whole order will go to ruin.[19]

Natural processes are a product of both structural and functional diversities that operate within Nature's ordered design, which cannot be seen, but instead are perceived as a kaleidoscope of ever-changing relationships. Nature's order is not understood because we are focused on the third story—the product—rather than the processes that are critical to maintaining the forest, of which the trees are only a result.

A Forest Is More than the Sum of Its Ingredients

In attempting to "manage" a system, the focus has been on the wrong end, whether the system is a forest, a grassland, an ocean, or society. "Management," after all, is only a metaphor to justify an impact on a system. The concept of management allows us to focus on the desired economic product rather than on the ecological processes that produced the product in the first place. In "forestry," therefore, focus has been only on the trees, and not the forest.

The relevance of focus can be explained in the following analogy. If you walk to the door of your living room and stop to survey the room, you see everything in the room in focus and in relationship (the forest). If you now walk to the coffee table, pick up the newspaper, and begin to read the headlines (the tree), your narrowed focus causes everything else to effectively disappear from view—to become out of focus.

This narrowness of focus brings us to "a lesson in a box"—a box of cake mix, that is! A critical ecological lesson lies inherent in making a cake from "scratch." Knowing what ingredients go into a cake, what they do, and why they are important allows us to understand why the cake comes out of the oven as it does.

Today, however, few people know how to make a cake from scratch. Instead, they buy a box of cake mix and erroneously think that they have purchased a cake. They have actually bought a box of some of the ingredients for a cake, but they do not know what those ingredients are, where they came from, their proportions or quality, who or what put them into the box, and whether or not everything is as it should be inside of the box. All of this is taken on faith.

If you buy a box of cake mix and put the contents into a bowl, do you have an instant cake in the bowl? No. What's in the bowl is a dry, powdery mixture of some of the ingredients for a cake.

If you read the instructions on the box, you will find that you must add additional ingredients and stir. Now do you have a cake in the bowl? No. You have the batter, which consists of all of the ingredients. Why don't you have a cake yet?

If you return to the instructions, you will learn that you must preheat the oven to 375 degrees Fahrenheit and then insert the batter, which you have been instructed to transfer to a cake pan and bake for 40 minutes.

When the 40 minutes have passed, you open the oven, take out the pan, and there is the cake. What happened in the oven that did not happen in the bowl? First, not all the necessary ingredients were in the box; you had to add some. Even then, something was missing. The heat of the oven caused the chemical interactions to take place among the ingredients, which in turn produced the cake. The heat was the catalyst that drove the chemical-physical processes that created the form and function of the cake.

The point is that you can understand a product, including what happens when an ingredient is omitted, only when you start from scratch, because only then can you see all the ingredients and their interrelationships before and after the chemical-physical interactions occur. A cactus, a grass, a shrub, and a tree are much the same as a cake. Each is but the physical manifestation of the chemical-physical interactions among them, in this case, a seed, soil, water, air, sunlight, climate, and time.

Thus, as a bakery produces cakes, so does a forest produces trees. As a cake does not make a bakery, however, neither does a tree make a forest. A bakery is housed in a building with electricity, water, ventilation, sewage disposal, and an owner, a baker, a salesperson, delivery trucks that maintain a supply of ingredients, and so on. Part of the profits from the bakery must be reinvested to ensure that it functions properly if it is to continue producing cakes. As a bakery is more than the sum total of the chemical-physical constituents that combine to make cakes, so too is a forest more than the sum total of the chemical-physical constituents that combine to make trees.

Suppose you enter the bakery as a customer with only one thing on your mind: to buy a cake. At that moment your thinking not only is linear but also is focused only on the product you want to purchase. However, if the bakery can produce only 75 cakes a day and if 76 customers show up, self-centered competition rules, tempers flare, and the polarization of duality sets in—mine versus yours, right versus wrong. If one person buys two or even three cakes, more trouble ensues.

So it is in a forest. Each timber company enters the forest as a customer intent only on securing as many trees for itself as possible, at the least cost, in order to maximize its own profits in the shortest possible time. Thus, while timber companies and "environmentalists" fight over who is going to get the last old-growth tree, society loses sight of the forest and the need to understand the complexities of its processes, because all the focus is on the tree.

What society has forgotten, however, in its linear, competitive drive to harvest trees, is that the quality and the quantity of the ingredients (seeds, soil, water, air, sunlight, climate, and time) and the interactions of the chemical-biological processes among them constitute the forest, which produces the trees. This is the "lesson in a box"—that focusing solely on the product and ignoring or even disdaining the processes that produced it is the major cause of extinctions worldwide, including that of forests.

ALL ELEMENTS ARE NEUTRAL

All things in Nature's forest are neutral; Nature assigns no values. Each piece of the forest, whether a bacterium or an 800-year-old Douglas-fir, is allowed to offer its prescribed structure, carry out its prescribed function, and interact with other components through their prescribed interrelated processes. None is more valuable than another; each is only different from the other.

Assigning a value to something in the forest begins to adjust that object in our focus, and bringing one thing into focus simultaneously forces almost everything else out of focus (Figure 17). For example, for many years rodents were poisoned in the name of forestry, because they were perceived as having only a negative

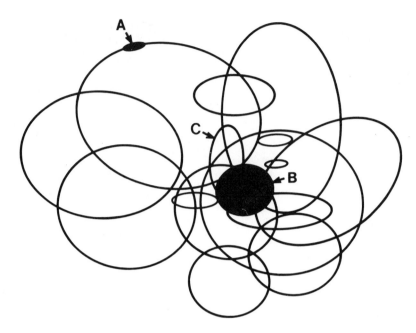

FIGURE 17   First we detect something we like in one part of a biological cycle (A) and assign it a value. Then we isolate it from the rest of the forest and, if possible, learn only those aspects that we want to know (B). When we focus on anything in Nature's forest and assign it a plus or minus (good or bad) value, we develop a myopic view by forcing it into an absolute value (A and B). Our view of its relationships to the interconnected cycles of other things then becomes distorted and out of focus (C).

value. They ate the tree seeds that grew into merchantable trees. Today, forest rodents are viewed differently. Some still eat tree seeds, but at the same time they disperse viable spores of mycorrhizal fungi, nitrogen-fixing bacteria, and yeast. The following scenario of the northern flying squirrel illustrates but one example of the dynamic interactions of small mammal–forest interdependence in the temperate coniferous forests of North America, Europe, and Asia, the high mountain forests of Argentina, and the eucalyptus forests of Australia.

The Squirrel

The northern flying squirrel (Figure 18) is common in conifer and mixed conifer–hardwood forests from the Arctic tree line throughout the northern conifer forests of Alaska and Canada, south through the Cascade Mountains of Washington and Oregon and the Sierra Nevada Mountains of California almost

FIGURE 18   A northern flying squirrel. Flying squirrels cannot fly; they can only glide downward from a higher elevation to a lower elevation. (USDA Forest Service photograph by J.W. Grace.)

to Mexico, the Rocky Mountains to Utah, and the Appalachian Mountains to Tennessee.

The seldom seen, nocturnal flying squirrel is primarily a mycophagist, a fungus eater. In northern Oregon, hypogeous fungi (those fruiting belowground) and epiphytic lichens (those growing in treetops) are the major foods of the flying squirrel. In northeastern Oregon, for example, hypogeous fungi were the principal food from July to December; from December through June, the lichen *Bryoria fremonti* (Figure 19) was the squirrel's predominant food and as well as its sole nest material. In southwestern Oregon, hypogeous fungi were the major food throughout the year; lichens were not important in the overall diet. Hypogeous fungi also are an important food for flying squirrels during all months that they are available in central Alaska, and the same is true for northern flying squirrels in the southern Appalachians of the eastern United States.

The Fungus

The term mycorrhiza, which literally means "fungus-root," denotes the symbiotic relationship between certain fungi and plant roots (Figure 20). Fungi

FIGURE 19 The lichen *Bryoria fremonti* forms the predominant food of northern flying squirrels from December through June in northeastern Oregon. The squirrels also make their nests of it. (A lichen is a fungus that houses an alga in a mutually beneficial symbiotic relationship; a lichen is thus two plants in one.)

that produce hypogeous sporocarps (belowground fruiting bodies) (Figure 21) are probably all mycorrhizal. Woody plants in the families Pinaceae (pine, fir, spruce, larch, Douglas-fir, hemlock), Fagaceae (oak), and Betulaceae (birch, alder) in particular depend on mycorrhiza-forming fungi for nutrient uptake. This phenomenon can be traced back some 400 million years to the earliest known fossils of plant rooting structures.

Mycorrhizal fungi absorb nutrients and water from soil and translocate them to a host plant. In turn, the host plant provides sugars from its own photosynthesis to the mycorrhizal fungi. Fungal hyphae (the "mold" part of the fungus) extend into the soil (Figure 22) and serve as extensions of the host's root system and are both physiologically and geometrically more effective for nutrient absorption than the roots themselves.

FIGURE 20 Mycorrhizae, the symbiotic relationship between certain fungi and plant roots. In this case the fungus forms a mantle (a covering over the root tip). (USDA Forest Service photograph by C.P.P. Reid.)

FIGURE 21   The belowground fruiting body (hypogeous sporocarp) of a truffle or false truffle (a fungus). The inner, dark tissue contains the reproductive spores and the light tissue is a tough outer coat. (USDA Forest Service photograph by J.M. Trappe.)

FIGURE 22   Mold-like threads of fungal tissue, called hyphae, extend from the fungal mantle around the root tip out into the soil, where they extract water and nutrients and translocate them into the roots of the tree. (Photograph by K. Cromack, Jr.)

Squirrel–Forest Relations

The most obvious squirrel–forest relations are those that occur on the surface of the ground, such as foraging. Even the nesting and reproductive behavior of these squirrels remains relatively obscure because of their nocturnal habits. In probing the secrets of the flying squirrel, however, at least four functionally dynamic, interconnected cycles emerge.

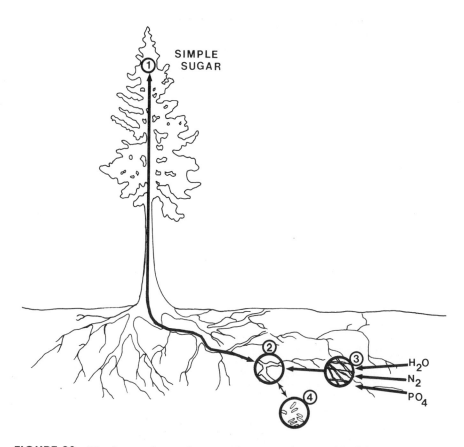

FIGURE 23 Nutrient exchange between the tree and mycorrhizal fungus. Tree (1) provides fungus with carbohydrates from photosynthesis. Fungal hyphae extend from the mycorrhizae (2) into the soil (3), acting as an extension of the tree's root system and assisting in uptake of water, nitrogen, and phosphorus. Inside mycorrhizae (2), fungus produces an exudate used as food by bacteria (4) that convert atmospheric nitrogen to a useable form, which may move through the fungus into tree roots, trunk, and crown (1).

The Fungal Connection

The host plant provides simple sugars and other metabolites to the mycor-rhizal fungi, which lack chlorophyll and generally are not competent saprophytes (a living plant that derives its nutrients from dead or decaying organic material) (Figure 23:1). Fungal hyphae penetrate the tiny, nonwoody rootlets of the host plant to form a balanced, harmless mycorrhizal symbiotic relationship with the roots (Figure 23:2). The fungus absorbs minerals, other nutrients, and water from the soil and translocates them into the host (Figure 23:3). Further, nitrogen-fixing bacteria (*Azospirillum* spp.) (Figure 24) that occur inside the mycorrhiza (Figure 23:4) use a fungal "extract" as food and in turn fix atmospheric nitrogen. (To "fix" nitrogen is to take gaseous, atmospheric nitrogen and alter it in such a way that it becomes available and useable by plants.) Nitrogen thus made available can be used both by the fungus and the host tree.

In effect, mycorrhiza-forming fungi serve as a highly efficient extension of the host root system. Many of the fungi also produce growth regulators that induce production of new root tips and increase their useful life span. At the same time, host plants prevent mycorrhizal fungi from damaging their roots. Mycor-rhizal colonization enhances resistance to attack by pathogens. Some mycor-rhizal fungi produce compounds that prevent pathogens from contacting the root system.

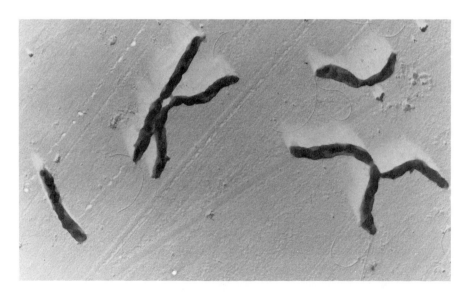

FIGURE 24   *Azospirillum* spp. is a bacterium that is capable of extracting nitrogen from the air and converting it to a nitrogen compound that plants can use. It is called a nitrogen-fixing bacterium. (Enlarged 4700 times.) (USDA Forest Service photograph courtesy of C.Y. Li.)

FIGURE 25   Northern flying squirrel–forest interaction. The northern flying squirrel nests and reproduces in the canopy (1); it descends at night to the ground to eat hypogeous fungal sporocarps (2).

The Sporocarp Connection

Sporocarps (fungus fruiting bodies) are the initial link between hypogeous mycorrhizal fungi and the flying squirrel. Flying squirrels nest and reproduce in the tree canopy (Figures 25:1 and 26) and come to the ground at night (Figure 27), where they dig and eat hypogeous sporocarps (Figures 25:2, 28, and 29). As

FIGURE 26   The northern flying squirrel both nests and reproduces in the treetops. Baby flying squirrels in their nest. (USDA Forest Service photograph by J.W. Grace.)

a sporocarp matures, it produces a strong odor that attracts the foraging squirrel. Evidence of a squirrel's foraging remains as shallow pits in the forest soil and occasional partially eaten sporocarps (Figure 30).

Sporocarps of hypogeous fungi contain nutrients necessary for the small animals that eat them. In addition to nutritional value, sporocarps also contain water, fungal spores, nitrogen-fixing bacteria, and yeast, all of which become important in the forest network.

FIGURE 27   The northern flying squirrel descends at night to the forest floor to hunt for its fungal food. (USDA Forest Service photograph by J.W. Grace.)

FIGURE 28   A northern flying squirrel digging in the forest floor for its predominant food, the belowground fruiting bodies of fungi. (USDA Forest Service photograph by J.W. Grace.)

FIGURE 29   A northern flying squirrel eating a belowground fruiting body of the fungus *Hysterangium* spp. (USDA Forest Service photograph by J.W. Grace.)

The Squirrel Connection

When flying squirrels eat sporocarps, they consume fungal tissue that contains nutrients, water, viable fungal spores, nitrogen-fixing bacteria, and yeast (Figure 31:1). Pieces of sporocarp move to the stomach (Figure 31:2) where

FIGURE 30   A small pit in the forest floor where a squirrel dug out the belowground fruiting body of a fungus and did not eat it. (USDA Forest Service photograph by J.M. Trappe.)

fungal tissue is digested, then through the small intestine (Figure 31:3) where absorption takes place, and then to the cecum (Figure 31:4). The cecum is like an eddy along a swift stream; it concentrates, mixes, and retains fungal spores, nitrogen-fixing bacteria, and yeast. Captive deer mice (Figure 32) retained fungal spores in the cecum for more than a month after ingestion. Undigested material, including cecal contents, is formed into excretory pellets in the lower colon; these pellets, which are expelled through the rectum (Figure 31:5), contain all the viable elements necessary to inoculate the root tips of trees with prolonged life (Figure 31:6).

The Pellet Connection

A fecal pellet (Figure 33) is more than a package of waste products; it is a "pill of symbiosis" dispensed throughout the forest. Each fecal pellet contains four components of potential importance to the forest: (1) spores of hypogeous

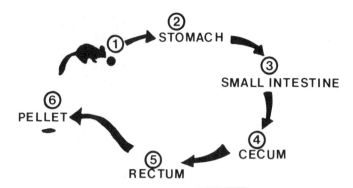

FIGURE 31   Passage of sporocarp through the northern flying squirrel: (1) sporocarp is eaten; (2, 3) fungal tissue is digested; fungal spores, nitrogen-fixing bacteria, and yeast propagules pass intact; (4) spores, bacteria, and yeast are concentrated and mixed; (5) fecal pellets are formed; and (6) fecal pellets containing viable spores, bacteria, and yeast are passed.

mycorrhizal fungi (Figure 34), (2) yeast (Figure 35), (3) nitrogen-fixing bacteria, and (4) like the yolk that feeds the chicken forming in the white of an egg, the complete nutrient component for nitrogen-fixing bacteria.

    Each fecal pellet contains viable spores of the mycorrhizal fungi, and each also contains the entire nutrient requirement for *Azospirillum* spp. The yeast, as

FIGURE 32   Deer mice are small, nocturnal denizens of the forest. (Photograph by M.L. Johnson.)

FIGURE 33 Fecal pellets of a northern flying squirrel act as an ecological pill that helps to inoculate the soil with live fungal spores, nitrogen-fixing bacteria, and yeast.

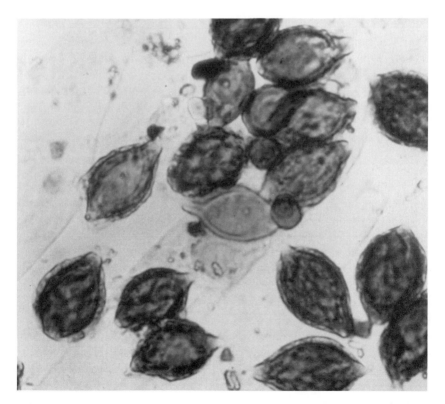

FIGURE 34 Fungal spores pass unharmed through rodent digestive tracts and germinate. (USDA Forest Service photograph by J.M. Trappe.)

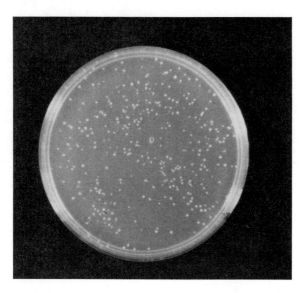

FIGURE 35   Yeast propagules (white spots) cultured from a fecal pellet of a northern flying squirrel. (USDA Forest Service photograph by C.Y. Li.)

a part of the nutrient base, has the ability to stimulate both growth and nitrogen-fixation in *Azospirillum* spp. Abundant yeast propagules may also stimulate spore germination because spores of some mycorrhizal-forming fungi are stimulated in germination by extractives from other fungi, such as yeast.

This is all incredibly complicated, but it is only a tiny glimpse of the total complexity of a forest. To continue, the fate of fecal pellets varies, depending on where they fall. In the forest canopy, the pellets might remain and disintegrate in the treetops, or a pellet could drop to a fallen, rotting tree and inoculate the wood. On the ground, a squirrel might defecate on a disturbed area of the forest floor, where a pellet could land near a conifer feeder rootlet that may become inoculated with the mycorrhizal fungus when spores germinate. If environmental conditions are suitable and root tips are available for colonization, a new fungal colony may be established. Otherwise, hyphae of germinated spores may fuse with an existing fungal thallus (the nonreproductive part of the fungus) and thereby contribute and share new genetic material.

The northern flying squirrel thus exerts a dynamic functionally diverse influence within the forest. The complex of effects ranges from the crown of the tree, down through the surface of the soil into its mantle where, through mycorrhizal fungi, nutrients are conducted through roots, into the trunk, and up to the crown of the tree (Figure 36), perhaps into the squirrel's own nest tree.

Such relationships are by no means confined to the northern flying squirrel

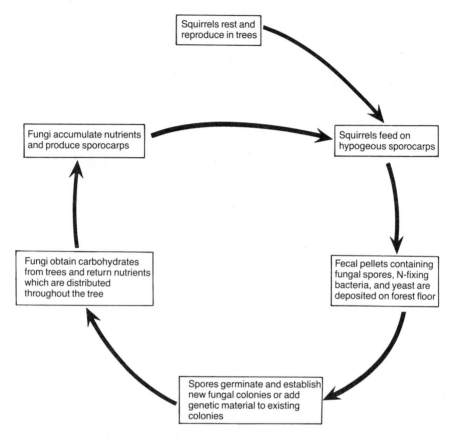

FIGURE 36 Major components of flying squirrel–fungus–tree mutualism.

or even to the North American continent. Many mammals, such as deer mice and white-footed mice, red-backed voles (Figure 37), chipmunks, mantled ground squirrels, Douglas squirrels (Figure 38), and western gray squirrels and other rodents, depend more or less on hypogeous fungi for food. In addition, this small mammal–hypogeous fungal–tree interdependence has been documented in rodents of the high, cloud forest of Argentina, the coniferous and mixed conifer–hardwood forests of southern Germany and southern France, and in small, marsupial mammals in the eucalyptus forests of Australia. (For those interested in the study of animal mycophagy, see Castellano et al.[20])

Today, however, both species and habitats are being lost because of a lack of sensitivity to how and why they are functionally interconnected. The organism and its habitat must be understood in relation to each other in order to understand the organism or its function within its habitat.

FIGURE 37   Western red-backed vole. (USDA Forest Service photograph by Douglas Ure.)

FIGURE 38   Douglas squirrel. (Oregon Department of Fish and Wildlife photograph by Ronald Rohweder.)

If the organism and its function within its habitat is not understood, how can the results of unexpected changes in the habitat when the organism is removed be understood? Can even a tentative, negative value be assigned to any organism in a forest without an understanding of its diversity of functions?

A FLEXIBLE, TIMELESS CONTINUUM OF SPECIES

As discussed, Nature operates in trends rather than absolutes. Further, Nature is neutral, assigning no values, making no predictions, and indifferent to the outcomes of constantly changing relationships. In Nature, what is simply is (Figure 39).

Nature's forest is a timeless continuum of species in which time is defined only as the measured or measurable period during which an action, process, or condition exists or continues, a nonspatial continuum in which events occur in apparently irreversible succession from the past through the present into the future. Assigning positive values to items in the forest allows them to be viewed as desirable products that must be harvested and marketed to make a profit. Time

FIGURE 39   A mixed-species stand of trees with western hemlock, western redcedar, and Douglas-fir. The forest is being simplified to contain almost exclusively one species of tree in any given plantation. (Photograph by L.D. Harris and the author.)

is used to measure efficiency in production of salable goods; hence the adage, "time is money."

Under present design strategy, time is perceived as a limiting factor on and against which management decisions are based. The artificial time constraints placed on diverse biological processes in an attempt to create fast, uniform, monocultural crops of trees serve to concentrate and magnify the perceived conflicts in time and space. What has yet to be learned is that any shortcut is inevitably the longest distance between two points.

For example, in focusing attention on a chosen product as a measure of the success of a management effort, what is forgotten is that a forest encompasses processes, which are cyclic and have no end. Short periods of time are arbitrarily chosen as frames of reference, usually based solely on economic considerations, within which to manage a normally long-lived forest for selected products; anything diverted to a different outcome is considered a loss. Such linear thinking ensures conflict and fosters the concept that anything in a forest not used by people is an economic waste.

The more intense forest management becomes, the less flexible it becomes. (The way the term forest management is used, it really means crop management.) The less flexible it becomes, the more stringent time constraints become. The more stringent time constraints become, the more varied, focused, and irreconcilable human conflicts become. This progression of events leads to competition and to human conflict: two or more persons want the same thing (such as old-growth trees) at the same time on the same acre for different reasons, and there is not enough to satisfy both.

The pivotal concept is single-species management under strict time constraints for a particular product, which forecloses options for other amenities that would require a different or greater mix of species and habitats with vastly different time regimes. For example, a mill owner wants to cut a stand of old-growth timber for the potential profit that can be made and then wants to streamline (redesign) the forest site to continually grow new trees faster so that more money can be made. People in a conservation organization want to retain the mixed-species stand of old-growth timber intact through time for a nesting pair of spotted owls. Both parties want the same stand of old-growth trees at the same time on the same acre, but for different reasons. Each perceives its purpose to be right and mutually exclusive.

LANDSCAPE PATTERN

Landscapes are designed by Nature as ever-changing relationships in time and space. We gaze on them for but an instant; yet past, present, and future are contained in that instant.

FIGURE 40   Coyote. (Oregon Department of Fish and Wildlife photograph.)

Nature designs landscapes with single-celled bacteria, coyotes (Figure 40), mosses, trees, slope, aspect, elevation, glaciers (Figure 41), fire (Figure 42), and volcanos (Figure 43). The variety and scales are infinite. Some landscapes are covered with forests and then uniquely molded with soils of different parentage, with insects and disease, fire and ice, drought and floods, and time.

Nature paints and repaints the landscapes with all manner of brushes and strokes, from fine and delicate to coarse and bold (Figure 44). Humans, on the other hand, lack Nature's imagination, skill, patience, and daring. Our landscapes, conceived piecemeal in planning rooms, are unimaginative, weak, and prone to unravel. The difference is that Nature designs over interrelated landscapes, whereas humans design a few isolated acres at a time. Nature designs with dynamic processes; we try to design by holding everything constant, including the dynamic ecological processes.

FIGURE 41   A large valley at Kiger Gorge, Steens Mountain, Oregon, carved out by a glacier during the last ice age.

FIGURE 42   A high-elevation forest in the Cascade Mountains, Oregon, following a forest fire. (Oregon Department of Fish and Wildlife photograph by Charles Burce.)

FIGURE 43   Nature often designs landscapes with volcanos, which spread lava over the earth. Note island around which the molten lava flowed.

A Lesson from Black Bear

A few years ago, I was asked to examine black bear damage to trees in the Mt. Hood National Forest. The wildlife biologist said the damage was severe, and it was indeed the most severe I had ever seen.

We spent most of the day in the field looking at 20- to 30-year-old plantations of Douglas-fir where black bear had been eating the cambium (inner bark) off the trees. These plantations were on densely forested steep slopes, on gentle slopes, and on flat ground with more widely spaced trees. The bear had climbed the trees on the steep slopes and had eaten the cambium on the open, downhill side about two thirds of the way up the trees. Bear had hooked their claws under the bark of trees on the gentle slope and pulled it off in great strips (Figure 45). They had then scraped off the cambium with their lower front teeth (Figure 46). On flat ground, the bear had to work hard to get around the low limbs and reach open areas of bark (Figure 47).

After four or five hours in the field, the biologist asked what we could do. "Cut trees differently" I said; "that will stop most of the bear damage. Let's go back to the office and I'll show you."

FIGURE 44   Nature designs landscapes with all manner of brushes and strokes, with ever-changing artistry.

FIGURE 45   Bark stripped off a young Douglas-fir by a black bear. The bear hooked the claws of its front foot under the bark and pulled it off. Arrows indicate some of the bear's claw marks on the inside of the bark.

FIGURE 46 Once the outer bark is removed, the bear scrapes the inner bark (cambium) off the stem with upward vertical strokes of its lower incisor teeth.

FIGURE 47 A black bear had difficulty maneuvering about the small limbs and an old wound from a bear that had injured the tree years earlier to eat the bark. This time the bear could only get at the bark on the scar tissue as it was healing over the old wound.

Once in the office, he got out all the maps and aerial photographs of the area. I then showed him that all the clearcuts were within one-quarter mile of one another and were all about the same age. The way the sale units had been laid out created much open bear habitat within an otherwise dense old-growth forest. The bear population had responded, as would be expected, to the boon of new habitat. As the plantations grew up, the expanded bear population suddenly found itself with a rapidly shrinking supply of food, and the bear were forced to supplement their diet with the cambium of plantation trees. The bears were "forced" because a bear population also existed outside the plantations, and those in the plantations could not move without having to compete with other resident bear.

With this understanding, a solution to the "bear problem" is simple: take the bears' habitat needs and habits into account when planning future timber sales and make appropriate adjustments. Bear habitat can either be created or minimized as desired.

Why Are Patterns Across the Landscape Important?

The transition from Nature's forests in the Pacific Northwest to our managed, economic plantations is marked by many changes and accompanied by grave ecological uncertainties. Disturbance was historically a common feature of the forests, many of which burned in massive fires in the Cascade Mountains of Oregon and Washington during the 14th century. A host of smaller fires have burned since then. Most of the forests in the Coast Range of Oregon date from a large fire that occurred during the mid-19th century.

In the interior forests east of the Cascade Mountains and extending into the northern Rocky Mountains, both fire and periodic outbreaks of insects appear to be part of the normal, historic, environmental pattern to which the organisms that comprise the forests were adapted. In toto, these are Nature's disturbances, and inasmuch as they represent the natural conditions that spawned the forests, they are a healthy part of a landscape's diversity, viewed over time.

Wildfire may be promoted by the trend toward homogeneous monocultures of young trees. The probability of a lightning strike in the Pacific Northwest turning into a wildfire depends on the amount, flammability, and both the vertical and horizontal distribution of fuels. Low branches and trees in the understory (termed "fire ladders" or "ladder fuels") provide a vertical avenue for the spread of a fire on the ground into the crowns of the overstory. Once in the crowns, the degree to which the crowns are "packed" (touching each other horizontally) comes into play, influencing the spread of a fire across the landscape. Like insect "pests," fire spreads most rapidly where it has an abundant, homogeneous source of "food."

Old-growth conifer forests are frequently considered to be vulnerable to catastrophic fire because of the large amounts of fuel on the ground (particularly since the days of fire suppression) and the sometimes-abundant ladder fuels. Although older forests in the northern Rocky Mountains (particularly those in transition from lodgepole pine to spruce and fir) are indeed the most susceptible to wildfire, the situation appears to be different in forests west of the Cascade Mountains of Oregon, Washington, and southwestern British Columbia. These forests appear to be most susceptible to fire during their first 75 to 100 years.

Here old-growth forests are less vulnerable to wildfire, because they frequently have a greater patchiness in the crown layer, which hampers the spread of flames. This is true except during regional "fire years" or during "super fire years," when everything is set to burn due to hot, dry, easterly winds. Although older stands do tend to develop fire ladders in the form of understory trees (particularly since the era of fire suppression began), such an understory, depending on its flammability, may or may not act as a fire ladder.

Recent fires in southwestern Oregon, for example, provided compelling evidence that stands with a component of hardwoods were less severely burned than those without. The reason is straightforward: the foliage of hardwood shrubs and trees is not very flammable and thus burns exceedingly poorly, except during regional "fire years" or "super fire years."

If the structure and distribution of young plantations west of the Cascade Mountains are increasing their vulnerability to catastrophic fires, then management practices are creating a fire-prone landscape, which is dominated by plantations with tightly packed crowns that are uniformly under 100 or so years of age. As long as dry summers prevail in the region, fires will periodically occur. Fires in such a culturally designed landscape can be very destructive and exceedingly difficult if not impossible to control. Managing for a mixed composition of species, including hardwoods, as well as for a horizontally heterogeneous distribution of the crowns may be the best way to protect plantations from catastrophic fires.

With the preceding in mind, it would seem that in making decisions about patterns across the landscape, it is important to consider the consequences of these decisions in terms of the generations of the future. Although the current trend toward homogenizing the landscape may make sense with respect to maximizing short-term profits, it bodes ill for the long-term, biological sustainability and adaptability of the land.

Here it must be noted that the economic and ecological systems are perceived to operate on different time scales, which means that the long-term, detrimental effects of decisions in favor of short-term profits are ignored. As a result, the long-term, nonmarket value of Nature's services is grossly underestimated.

When considering diversity and stability, it is important to remember that it is relationship of pattern, rather than of numbers, that confers stability on ecosystems. To understand this, according to soil scientist David A. Perry, it is helpful to think of an ecosystem as somewhat analogous to language. The raw material for languages is words, but its richness is imparted by syntax and context, the way words are arranged into sentences and sentences into phrases. In an ecosystem, stability flows from the patterns of relationship that have evolved among the various species. Creating a sustainable, culturally oriented system, even a very diverse one, must attend to these co-evolved relationships or it has about as much chance of succeeding as has a sentence made out of randomly selected words.[22]

THE LONGEST LIVING BEING ON EARTH

Although Nature designed forests in the Pacific Northwest to live 1200 years and sometimes even more, 400- to 500-year-old forests are more usual. Because people do not live as long as trees, it is difficult to conceive the complexity of the life cycle of a tree. If a Douglas-fir (Figure 48) lives for 800 years and then takes another 400 years to decompose and recycle into the living forest, one third of its useful existence occurs after it has died—as a standing, dead tree (snag)

FIGURE 48 Old-growth Douglas-fir. Note that the tree was partly blown over by the wind. (USDA Forest Service photograph by J.M. Trappe.)

FIGURE 49 Dead, standing old-growth Douglas-fir. (USDA Forest Service photograph by Tom Spies.)

FIGURE 50   Old-growth Douglas-fir and western hemlock forest with numerous fallen trees. (USDA Forest Service photograph.)

(Figure 49), a fallen tree (Figure 50), or both, which are merely altered states of the live tree.

An 800-year-old Douglas-fir could be used to further illustrate the longevity of forests, but instead a 500-year-old tree will be used. A 500-year-old Douglas-fir that blows over in 1994 and takes 250 years to decompose and recycle into the forest represents 10 lifetimes of a 75-year-old person and 25 working careers of 30 years each. Further, that 500-year-old tree would have germinated in 1494, and it would lie on the forest floor decomposing until 2244.

Another way of looking at this is to consider a person who is 50 years old in 1994; the person was born in 1944. Subtracting 50 years from 1944 takes us to 1894; adding 50 years to 1994 takes us to 2044. Thus, three lifetimes of a 50-year-old person (150 years) span from 1894 to 2044. Think of the monumental technological advances that occurred in that 150 years (three lifetimes of a 50-year-old person) compared to the little that has been learned about the ecological processes that maintain the forests. Consider further that subtracting 150 years from 750 years (the length of time it took the tree to complete its cycle) leaves 600 years unaccounted for—600 years of intense interactions of which we know next to nothing.

Our intent in redesigning the forest is to permanently and drastically shorten the life of the trees. In so doing, the entire dynamics of how the forest functions is being altered without the slightest idea of (or any data to show) how such actions will affect the sustainability of the forest. To *maximize* the harvest of wood fiber today, we gamble with the existence of the forests of tomorrow.

A UNIQUE ENTITY

Over 40 genera of woody dicotyledons (plants with two seed leaves rather than one) occurred from Oregon north, through Alaska and Siberia, to Japan during the early and middle Miocene (18 to 28 million years ago). A pure coniferous forest, mostly fir, spruce, and hemlock, existed only above 1625 feet in Japan and above 2275 feet in Oregon. These forests were therefore widely separated in northwestern North America and northeastern Asia during the early and middle Miocene. However, they began to occupy large areas by the late Miocene (12 to 18 million years ago), when fir, spruce, and hemlock spread through the middle elevations of the western United States and a continuous coniferous forest extended northward for the first time from Oregon, through British Columbia, into Alaska.

A rich northern forest of spruce, pine, and hemlock (with some larch, fir, beech, oak, and elm) became established in northeastern Siberia during the late Miocene or early Pliocene (10 to 12 million years ago); a similar trend occurred in Alaska during the same time interval. West of the Cascade Mountains in Oregon, however, an impoverished deciduous hardwood forest of hickory, elm, and sycamore remained during the early Pliocene. As these forests began to shift from deciduous to coniferous trees during the late Pliocene, more deciduous trees became extinct than during any period since.

By the early Pleistocene (about 1.5 million years ago), before the major glaciation, the forests of the Pacific Northwest contained the extant species, but it was not until 10,500 years ago that the Douglas-fir forests were established much as they are today.

The coniferous forests that clothed the landscapes from northern California to the panhandle of Alaska were unrivaled in both size and longevity of individual trees and in the accumulations of living matter (biomass) of individual forest stands. Many of the 25 species of coniferous trees represented the largest and often longest-lived of their genera (Table 1). Extinction of the deciduous hardwood forest of the Pacific Northwest probably was related to climate that favored the conifers. Major components of the original widespread deciduous hardwood forest still persist in Japan, China, Europe, and parts of the eastern United States.

Redesign of the coniferous forests of the Pacific Northwest is based largely

TABLE 1   Typical and Maximum Age and Diameter at Breast Height
Attained by 16 Species of Coniferous Trees on Good Growing Sites
in Forest of the Pacific Northwest

| | | | | Typical | |
| | | Maximum | | | |
Species	Age (years)	Diameter (inches)	Height (feet)	Age (years)	Diameter (inches)
Silver fir	>400	36–44	143–179	590	82
Noble fir	>400	40–60	146–228	>500	108
Port-Orford-cedar	>500	48–72	195	—	144
Alaska-yellow-cedar	>1000	40–60	98–130	3500	119
Western larch	>700	56	163	915	93
Incense-cedar	>500	36–48	146	>542	147
Engelmann spruce	>400	>40	146–163	>500	92
Sitka spruce	>500	72–92	228–244	>750	210
Sugar pine	>400	40–50	146–179	—	122
Western white pine	>400	44	195	615	79
Ponderosa pine	>600	30–50	98–163	726	107
Douglas-fir	>750	60–88	228–260	1200	174
Coast redwood	>1250	60–152	244–325	2200	200
Western redcedar	>1000	60–120	>195	>1200	252
Western hemlock	>400	36–48	163–211	>500	104
Mountain hemlock	>400	30–40	>114	>800	88

Source: Modified from Waring, R.H. and J.F. Franklin. 1979. *Science* 204:1380–1386.

on a single-species, short-rotation monoculture that emphasizes the production of
wood fiber, rather than the sustainability of a species-rich forested landscape.
How can we expect to improve in less than 100 years the socially valuable
landscape that Nature took 28 million years to design?

SELF-SUSTAINING, SELF-REPAIRING

That Nature designed a forest to be self-sustaining and self-repairing is
evident from the exceedingly rich forests in the Pacific Northwest. It is also
evident, however, that Nature designed a dynamic, timeless forest. Indeed,
forests have been burned, ravaged by insects, frozen by glaciers, and changed by
time, but they have always been allowed to heal and to replenish themselves.

In viewing forests as a commodity and by not allowing them to repair themselves, we miss the lesson which Nature tries to teach us:

> "When someone is seeking," said Siddhartha, "it happens quite easily that he only sees the thing that he is seeking; that he is unable to find anything, unable to absorb anything, because he is only thinking of the thing he is seeking, because he has a goal, because he is obsessed with his goal....in striving towards your goal, you do not see many things that are under your nose."[22]

So it is with our management of forests. We are so obsessed with our small goals that we do not see the forest and do not recognize Nature's warning about gross simplification seen in outbreaks of the Douglas-fir tussock moth, western spruce budworm, barkbeetles, laminate root rot, blackstain root rot, red ring rot, and other signs. As long as we view the forest as a commodity (Figure 51), we are blind to the fact that forests cannot be repaired with fertilizers, herbicides, or pesticides. Forests can only be healed with humility, love, understanding, and patience.

This last point is important and can be illustrated as follows. Let's say that

FIGURE 51   The way we treat the land is a result of how we think about the land. Clearcut at Olympic Peninsula, Washington.

you break your leg in a skiing accident. The doctor informs you that your leg will require three metal pins and two screws to hold the pieces of bone in place so they can heal. You allow the doctor to insert the hardware into the shattered bone of your leg and to put a cast on it. You ask how long you will have to wear the cast. "A minimum of eight weeks," the doctor replies. "It will take at least that long for the bone to knit and your leg to heal." "Can't you fix it any faster?" you ask. "If it takes that long, I'll miss the rest of the skiing season." "Look!" says the doctor, "I can only set your leg. I can't heal it for you. Nature and time must do that. I can tell you this, however; if you get impatient, you will have complications, and your leg won't heal correctly, in which case, I may have to re-break it, and you will have to wear that cast much longer. The choice is yours."

The word *heal* is the crux of the matter. A forest is a living entity. As such, we can break it and we can disrupt its processes, but we can neither fix it nor heal it by managing it. We can only impose a "treatment" and hope that it heals well through a succession of stages. Thus, while we can help to some extent, we must then step back, because all we can do is allow it time to heal.

We thus revisit the element of time and a tendency toward shortcuts, particularly in an effort to maximize profits or results. Namely, there never seems to be enough time to do something right initially, but there always seems to be time to do it over and sometimes over again—at a tremendous cost, both biologically and monetarily. Coupled with this is an oft-observed outcome: anything fixed out of impatience usually takes longer and costs more than anticipated and, when fixed, does not work very well.

Part II

AS WE THINK,
SO WE MANAGE

All of nature's systems are closed loops,
while economic activities are linear and assume inexhaustible
resources and "sinks" in which to throw away our refuse.

Kenneth Boulding

The unleashed power of the atom has changed everything,
save the way we think
and thus we drift toward unparalleled catastrophe.

Albert Einstein

"As we think, so we manage" bears the germ of our understanding and social consciousness about Nature. Nothing was known about forestry when North America was settled. Trees were considered impediments to agriculture and were removed in order to expose virgin soil for crops. The early view of the vast forests was largely negative in that they had to be conquered. They were then viewed as a free, inexhaustible resource. How is their value perceived today?

Vast areas of the world's forests are still being cleared and changed today, but for a different reason: profit. Forests are no longer cut to be conquered; now they are cut as a commodity to be used, or else they are considered an "economic waste." The result, however, is the same. Vast areas of the world are being deforested, and the case is made that it is being done for the "right" reason. In fact, the argument is the same, *short-term economic expediency.*

Short-term economic expediency, one of humanity's earliest thought and behavioral patterns, was seemingly harmless as long as resources were abundant and free for the taking by a relatively small human population. Under such circumstances, the resources that people exploited eventually were restored by Nature. When human use surpassed Nature's yield, the concept of short-term economic expediency resulted in increased biological simplification and loss of vigor within the world's ecosystem.

Civilizations have fallen into ruin through the millennia, suffering the consequences of this same decision. Short-term economic expediency stems from two roots: one private and one public. In the first case, a person may own a resource, but does not value it sufficiently to maintain its quality or quantity. The second case involves a public resource that everyone owns and yet no one owns, so everyone exploits it. The attitude is "If I don't get my share now, someone else will." Short-term economic expediency, if not the impetus, at least describes the accelerated technological developments in the unrelenting conversion of natural resources into money.

2

FORESTRY
PAST, PRESENT, AND FUTURE

When Europeans invaded the New World, they viewed the land and its products as unlimited commodities to be exploited for short-term gain. Rapid liquidation of inherited natural resources pays handsomely because the liquidator has nothing invested in producing the goods. Conservative use of such resources is not attractive to the liquidator because it reduces short-term profits. The Europeans viewed the land and its products only in terms of immediate conversion into money, i.e., its "conversion potential." Thus, in less than 400 years, European Americans have dominated the land, squandered its resources, slaughtered its indigenous people and commercially exploitable wild animals, and polluted its soil, water, and air.

"Profit first" remains the guiding rule today in justifying the worldwide liquidation of native old-growth forests, because in large measure the traditional foresters' view of a forest has influenced our national/social image of not only forestry but also of what a forest is. The result is seen in economically designed, crop-like plantations of young trees, theoretically to be harvested over and over into the distant future like fields of corn. Trees, however, are only one part of a forest system, the part to which the primary potential of conversion into money is assigned. The rest of the forest is largely destroyed, its soil impoverished, and its myriad organisms and processes dismissed as useless.

Linear, product-oriented thinking leads to the concept that an old-growth forest is an economic waste if it is not converted into money. Such notions stimulated professor Garrett Hardin to observe that "Economics, the handmaiden of business, is daily concerned with 'discounting the future,' a mathematical operation, that under high rates of interest, has the effect of making the future beyond a very few years essentially disappear from rational calculation."[23]

The potential conversion of resources into money is deemed so important because the economically effective horizon in most economic planning is only about five years away. Thus, in traditional linear economic thinking any merchantable tree that falls and reinvests its nutrient capital into the soil is considered an economic waste, because it has not been converted into money.

Forests are being decimated the world over because conversion potential dignifies with a name the erroneous notion that unharvested resources have no intrinsic value and must be converted into money in order for any value to be assigned. For example, Clyde Martin, of the Western Pine Association, wrote in the *Journal of Forestry* in 1940: "Without more complete and profitable utilization we cannot have intensive forest management....When thinnings can be sold at a profit and every limb and twig of the tree has value, forest management will come as a matter of course."[24]

Martin's notion still predominates: anything without monetary value has no value, and anything with immediate monetary value is wasted if left unharvested. Short-term economic profitability is always the bottom line.

This thinking persists not only in industrial circles but also in university colleges of forestry. Colleges of forestry, often the recipients of industrial monies, deprive their students of a sense of wonder about a forest. Such sense is replaced with the shortsighted economics of wood fiber production. The real issue, therefore, is the protection of decades-old false economic dogma vs. telling the ecological truth about forestry.

Simply put, traditional forestry is the exploitation of trees for their economic value. It is founded on growing and harvesting trees as rapidly as possible to maximize short-term profits and minimize capital outlay. Everything but immediate profit is ignored when forest economists assume that soil, water, air, sunlight, and climate are nondegradable constant values.

If the 20th-century exploitation of forests is to be replaced by a 21st-century healing of forests with a view toward biological sustainability, forestry as a profession must be founded on documented ecological truth. Universities must have daring leaders with a vision of imaginative possibilities with which to inspire future foresters.

SOME CAUSES OF TROUBLE IN THE FORESTRY PROFESSION

Ignorance might be excused in the absence of information, but to act in defiance of documented knowledge is inexcusable. The forestry profession is in trouble because of the resistance of many traditionally educated foresters to alter their thinking in terms of the world today. Five major causes of trouble in the profession of forestry are (1) the economic myth of forestry, (2) dogmatization of forestry, (3) limitations of science, (4) informed denial, and (5) university training.

The Economic Myth of Forestry

The practice of forestry began with the idea that forests are perpetual producers of commodities. A disciplined economic rationale was needed in order to capitalize on the harvest of such commodities as a way of life. The "soil-rent theory" became that rationale.

The soil-rent theory—a classic, liberal, economic theory—is a planning tool devised by economist Johann Christian Hundeshagen in the early 19th century to maximize industrial profits. Since its unfortunate adoption by foresters, it has become the overriding objective for forestry worldwide. However, the theory is based on six flawed primary assumptions: (1) the depth and fertility of the soil in which the forest grows is a constant, (2) the quality and quantity of the precipitation reaching the forest is a constant, (3) the quality of the air infusing the forest is a constant, (4) biological and genetic diversity are nonessential, (5) the amount and quality of solar energy available to the forest are constants, and (6) climatic stability is a constant.

Falsely assuming that all ecological variables can be converted into economic constants leads to the misconception that all that is needed to produce an economically sustainable tree farm is to calculate the species of tree, the rate of growth, and the age of harvest that will give the highest rate of economic return in the shortest time for the amount of economic capital invested in a given site—the soil-rent theory.

Dogmatization of Forestry

Once the soil-rent theory was set as the foundation of forestry, it had to be institutionalized in order for forestry to become a controllable discipline. The Germans and the British led the way by institutionalizing the practices, which sufficed to protect the discipline for many years.

Eventually, however, it became evident that cumulative ecological damage resulted from this flawed economic basis of forest management. Ecological damage threatened not only economic interests but also reputations of the idealized founders of the profession. Nevertheless, dogma continued to supplant new information because dogma explicitly forbids taking new facts seriously and recognizing old mistakes by making such information appear to be unsubstantiated and confusing. Defeating challenges to forestry's dogma tends to protect members of the profession from perceived ecological malfeasance and liability for embracing false assumptions.

Once the soil-rent theory was cloaked in dogma, vested interests had effectively institutionalized denial of their economic motives while simultaneously alleging scientific legitimacy. However, such legitimacy can be claimed only if it is protected from questioning and from testable scrutiny. As a result, students are unknowingly trained in the dogma of the soil-rent theory, which prevents recognition of the original economic errors in the foundation of today's so-called scientific or "new" forestry.

What is taught therefore perpetuates the original economic falsity. So long as students uncritically accept what they are taught, the original economic falsity will remain the foundation of today's forestry practice.

To help ensure the persistence of dogma, one is encouraged, through a system of rewards and punishments, to accept proffered dogma unquestioningly. A working forester who does exhibit the courage to challenge dogma is initially met with anger and then by a wall of denial and resistance.

Criticism, rejection, and shunning are powerful motivators for keeping a person's questions within prescribed boundaries. Dogma lives on fear of losing approval and acceptance, both of which can be expressed through losing affiliation with colleagues, or livelihood, or both. The power of fear keeps people too frightened to openly and freely voice their intuitive beliefs and even prevents people from feeling and knowing their own inner truths.

Increasingly, people in the profession speak of "progressive forestry," "scientific forestry," and "new forestry." What do they mean? Forestry based on faulty premises decades old is no more progressive, scientific, or new today than in the past. How can this be with all the ongoing scientific research and with all of the research information at hand?

Limitations of Science

The true goal of scientific endeavor is the pursuit of pure knowledge, which, by definition, demands that the scientist be unencumbered and forever open-minded in thinking. The greatest triumphs in science are not triumphs of facts, but rather of new ways of seeing, thinking, perceiving, and asking questions. At least five roadblocks stand in the way of the search for scientific truth: (1) the lure of grants aimed at predestined results, such as monies from the timber industry aimed at genetic engineering of tree seedlings for faster growth; (2) attachment to a single hypothesis, such as the necessity of fertilization to improve soil productivity or of herbicides to control unwanted vegetation; (3) the fact that science can only disprove information, which means the timber industry cannot prove its dogma through science, but science can be used to disprove industrial dogma; (4) the fact that science began as an endeavor to seek knowledge for curiosity's sake, but too often has become a problem- and method-oriented means of determining politically correct answers to safeguard established dogma; and (5) objectifying human participation with Nature, which denies the subjectivity with which a timber company executive and an ecologist might view the same information.

The Lure of Grants

Most research institutions depend on grant money to help defray the costs of operation, and researchers are strongly encouraged to seek grants. Such encouragement sometimes amounts to administrative coercion in that promotions and even tenure are linked to the amount of grant money secured by an individual. Researchers therefore tend to focus on those areas within their disciplines for which they have the best chance of getting money. The questions to be asked and the answers to be derived may thus be controlled by the allocation of funds. Dogma can also thus be protected, because a researcher may be prevented from asking an unrestrained question.

Attachment to a Single Hypothesis

To keep the search for truth on a credible track, it must be noted that a person not only tends to form a single hypothesis but also tends to become so attached to it that any criticism or challenge raises defense mechanisms. The moment a person derives what seems to be an original and satisfactory explanation for a phenomenon, the attachment to his or her intellectual "child" springs into existence. The more this explanation grows into a definite theory, the more near and dear it becomes. It is then massaged (as it is often called in government agencies) to make the theory fit the information and the information fit the theory, which is just another way of making sure it all fits within accepted dogma, i.e., the current level of harvesting timber can be increased by ten percent because someday it will be possible to grow ten percent more trees that have been genetically improved.

Science Can Only Disprove Information

The truth about scientific research is that nothing can be proven—only disproven; nothing can be known—only unknown. Thus, we can never "know" anything in terms of knowledge but only in terms of intuition—the knowing *beyond* knowledge, which is not admissible as scientific evidence. Scientific truth, whatever it is, can only be intuited and approached; it can never be caught and pinned down.

Although science can prove nothing, it can *disprove* dogma, by posing unrestrained questions dangerous to economically vested interests. Hence, those who resist change do so by demanding "conclusive proof" that change is necessary or even desirable.

Method-Oriented vs. Problem-Oriented Questions

People tend to be "method-oriented" rather than "problem-oriented" in their thinking and in many of the questions they ask. A method-oriented question might be: How can trees be planted to make them grow fastest? A problem-oriented question might be: How much organic material in the form of large woody debris must be left on site following logging to ensure continued productivity of the soil?

Method-oriented questions are invariably aimed, unconsciously perhaps, at strengthening dogma. Nevertheless, we believe that through method-oriented experiments we can learn the truth about Nature, when in fact we are learning only about our methods, which are our experimental designs, assumptions, and expectations about the outcome.

Objectifying Nature

Nature cannot be accurately represented through science. What is observed and interpreted in Nature is a product of the personal lens through which a scientist peers. Although one can attempt to detach himself or herself from Nature in the name of science and try to become "objective," this is not actually possible if for no other reason than he or she is part of Nature and must interact with Nature in order to study it. In addition, the very act of observing something changes its relationship to everything else and hence its behavior, because in the act of observing a *new* relationship is established.

The insistence in science to neutralize human subjectivity is an attempt to objectify Nature and thereby to deny participation with Nature in any form, which is impossible. Nevertheless, the attempt to objectify Nature makes scientific theories and facts no more than social constructs. Instead of scientific consensus being achieved when the facts reach the state of speaking for themselves, scientists come to consensus only when the political, professional, and economic costs of refuting them make further negotiation untenable.

Informed Denial

Those opinions most strongly defended usually hide the root of a lie. For example, overcutting old-growth forests is staunchly defended by the timber industry through such statements as: "We can increase the timber output today, because we are planting genetically superior trees that will grow faster tomorrow." Defending such opinions protects a person from painfully awakening to the truth. Only truth, even the most uncomfortable, endows a thought and its action with the power to change society, which denial can never do.

It is interesting, therefore, that so much is ostensibly devoted to gathering scientific information with which to inform the public, while a similar effort is devoted to preventing the spread of this information by diverting public attention to other subjects. Such displacement of focus prevents the public from benefiting from whatever truthful information they may have been given.

Diversion of attention, even by some scientists, is based on fear of the truth. It is easier to scrutinize available information, find it to be inconclusive, and then, based on this "emotionally safe" determination, deny that there is a problem of sufficient magnitude to act.

Informed denial serves the purpose of not having to acknowledge the perceived truth of a situation, of rendering it invisible, or at least of describing it as purely "subjective." In this way, voices are silenced, and the most "objective" truth based on the best and/or most recent knowledge remains hidden.

University Training

A curriculum can be purposefully designed and used to liberate a student's imagination to question ideas and soar with new possibilities, or it can be designed and used to imprison that imagination in the rut of worn-out dogma or "business as usual." This is possible because power is free of value. In the hands of any special interest, it can be used to liberate or to confine. Unfortunately, the selection of courses offered in most forestry curricula is academic ritualism aimed at confinement and compliance with the decades-old flawed economic dogma on which today's forestry is based.

By exposing a student to selected information through the courses both offered and required, a mind is molded in the shape of the prevailing dogma. At the same time, the opportunity to examine information opposing desired dogma is not required or may not be readily available.

As a case in point, I was a guest lecturer in a forest management class in which I discussed large woody debris, small mammals, mycorrhizal fungi, nutrient cycling, and the effects of gross habitat alterations in coniferous forests. When the class was over, an angry young student approached me. "I'm a senior," he almost shouted, "and I'm going to graduate in a couple of weeks. How come this is the first time I've heard any of this? I've just spent four years in what they call forest management! You just showed me that I don't know a damn thing about how a forest works! And now I'm supposed to be a forester! What in the hell am I going to do out there?" This student was astute enough when given opposing information to let his intuition speak and thereby penetrate dogma's armor and see the economic lie of forest management. In terms of his university training, however, the truth came late.

Another way dogma is perpetuated through a curriculum is to offer a course, such as Forest Entomology, in which only a few species of insects are selected and treated as economic pests that threaten the survival of the forest. By thus teaching one facet of a broad subject, dogma is again protected and spread.

Now is the time, however, to ask where current teaching and management practices are headed, what kind of forest landscape is being created, and how susceptible this future landscape is to outbreaks of forest insects. Once the management landscape pattern is established, it will be too late, something that is seldom considered in teaching forest entomology. Consider, for example, that "foreign" disturbances (those introduced by humans, such as converting diverse forests into monocultural tree farms, which in one way or another weaken rather than strengthen the adaptive mechanisms that confer stability on ecosystems) are potentially destabilizing.

Before discussing "pest" insects in the context of landscape patterns in forested areas, it must be clearly understood that, since biblical times, most insects that feed in one way or another on plants have been considered to have only negative effects on the resources that are valued for use by humans. The term *pest* reflects this traditional bias and perceived battle for control of the resources. Only within the past decade or so has evidence emerged to show that many of the so-called pests provide largely unrecognized benefits to the forest, even during apparently destructive epidemics.

Patterns on the landscape influence populations of insects in three general ways: (1) through the degree of diversity in the distribution of their sources of food, (2) through the quality and quantity of habitat provided for their predators, and (3) through influencing how insects move across a landscape. Not surprisingly, insects multiply and disperse much more effectively when suitable food plants are uniformly distributed. "Suitable" in this sense may mean a given species of plant or a group of species of plants of a certain age or size.

The implications of "homogenizing" forested landscapes as related to insect activity are interesting and instructive, but seldom discussed in the classroom. Taking a landscape of diverse, native forest and homogenizing it through clearcutting and planting single-species monocultural plantations has the effect of eliminating predators and such physical barriers to insect dispersal as fire-maintained habitat diversity. Loss of such habitat diversity increases both the survival of forest-damaging insects and the likelihood of regionwide outbreaks.

Old-growth forests in the Cascade Mountains support far more predatory invertebrates (mainly spiders) than do plantation "forests." This raises the question of how much help in controlling insects foresters can expect from invertebrate predators, which originate in old-growth stands. Although the consequences of reduced numbers and kinds of predators are impossible to predict with any certainty, it seems likely that severe epidemics of plantation-damaging insects will become increasingly frequent.

How might this work? First, the success of plantation-damaging insects increases when the landscape is intersected with roads and planted in young, single-species monocultures of trees, thus decreasing the average size of the trees and the diversity of age classes. This makes it easier for the insects to find suitable host trees by removing the confusion factors inherent in a diverse habitat and thereby reducing the time it takes the insects to locate food. Second, simplification of the forest also reduces the diversity of habitats and the variety of species of prey that are necessary to maintain the populations of generalist, opportunistic predators, such as spiders and birds. These predators are more important in preventing outbreaks of insects than are host-specific predators, such as parasitic wasps, which are dependent on finding a particular species of prey to exploit.

Patterns of the forested landscape also influence the habitats provided for birds and mammals that prey on defoliating insects, a subject that must be integrated into forest entomology. For example, bird communities in coniferous forests of the west are dominated by species that feed on insects in the foliage. Roughly 80 percent of the food consumed by northwestern birds is animal prey, mostly foliage-feeding insects. It would cost about $1800 per 1.5 square miles per year in insecticides to kill the same number of spruce budworms that are eaten by birds in the forests of north-central Washington, and this does not even count the predaceous ants that complement the birds. These insect-eating birds and mammals, such as bats (Figure 52), depend on forests in certain age classes for

FIGURE 52 Long-legged bat.

nesting and roosting, and their numbers are declining where the landscape no longer contains the required habitats.

Thus, old-growth forests in the Pacific Northwest, with their complex array of species of both trees and predators, large size of stands, and high diversity of age classes are less vulnerable to epidemics of forest-damaging insects than are the simplified plantations that are being created over vast areas and through which the landscape pattern is being altered.

Circumstances may be somewhat different, however, in forests outside of the Pacific Northwest. Some old-growth forests in the Rocky Mountains, for example, are epicenters in which insects build in numbers, but these same forests survive the damage of the ensuing epidemics.

It is thus necessary for professors to broaden their horizons and integrate their disciplines so that students are exposed to such systemic questions as: Can plantations, forests, and landscapes be designed in such a way that problems of forest insects are minimized? The answer to the question is yes, but it will be neither simple nor easy, because different insects respond differently to a given landscape pattern. In turn, a pattern that reduces problems with one insect may well create problems with another. Further, air pollution and change in the global climate will stress some forests and further stress some plantations of the Pacific Northwest, and stress often translates into increased problems with forest-damaging insects.

In addition, global influences extend in unforeseen ways beyond those mediated by the atmosphere. For example, the most serious threat to insectivorous birds in the Pacific Northwest may not be the loss of their habitat once they reach the Northwest, but rather the loss of their habitat in Central and South America, where they spend the winter. Roughly one half of the species of insectivorous songbirds in the Northwest are migratory and spend the winters in tropical forests. The large-scale destruction of the tropical forests means lower numbers of insectivorous birds in the temperate forests where they return for the summer.

Thus, insects, even those causing damage to trees, are not only natural components of the forest but also are necessary to the long-term health of the forest, because they are part of a whole complex of organisms that forms the forest's "immune system," which includes diseases, parasites, and a variety of predators, such as spiders, bugs, beetles, flies, wasps, ants, birds, and bats, to name a few. Nevertheless, damage-causing insects are still viewed solely as pests in the narrowly focused values of industrial forestry, which are still taught in most North American universities.

In fact, the great majority of courses required for a degree in forestry at today's universities are aimed at producing wood fiber with the economic underpinnings of rapid tree growth, genetic "improvement," economically efficient harvest, cheap transportation, and maximum utilization of wood. The prime objective is the greatest short-term profit for the least outlay of economic capital. The forest as an ever-evolving living organism is discounted as having no intrinsic economic or ecological value over the long run.

Focus on the economics of trees is so narrow that we lose sight of the forest. Colleges of forestry are training "logging technicians," "industrial tree production managers," and "plantation managers," but not foresters in the sense of nurturing a forest as a whole living organism and educating the general public about the wonder of it all.

Forestry is no longer just a matter of cutting trees, or even of cutting and planting trees. It has progressed far beyond solely the economic—or even the ecological—aspects of forests. It has become a profession of people working with other people who want many different values from forests, especially national forests, which are owned by the public. Yet communication skills, such as writing and public speaking, the most important tools of the trade, are all but left to chance in many forestry programs. Little or no emphasis is placed on learning the skills of interpersonal relationships, how to conduct public meetings, or about professional ethics.

In short, the colleges of forestry are years out of date because those courses needed by current and future foresters are seldom required or even available in the core forestry curricula. If colleges of forestry refuse to change, acquiring these courses may necessitate a fifth year of training outside of the college of forestry.

FORESTERS OF YESTERDAY, TODAY, AND TOMORROW

Yesterday's foresters were trained as utilitarian servants of industry. Their job was to protect the commercial value of the timber from fire and insects and to get out the maximum cut. Today's foresters, while receiving essentially the same training, are beginning to question the philosophical premises of the old school. What about the foresters of the 21st century?

The 21st-century forester must be a bridge between the ecology of the forest in relation to the dynamics of the landscape and the cultural necessities and human values of society. Tomorrow's forester is the guardian of options for both the forest and the future, because when all is said and done, the only gift we can give future generations is the right of choice and some things of value from which to choose.

In 1968, for example, forestry professor Richard Plochmann described the painful dilemma faced by German foresters in trying to respond to changing cultural conditions: "Our [German] forestry will be carried on even under bad economic situations. We could better the return if we would be willing to give up the high intensity now maintained or if we gave up the principle of sustained yield. We cannot do both and we do not want to do either. The first seems imperative for the multiple uses of our forest and the second for the benefit of following generations...."[25]

By 1989, however, Plochmann could report a truly remarkable response by the Central European forestry profession to the pressures of the European people:

> Many writers have described the poor state the [Central European] forests were in, often in quite drastic terms. One writer claimed that on 10,000 acres of a certain forest district, no tree could be found strong enough to hang a forester on....It was in the 1960's that forestry came under public criticism. This boiled down to the general reproach that forestry solely oriented towards the maximization of profit can no longer meet the expectations and needs of society....Forestry by itself came to the conclusion that a new concept of management should be found to meet future demands and developments....The realization of our changed view will take a long time. After a century of forest rehabilitation, we have now another one of conversion ahead of us. We are convinced that the new concept is a fair compromise of ecological, social, and economic goals and therefore can be jointly carried out by a large majority of our public. We started it on its way. We hope too it will be a success.[26]

To accomplish this kind of philosophical revolution in North America, university colleges of forestry must be imaginative and daring. They must lead the profession of forestry toward a new vision, one that harmonizes human culture and Nature's forest beyond the dogma of traditional forestry and the economics of commodity extraction. This can best be done by helping students realize their dreams of finding a nurturing philosophy of life within the profession as trustees of forest health and productivity.

What will the 21st-century forester need in his or her training to fulfill the philosophical goal? To be equipped for the 21st century, the forester of the future will need to:

1. Understand the parts of a forest in relation to its organic whole and the whole in terms of its parts by learning to sit with and in the forest and hear its voice.

2. Understand the science of each component and subsystem of the forest, i.e., to be able to evaluate and openly question new information in terms of the old, old information in terms of the new, and both in terms of existing ignorance.

3. Understand enough about the ecological linkages among forest components and subsystems to be able to anticipate effects of management stresses.

4. Understand and work with a forest as a dynamic component of an ever-evolving, culturally oriented landscape while simultaneously honoring the integrity of the blueprint of Nature's processes and patterns.

5. Understand that the only sustainability for which we can manage is that which ensures the ability of an ecosystem to adapt to evolutionary change, which means managing for choice (maximum biodiversity).

6. Understand how to manage for a desired condition of the landscape and abandon the unworkable notion of sustained ever-increasing yield of natural resources.

7. Understand how to manage for the connectivity of habitats to help ensure the ecological wholeness and the biological richness (biodiversity) of the patterns they create across the landscape.

8. Be able to abstract, simplify, synthesize, and generalize information about complex systems so that his or her "intuitive mind" can act as the final reality check of relevant information prior to making decisions.

9. Be able to articulate ideas effectively, clearly, and accurately both in writing and in public speaking.

10. Be able to work openly and skillfully with people with sufficient knowledge and in sufficient depth to validate their concerns and to give them the critical understanding and trust of the professional rationale behind a decision, even if they are opposed to it.

If university colleges of forestry in North America can find the courage within the decade of the 1990s to do their jobs with the professional truth, openness, and excellence foresters both deserve and need, the next 100 years can indeed become the first century of healing forests both at home and abroad. If the 21st-century forester can stand in the forest of which he or she is the trustee and can truly say in his or her heart, "I feel good about what I'm doing here today for the forest and for the people of tomorrow," then forestry as a profession is on a moral path into the future.

The rest of this book is dedicated to the proposition that the time for articulating a bold new vision is at hand. Some of the concepts that are necessary components of any vision for sustainable forestry in the next century will be discussed.

To begin, we must look at how an agency, such as the U.S. Forest Service, comes into being and functions. Keep in mind that the cycle of a professional society, such as the Society of American Foresters, is essentially the same as that of an agency.

3

THE CYCLE
OF AN AGENCY

As you read that what we are doing to our environment in the name of management is really a peek into our human inner workings, keep in mind that the cause behind our environmental crises is brought about largely by our resistance to change. Whether we like it or not, we now have the technological capability to disarrange and disarticulate the entire global ecosystem. Thus, for human society to survive as we know it, we must face ourselves, as uncomfortable as it may be, and help one another to confront our human failings and blind spots. We must be willing to risk changing our thinking and our behavior and get back in touch with our repressed feelings, our exiled intuition, and our lost spirituality.

Although one individual cannot change what history has already written, that person can change himself or herself and thereby influence what may be written in history when the present becomes the past. In all likelihood, this change does not occur over night, but rather, like the man who eventually moves the mountain by carrying away small stones, it begins with one step.

Change is an immutable law of the Universe and is an ongoing process that is always in the present tense. It is an inner journey, without end and without distance, a journey in which our thoughts and actions create our reality. The choice of reality is ours because we have free will, which allows us to choose how and what we think and do.

Each individual has the power of choice, the power to effect change in any arena of life, including an agency. We take ourselves with us wherever we go, and we are the common denominators—the threads—running throughout the entire tapestry of our social structure. The irony is that as a species we must actually study ourselves so that we can survive with ourselves.

If the individual is an extension of the family, and through the individual the agency also is an extension of the family, and society is an extension of the agency, then an individual who is willing to change can change society. If this is correct, then the dynamics in a family is similar to the dynamics in an agency. Thus, an understanding of one may help to understand the other, because each is composed of individuals.

By our thoughts we privately define and by our actions we publicly declare who and what we are. Keep in mind, therefore, that a word spoken is but the manifestation of a thought, positive or negative, and once spoken it can never be withdrawn.

This is written with the hard-won realization that, of all species on earth, humans are both blessed and cursed with the greatest of powers: the power to consciously change ourselves, to struggle toward an ideal of being, and to frequently fall short of that ideal. In struggling, however, we must understand and remember that anything worth doing well is worth doing poorly for a while.

Humans have been granted the power to create. We create ourselves with our thoughts, and with our thoughts we create the society and the environment in which we live, survive, or become extinct.

We have control over what we choose to think and do. The outcome is our choice and therefore our responsibility. This being so, it is within our creative power to change ourselves, one by one, from collectors of the society's psychological garbage to trustees of one another's dignity, and it is within our power to transform the world from a toxic waste dump into a heavenly garden.

Today, we are at a spiritual/ecological threshold of social survival, a threshold at which humanity has never before stood, and we must now come to grips with ourselves. These pages are therefore written within the context that each individual has the power to change and in so doing to move the agency and human society in a new and vital direction.

THE INCEPTION AND FUNCTION OF AN AGENCY

An agency is simply a business or service authorized to act for others. Public land-management agencies will be the main focus here, not only because their function is one of primary service to the public but also because my experience is with such agencies.

When speaking of an agency, the possessive is often used, such as "the Forest Service's point of view," as though it were an individual, which in a way it is. An agency is a collection of people, much like an extended family. Therefore, its benefits and deficits are the collective contributions of both its past and present members.

An agency during its inception not only gives form and substance to a mutually held ideal, which is projected into the future, but also gives the individual a sense of participation, achievement, and pride in the furtherance of that ideal, something no one can accomplish alone. Another opportunity an agency offers to its employees and volunteers is the potential for a "team effort," which makes the whole greater than the sum of its parts.

Although many people tend to be cynical about our public agencies, I believe each began with noble ideals of service, worthy of their time in history and worthy of the effort people invested in them. Somewhere along the way, however, we lost sight of ourselves and our ideals.

WE ARE THE AGENCY

A brief overview of my experience and observations in government agencies will be given here in order to better discuss the notion that as we change ourselves, we can change an agency. It is not the intent, however, to delve into a technical discussion of the psychology of an agency.

Before we can see how to change an agency, we must understand that there is nothing sacred about it. An agency is merely a collection of people—as good as its best, as bad as its worst, and as mediocre as the average. Further, individual motives and conduct give the agency whatever meaning, purpose, integrity, foresight, or credibility ascribed to it. In essence, an agency is an extension of one's family of origin in that individual reactions to coworkers and authority figures are often symbolic not only of one's stage of personal development but also of the dynamics of one's relationships with significant family members, especially in relation to unresolved issues.

Those who work in an agency—who set its standard, who determine its budget, who oversee it, and who vote—*are* the agency. Because we are all the agency, we also share its destiny, and because it is in our power to be authentic individuals, it is also in our power to heal the current dysfunction in our agencies.

STAGES IN THE CYCLE OF AN AGENCY

I am not aware of any reference that describes the developmental stages of an agency, as has been done for an individual and a family, and I have not taken part in the formation of an agency. I can, however, think of four generalized stages of development, which I have personally experienced while working on large projects within an agency. Although the projects were undoubtedly simpler than the creation of an agency itself, some were incredibly exciting, and it is from participating in one of these projects that I can imagine what it must be like to take part in the creation of an agency like the U.S. Forest Service. To this end, I will briefly share my experience of helping to shepherd a large project from its inception through its completion, because I perceive a similarity between the life of a large project and that of a public agency.

Stage one.   The inception of an agency is based on a perceived need that is in the public interest. This perception revolves around one person or a small nucleus of people with a vision, as clearly stated by Gifford Pinchot in his book *Breaking New Ground*. He saw the "conservation policy," which he helped to forge, as the guiding principle of the U.S. Forest Service, of which he was the first chief from 1898 through 1910:

> The Conservation policy...has three great purposes.
> First: wisely to use, protect, preserve, and renew the natural resources of the earth.
> Second: to control the use of the natural resources and their products in the common interest, and to secure their distribution to the people at fair and reasonable charges for goods and services.
> Third: to see to it that the rights of the people to govern themselves shall not be controlled by great monopolies through their power over natural resources.[27]

A letter written by Pinchot on March 5, 1905 concisely stated his mission, the guiding philosophy on which the early U.S. Forest Service was founded: "the greatest good for the greatest number in the long run." Yet even with the clearly perceived need for a guardian of the public interest, the formation of the Forest Service was no easy task. There were many bitter, political battles to be fought with men who wanted all the land put in private ownership for their personal gain.

Although the ideals, which are seminal in the inception of an agency, may be clearly defined for their time and place in history, such as those of the U.S. Forest Service, we now look back and wonder exactly what was meant. In this sense, the following question has been raised several times in recent years in the struggle to meet today's perceived "needs" from our national forests: What exactly did Pinchot mean by "the greatest good for the greatest number in the long run?"

While one may not agree with his perceived motives as attested by the political actions taken, it must be noted that whatever the ideal with which the Forest Service was founded and whatever actions were taken to implement that ideal, it was new and daring in its time, and it was meant to be a service held in trust for all the people both present and future.

It is easy to consider the vision of the pioneers of the past as simplistic today because of a greater biological knowledge about forests, a different perception of desires and necessities from forests, and a different perception of society. It is important to remember, however, that those pioneers did the best they could with the knowledge available and the vision they had. It is also important to keep in mind that today we are the pioneers of the future. Are we doing the best we can with what we know? Will the agencies in which we now serve or which we now create fare any better than those of the past?

Stage two.   After its inception, an agency goes through a period of false starts and misfires in searching for an identity. Out of this fumbling can come growth and joining together to fulfill the vision, provided the vision is clearly stated and firmly agreed to in the first place:

> Pinchot was a great and electric leader by any standard. Stewart Udall called the Forest Service's Washington headquarters during Pinchot's regime "the most exciting place in town." And it was, as Pinchot and Teddy Roosevelt successfully conspired against private western ranching, timber, and water interests to set aside 148 million acres, three-fourths of today's system, as national forests.[28]

One can only imagine what it must have been like in Pinchot's day. I was privileged to experience this type of electric excitement first-hand during the years that I worked in La Grande, Oregon, at the Forest Service research laboratory. We were putting together a book entitled *Wildlife Habitats in Managed Forests: The Blue Mountains of Oregon and Washington* (referred to simply as the "Blue Mountain Book") and were frequently "in over our heads." Part of the time we spun our mental wheels as we struggled to say what we thought we meant and at the same time be sure that we really meant what we thought we were saying.

This is the excitement of growth not only in a project with clearly articulated vision, goals, and objectives but also in an agency where the people are clearly committed to and empowered to follow the vision of public service for which the agency is being created. "Ideals," said journalist Carl Schurz, "are like the stars, we never reach them, but like the mariners of the sea, we chart our course by them."

Stage three.   Growth, with the stimulation of its unknowns, its gropings, its many false starts and surprising insights, eventually gives way to maturity in which the outcome seems assured. Again, I gained some sense of what this stage must be like while working on the "Blue Mountain Book."

During the first year and a half of struggling to put the ideas coherently on paper, we had many false starts. Then the pieces started coming together, slowly at first and then faster and faster. The wonderful thing was that everyone from every necessary discipline seemed to materialize "out of thin air" just when we needed them most. There was an ideal mix of people, which kept the interest charged and the enthusiasm crackling.

By the end of the second year, we saw where the project was going, and we knew what the product would be. At that point, the project came into its maturity, and everyone seemed to know it. From then on, it was almost impossible to keep up with the requests to put on workshops explaining the concepts embodied in the "Blue Mountain Book."

Agencies reach maturity in a similar manner. They have a function to perform, and after many false starts, they begin to accomplish what they are designed to do. They mature into their stated mission, and once this has been accomplished, the turning point has been reached.

Stage four.   Thus, the beginning of stage four is pivotal to the life of an agency, for here is where its direction is ultimately determined. As the "Blue Mountain Book" project reached maturity, we had to recognize and accept that it was time to begin thinking about the next project. By the time the "Blue Mountain Book" was finally published, however, it was like an old friend, and going on to something else was extremely difficult because it felt like the death of a relationship. Even my working relationships with my colleagues of five wonderful years were forever altered. It was time to move on, to risk major change, to grow.

An agency reaches the same point. Having fulfilled its original charge, it must be reenvisioned, reoriented, rechartered, and revitalized *or* disbanded. If left solely to its own devises, senescence creeps in. This is the point at which an agency becomes dysfunctional.

The now-declining agency tries to "hang on," to live as in the past. In reality, however, it becomes a self-perpetuating machine, which, having "forgotten" its original charge and having outlived its function, becomes dysfunctional, looking out for its own survival at any cost.

DYSFUNCTION

Each individual represents the behavioral strengths and weaknesses of her or his upbringing. Behavioral patterns, whether functional or dysfunctional, tend to be repeated over and over again unless the cycle is consciously broken. Dysfunction means that a given system is impaired in its ability to function harmoniously.

There are three basic instances in which an agency becomes dysfunctional. The first is when a person (or persons) takes over control and listens only to vested interests at the expense of the agency's mission, its people, and the public at large. The second is when people in an agency created for one purpose try to make the agency into something it was not designed to be, but are either unwilling or unable to change its perceived mission and infrastructure to accommodate its new identity. The third is when an agency has outlived its original purpose or has simply lost sight of it and becomes a self-serving, institutionalized machine to perpetuate its own survival.

In any case, the agency no long serves the constituency it was designed and empowered to serve, but instead compels the constituency to serve the survivability of the agency. Here it must be remembered that individuals are the agency. Therefore, the more dysfunctional the leadership, the more dysfunctional the agency; the more dysfunctional the rank and file members, the more dysfunctional the agency. Although the dysfunction of an agency is a reflection of the dysfunction of its individual members in the collective, especially the leadership, there is more to it than this statement suggests.

Complex organizations, such as the Forest Service, the Bureau of Land Management, or the Environmental Protection Agency, are typically composed of people who fall into three major categories, which are prone to becoming subcultures:[29]

1. **Politicians** are those people who expect their futures to lie in appointed or even elected office at large and therefore concentrate on maintaining and expanding their career opportunities outside of the agency. Today they are often appointed as heads of agencies about which they know little or nothing.

2. **Careerists** are those people whose rewards come largely from inside of the agency of employment. They therefore define their primary goals and loyalties within and up the agency's hierarchical career ladder in a way that maintains the stability of the agency and their own position within it.

3. **Professionals** are those people who relate to norms set by others in the same profession, say forestry and wildlife management, that generally cut across organizational boundaries. A professional's rewards come independently from peers within the professional society and not from within the agency of employment. The professional role is seldom the most powerful in an agency.

Although an individual often plays two or more roles, one is dominant. It is not surprising, therefore, that people in these three categories within a single agency often come into conflict when they are brought together to deal with a complex, integrated issue, because each belongs to a different subculture with different supporting structures and different systems of ideals, rewards, and incentives. This is especially true if an agency tries to shift its identity in a way that also shifts the perceived basis of power from one subculture to another.

Adding to the confusion is the fact that while the formal goal of an agency may be sound forestry, its informal operating goals may differ substantially among the three subcultures and may be based on differing senses of idealism, integrity, and objectivity. Thus, each subculture not only perceives a given situation differently but also defines the problem and envisions the solution differently.

Most professionals want to keep two professional norms inviolate: *personal integrity,* which is a sense of authenticity based on sound, sincere, personal, and professional principles, and *professional objectivity,* which means dealing with ecological issues with as little personal and political bias or prejudice as humanly possible. The way we organize ourselves into agencies can either encourage and nurture professionalism or stifle it.

It takes more than cooperative behavior and a sense of good will to protect integrity and objectivity. Such cooperation and good will must be nurtured by the appropriate organizational design, one which is "safe." In addition, the organizational design must support and nurture integrity and its creative source, as well as differing ideologies, through positive rewards and incentives.

Finally, the U.S. Forest Service, now over a century old, is an example of a public agency that has lost sight of its original charge and has become dysfunctional with age, which points out that our present form of exploitive capitalism is not working for the good of the American people, especially the children, who will need healthy forests in their future. When a dysfunctional agency serves the timber companies, who is it then that serves the forests and the people of the United States, who by and large have no vested profit motive in timber from public lands?

The same question can be asked about the Bureau of Land Management, which serves the special interests of a small portion of the livestock industry and the mining and oil and gas industries at the expense of public lands. Again, the American people by and large have no vested profit motive in grazing livestock, mining, or drilling for oil and gas on public lands.

Dysfunction tends to creep into an agency when the production of a product, coveted by a politically strong special interest group, becomes more important than the process of human interactions that fulfill the agency's original charge. The product is more important than people and human dignity.

What happened? Rather than recognizing that the people of the nation are both the bosses and the customers of the agency and treating them accordingly, the agency bows to corporate interests through political pressures and serves the corporate/political bosses. Again, our political system serves the private power base, not the people. Thus, the main enemy to a dysfunctional agency is the very *public* it is meant to serve.

When such dysfunction is pointed out in news stories and court cases, the "views and policies" of the agency become the backbone of the agency's homeostatic defense, a defense that is usually upheld by the courts if legal procedure has been followed, regardless of the consequences.

HOMEOSTASIS

A family is a system governed by a set of rules, which determine and control the interaction of its members in organized, established patterns. The rules are a prescription of what shall and shall not occur within and outside of the family. Homeostatic mechanisms maintain the ongoing arrangement among members of the family by activating the rules defining each member's relationship to the whole. Although homeostasis is the maintenance of a dynamic equilibrium within a system, such as an agency, it is also a mechanism through which an agency's name is kept pure.

Once an agency has become dysfunctional, it begins to perform institutionalized rituals to ensure its survival and hide its dysfunction. Like a family, an agency is a system governed by a set of rules that determine and control the interaction of its members in organized, established patterns.

In today's dysfunctional agencies, therefore, those ideas contrary to the established view are termed "heresy," and those that implicitly support the established view are termed "policy." Policy is thus synonymous with homeostasis, which determines what shall and shall not occur within and outside of the agency.

Homeostasis is thus a mechanism through which an agency's inner workings are kept hidden from "outsiders." This is a very important concept. In fact, one man's intuitive sense about my lack of unquestioning "loyalty" to and for the good of the Bureau of Land Management (commonly referred to as BLM) almost kept me from being hired as a full-time, permanent employee.

When my initial six-month probationary period was over and I was eligible for permanent status, this man said that he had grave concerns about my being a permanent, full-time employee, because I was not a "BLMer," and he was right. I did not worship the agency and did not consider its regulations and policies to be sacred.

The Bureau of Land Management, like any other agency, is only a collection of people who have come together to serve the public—not to make the public serve them. For a blindly "loyal," unthinking, unquestioning agency employee, however, this is an intolerable point of view, because the homeostasis of the agency is always in jeopardy from the unknown, impure, questioning, and, therefore, potentially dangerous actions of a thinking "outsider" who is within the agency.

Such concern is, to some extent, reasonable in that to function, even within its original charge, an agency must work as a team, and one uncaring maverick can disrupt the whole effort. On the other hand, homeostasis within a dysfunctional agency hides corruption.

To safeguard its "good" name, a dysfunctional agency becomes a self-serving machine that often defines more and more narrowly and more and more rigidly the job descriptions of its employees and thereby controls them. In this case, the description of an employee's job, which is available to the public, directs the employee to do something as an employed professional, but the dysfunction of the agency requires that the employee must be prevented from carrying out an assignment in a 100 percent professional, questioning manner if the agency's deception of the public is to be kept under wraps.

Keeping the dysfunction under wraps is a test of blind loyalty to the agency. Failure to pass the test usually has severe consequences. The situation becomes even more ensnarled when a person's job is specifically to uncover and report waste, fraud, and misuse of government property, and in the professional performance of that job a person runs into the homeostasis of a dysfunctional employer. Consider, for example, the following story about the military in which a man did exactly as he was supposed to do within his job description and also did exactly as he was supposed to do as an ethical, professional employee:

> An Army officer who directed an inventory that uncovered improperly diverted explosives at Fort Lewis [Washington] is trying to win another command, after he resigned his previous post because of conflicts over the incident....
>
> Flick said his role in identifying the mismarked and diverted munitions has cost him his career because he broke a bond of trust by going over his commanding officer [the immediate source of dysfunction] to point out the irregularities.
>
> "They feel that I broke the bond of trust, even though the part I did was lawful [within his professional job description and Army regulations] and what they did was unlawful...."[30]

Few people who have the courage to stand up and speak out for what they know is right are rewarded for their honesty and their loyalty to their ideals as professionals, because preventing these kinds of "leaks" is what homeostasis is all about. In some cases the homeostasis fails and change is affected, but in most cases it succeeds, and an untold number of truly loyal people die a slow, unknown "professional death" for their integrity and courage as human beings and professional employees.

A key to homeostasis is the job description. Although a job description is necessary and seems harmless in itself, it can be used either to relate the work of professionals within an agency to one another and among disciplines or to isolate professionals and disciplines from one another. The danger lies in the isolation of individuals by increasing professional specialization as manifested through the purposeful narrowing of the interpretation of job descriptions and through the careful, rigid, and absolute control of those descriptions.

The use of job descriptions as an isolating mechanism is one of the ways an agency protects its members from "unfavorable" information. The institutionalized, internal policies of the agency shape the perceptions and beliefs of uncritical employees in ways that protect the agency's dysfunction, even when catastrophic outcomes are involved.

A distortion of information is not limited to willful deceit on the part of an individual who is perceived as loyal to the agency. Even honesty within a dysfunctional agency is insufficient to prevent the widespread distortion of information. The weakness lies within the job description itself and with an individual's acceptance of completing an assignment without thinking about or questioning the consequences.

Although this hardly sounds untrustworthy, much less dangerous, it is just this "functionary" behavior that allows systematic, homeostatic distortions to occur. This is why decisions are often difficult to deal with in agencies, because one is seldom sure who makes them. Almost everyone seems to be afraid of taking a risk; therefore, if more than one person is responsible for a decision, especially a miscalculation, no one will be at fault.

One thus becomes a functionary by limiting one's inquiries only to questions of how best to accomplish an immediate assignment and allowing the agency to shape one's perceptions. The fault, therefore, lies neither in the job description nor in the assignment. The fault lies in not accepting personal responsibility for the outcome of the assignment. This is the seed of dysfunction, as well as the birth of the machine through the loss of identity, individuality, and human dignity.

Simply taking orders without thinking about or questioning them is personally safe but environmentally and socially risky. On the other hand, it is often personally risky—if you want to keep your job—to question orders, but questioning the orders is both environmentally and socially responsible.

Most professionals in land-management and regulatory agencies are told what level of professionalism to practice in order to stay employed. People therefore trade their dignity and professional ethics for job security. Unfortunately, these frightened employees are often judged harshly for being functionaries, but this judgment is usually rendered by people outside of the "system" who do not have the benefit of understanding how it works.

Another aspect to the homeostasis of public land-management agencies has to do with such laws as the National Environmental Policy Act of 1969 and the Rare and Endangered Species Act of 1973. Both of these laws were passed in an effort to control abuses of how land is used. Outside control, however, is the fear of a dysfunctional agency and not a cure.

The problem with such laws as the Rare and Endangered Species Act is that to a dysfunctional agency they are like an unpredictable disease, which can strike at any time and totally disrupt the homeostasis. Take, for example, the ongoing battle over the spotted owl and its habitat, the ancient forest. Both the U.S. Forest Service and the Bureau of Land Management are required by law to pay attention to the spotted owl and its habitat requirements, yet people in both agencies have tried to ignore this responsibility with respect to their timber sales, as did the U.S. Fish and Wildlife Service in listing the owl in the first place.

To ignore the law, however, an agency must set up a "spotted owl S.W.A.T. team" (Special Weapons and Tactical team) whose sole charge is to maintain the agency's homeostasis at any cost. Unfortunately, not even Congress is immune to this dysfunctional, S.W.A.T. team mentality:

> [Sen. Mark] Hatfield [R-Ore.] said thousands of jobs will be lost ... if environmentalists succeed in getting the spotted owl listed as an endangered species.
>
> "It is an attack on all Oregonians, because they threaten our jobs, our roads, our schools, our ability to meet Oregon's diversification needs for the future," Hatfield said.[31]

Senators and representatives often forget that these are public forests, which belong to *every* citizen of the United States, as opposed to their own private, industrial forests. Further, court appeals are the legal right of American citizens, thanks to such far-sighted laws as the Rare and Endangered Species Act. When the appeals are upheld, they are justified, and the law to protect a major component (diversity, both biological and genetic) of the long-term biological sustainability of the national forests is working as it was intended.

The S.W.A.T. team mentality and the congressional tinkering with morally correct, as opposed to politically correct, laws is founded solely on trading political support for short-term economic favors based either on abysmal ecological ignorance or informed denial of the problem. During my 20 some years as a research scientist, I have repeatedly noticed that at least 90 percent of what we know ecologically about our forests, ranges, deserts, arctic, and oceans is not allowed to be applied to their management. It is curious that the vast majority of our hard-won ecological knowledge about the way in which our planet functions is summarily dismissed (in court and in Congress) if that knowledge interferes with a private interest in short-term profits, political support, or both.

By the same token, if Congress passes a law to create a legal principle of reasonable governance of social behavior for the long-term good of society, the lawmakers want others to obey that principle but want to exempt themselves from that same edict in their home districts or states. They want to be a "special case" (above and beyond the law) to protect their own special interests. The danger of such irresponsible short-sightedness is that the course of planetary evolution is being directed from a position of greed, ignorance, and illegality. Their dysfunction has obliterated the boundaries of the morally correct social behavior on which our democracy was founded and the biological sustainability of our forests depends.

BOUNDARIES

Boundaries can be thought of in a manner similar to the home ranges and territories of animals. A home range is that area of an animal's habitat in which it ranges freely throughout the course of its normal activity and in which it is free to mingle with others of its own kind. A territory, in contrast, is that part of an animal's home range that it defends, for whatever reason, against others of its own kind. This defensive behavior is most exaggerated and noticeable during an animal's breeding season.

The same dynamic is in effect within your home. How well you know someone determines how freely they may use your house, and the place you are the most particular about (defend) is your most private space.

Although boundaries are frequently blurred and abused outside an agency, the dynamic becomes even more complicated, less clear-cut, and often partially or largely breaks down within a dysfunctional agency. It may function in a "normal" manner with politically important outsiders, but among the members of the agency, personal and professional boundaries may be so blurred that one's private space (including one's physical body) is frequently violated.

Even at the highest national level, one must be aware that unconscious, narrowly focused people in the U.S. Congress who lack a sense of personal or professional boundaries often become role models for a dysfunctional agency. So how does one cope with all of this?

COPING MECHANISMS

The main "product" of an agency is information, which is translated into laws, policies, directives, management plans, and public relations. Therefore, all products are really translations of the information and the system that produces it. The way in which that information is used determines whether or not an agency remains functional or has become a dysfunctional machine.

There is a vast array of coping mechanisms, first deciphered and named "defense mechanisms" by psychiatrist Sigmund Freud. Coping mechanisms begin as thought processes that we devise to protect ourselves from that which we deem dangerous to our well being. What begins as a thought manifests into behavior when we are confronted with a perceived life-threatening circumstance from which the thought process was devised to protect us. If the combination of thought and action is successful, then a functional mechanism of survival, a "coping mechanism," has been devised, which is reinforced by a feedback loop every time it works as expected. As it is repeatedly used, the thought process is relegated to the subconscious, and only the behavioral pattern is manifested.

Coping mechanisms therefore become the unconscious behavioral devices we learn to use to help us retain or regain control in uncomfortable circumstances. This really means we are trying to cope with a Universe in the process of constant change.

Coping mechanisms as a strategy for survival may be functional, positive, and entirely appropriate for a given circumstance when first developed, but eventually they can and often do become outmoded and dysfunctional as circumstances change. Clinging to dysfunctional coping mechanisms when they fail to meet current or new situations in life can stifle inner growth and development.

With this in mind, six examples of the more common coping mechanisms will be briefly discussed here. Before continuing, however, it is important to reiterate that perceived enemies are likely frightened people who may have shirked their responsibilities as professional employees and, having done so, forfeited control of their lives to an ever-growing dysfunctional machine, which uses homeostatic strategies to further dehumanize them in order to maintain itself.

Anger and Aggression

Anger and aggression are discussed together because anger is the emotion that triggers aggression as an act. Anger, a feeling of hostility toward someone or something, is based on fear, which is violently projected outward, and serves to isolate us from the facts, ourselves, and one another.

People are not angry for the reason or at the person or thing they think they are. People become angry when they feel out of control of circumstances, which has nothing to do with the person or thing toward which the anger is directed. Unfortunately, this realization all too often follows an outburst, which has been projected onto someone or something else. Anger is a means of not having to deal with the feared circumstance.

Anger often translates into aggression, which is manifested as hostility. If I show enough aggression toward the person I think I am angry at, I am coping with my fear by causing that person to back away from me. Through aggression, I think I can avoid having to deal with the circumstance over which I have lost control and of which I am afraid. All I have really accomplished, however, is to isolate myself from any understanding of the information and from the person who is presenting it.

Appraisal

Appraisal is the act of evaluating something; of estimating its quality, amount, size, and other features; of judging its merits. As such, appraisal is an interesting coping mechanism in that it effectively prevents forward motion, and it is used as a form of resistance (another coping mechanism) to change.

Some scientists are so afraid of making a mistake and of being criticized that they get bogged down appraising the scientific details in an attempt to cover all of the bases and thus never really accomplish anything. It often seems that no one wants to take the responsibility for making a clear-cut decision.

Appraisers cope with their fear of criticism by checking, rechecking, and triple checking the data and are seldom willing to make a decision for which they are accountable. Thus, when in doubt, conduct another study, but refrain, at any cost, from saying or doing anything until all the data are in and have been carefully and properly analyzed. This, of course, will never happen, because even if all the data could be collected, the appraiser would continue the analysis indefinitely. Although appraisal is ostensibly appropriate in analyzing scientific information, it is too often used to ignore new and potentially threatening data, which all new data seem to be if they do not conform to current policy

Defensiveness

To be defensive means to protect that which already is, to resist a new view, to resist the possibility of change, or to resist the truth about oneself. Defensiveness limits growth in that the status quo is maintained rather than being open to a new possibility. People become defensive because at some level they know that there is some truth in what is being said. The truth is frightening because it means having to act, which means having to change, which is also frightening. It thus fosters an obligation to defend a comfortable, known entity.

Defensiveness is a coping mechanism that comes rapidly to the fore when we feel unsafe, when we are losing control of a circumstance, and when, in order to retain control, we have hidden agendas motivated by fear. Defensiveness is widespread in our public land-management agencies.

Defensiveness about one's job or what one is doing is a self-centered mode of fear, which not only robs the individual of dignity but also isolates the individual from peers and fellow workers. It creates a sense of distrust that can spread like an epidemic throughout an agency.

Denial

Denial is a refusal to recognize the truth of a statement; it is a contradiction, a rejection of what is. Although denial as a coping mechanism is part and parcel of almost all the other coping mechanisms and one of the most pervasive, it is also an entity unto itself.

We isolate ourselves when we do not accept change. We become defensive, fearful, and increasingly rigid in our thinking; we harden and close our minds.

If I become defensive about anything, if I start to form a rebuttal before someone is finished speaking, and if I filter what is said to hear only what I want to hear, I am in denial of what is being said.

The oldest argument of denial I encountered while working with the U.S. government and in court with the timber interests was "informed denial": It is inappropriate to do anything until we have all the data or facts, which of course we will never have. Thus, deny there is a problem until someone dies or there is an environmental catastrophe, and then it is too late. Consider the following example:

> "We don't see any basis for taking (Alar) [a chemical spray for apples] off the market," Uniroyal [Chemical Co., of Middlebury, Conn.] spokesman Yanis Bibelnieks said. "All of the studies that have been done since 1985 have supported our position that Alar poses no significant health risk." [Who defines whose health and life as significant?]
>
> He said it was "inappropriate to make any judgment based on interim results" cited by the EPA.[32]

Filters

A filter is a device through which a substance (such as light, water, or thoughts) is passed to remove what we define as "unwanted impurities." In the sense of a coping mechanism, unwanted information is filtered out, and we can accept and understand whatever we want to. Have you, for example, ever tried to explain something to someone and had them hear only part of it, the part they *wanted* to hear?

I often find this to be the case when speaking to a group of people comprised of the timber industry, environmental organizations, and land-management agencies. They each hear what they want to hear and address these different aspects during the question and answer period. The more polarized the audience, the more predictable are the questions they are likely to ask.

At times we live as though we are in a giant "safe" with filters to control what we see, hear, and feel. We can accept and understand that which we choose and do not have to step out of our comfort zone and be accountable.

Filters in an agency are almost always fully engaged and can be very frustrating for the person who is trying to communicate with someone who does not want to hear what is being said. Yet we all filter information simply because we have different frames of reference.

Another interesting and transparent filtering maneuver is to hear information presented and consider it to be the "presenter's data" and therefore to summarily dismiss it as that person's unproven hypothesis, even though it may be the collective data of a number of scientists who are in agreement. By filtering the information and hearing only what fits into the homeostatic pattern, it can be considered someone's unproven hypothesis and discounted, remaining safe from having to confront the machine by changing policy to accommodate these new data. This type of filtering system to ward off new, imaginative data cannot continue indefinitely, however, as observed by the poet William Blake in 1793: "What is now proved was once only imagined."

Rationalization

To rationalize in the sense of a coping mechanism is to devise self-satisfying but incorrect reasons for one's behavior. For example, I have been told to do something in my job with which I disagree ethically, but if I do not comply, I will lose my job. So I rationalize that I can do more good by keeping my job and working unobtrusively for change on the inside of the agency than I can by getting myself fired for sticking openly to my beliefs. In so doing, I intellectually rationalize the situation and comply with it, but I have simultaneously repressed the honesty of my feelings.

A few years ago, I put on a workshop for the Forest Service fuels managers who, simplistically stated, are those people who help control and prevent forest fires. We talked about forest ecology, fire ecology, Forest Service policy with respect to the management of ecosystems and fire, and human dynamics.

Finally, toward the end of the day, I said, "Now turn off your minds. I don't care about what you *think*. I want to know how you *feel* about what you're doing." Silence. Then an older gentleman got up and said, "I'm going to retire in about six months, and I don't feel good about what I'm doing anymore. I haven't for some time now." This turned out to be the general feeling of the audience. I then asked: "If you don't feel good about what you're doing, why are you doing it?" The answers had the tenor: "I do it because it's Forest Service policy. I can rationalize its being okay in terms of policy and politics if I don't examine it too much. It may sound good, but now that you've asked, it doesn't feel good."

Our truth is how we *feel* about something, not what we think about it. Those workshop attendees could no longer just think—rationalize—that something was okay when in fact it was not. From that day on they would have to check in with their feelings at some level, and if they were untrue to their feelings, they would have to deal with a moral and ethical crisis.

The Upshot

All personal coping mechanisms, in one way or another, become the collective coping mechanisms of the agency, because *we* are the agency. My feeling is that the industrial/political machine has so completely and so insidiously taken over public land-management and regulatory agencies that many people have lost touch with themselves and no longer know the difference between the truth and their coping mechanisms. I see more and more frightened people being chewed up and spat out by the dehumanizing machine, which seeks short-term profits at the tremendous cost of truth, trust, human dignity, and public service.

BREAKING THE DYSFUNCTIONAL CYCLE

To break the cycle of a dysfunctional agency, we must deal with all of the pieces in context with the whole, which is something we seldom do. We try instead to fix individual pieces in isolation, which cannot work. An agency can be restructured, for example. I survived several restructurings, but that did not fix the cause of the dysfunction in that it did not heal the people who made up the agency—beginning at the top.

An agency can be given a new charter, but that does not fix the dysfunction, which once again is caused by the people who make up the agency. *We* are the agency; therefore, *we* are the problem, the dysfunctional components; therefore, *we* are also the solution.

The main problem is the constant struggle to retain one's dignity and integrity in an agency that has become a machine intent primarily on its own survival by maintaining the status quo through putting out "crisis fires." Such an agency is indifferent to its real mission and to those who want to carry it forward as it was originally intended.

It is up to the individual to learn how to work within the agency while, at the same time, maintaining personal and professional integrity. We cannot blame externals for our failure to maintain our integrity, and we must beware of using statements that seem to absolve us of our responsibility for our personal behavior, because they can only lead to personal powerlessness. Whatever the external obstacles, it is our personal responsibility to face them and retain our empowerment.

Any kind of professional impotence is a condition that feeds on itself. When we, as professionals, abdicate our own power, we assume the role of victims and develop the cynical attitude that "we can't fight the system." We thus justify the prophecy that all of our efforts are doomed and that nothing we do matters or makes a difference, and herein lies the seed of the problem.

When we surrender our power by placing all the responsibility for the failures of our efforts outside of ourselves, we are in jeopardy of having our work devitalize us. If, on the other hand, in spite of the obstacles, we assume responsibility for the professionalism of our work, it can vitalize us.

We simply cannot get away from ourselves as individuals. We all bring ourselves as extensions of our familial backgrounds into the agency. We must therefore recognize that, in a sense, we have become part of an "extended family" filled with "relatives," all of whom are doing their best to cope with their ever-changing experiences of themselves and of one another.

One of my more vivid lessons about how to break an unwanted cycle came from Ol' Red, a huge, male red diamond-backed rattlesnake I took care of in graduate school. Once a week I fed him wild mice. When I put the mice into his cage, they immediately exhibited their instinctive fear response to the odor of Ol' Red by running to the farthest corner of the cage and trying to climb out or to hide. Ol' Red in turn reacted to the expected fear response, which in turn elicited his normal strike response.

I once failed to catch "wild" mice. To amend my dereliction of duty, I went to the animal laboratory and got "lab" mice, which had been bred in the lab for at least fifty generations and had lost all fear of snakes. When I put two of these mice into the cage, they approached Ol' Red, nipped him on the nose, climbed up on his back, and chewed off some of his magnificent rattles. Ol' Red in the meantime put his head beneath his coils and refused to move. I left the mice with Ol' Red for two to three days, but he would not touch them.

The lab mice had broken the stimulus–fear response cycle and Ol' Red did not know what to do. The fearless mice had simply rewritten the rules and in so doing had changed the game. Ol' Red no longer knew how to play the game he had inherited from his long line of ancestors. Unlike Ol' Red, who simply hid his head, we have the ability to make a conscious choice to break the old stimulus–response cycle in favor of something more desirable.

To break any cycle, we must first be aware of it and how it functions. Admitting, owning, and accepting the problem brings it into the light of consciousness, where it can be recognized and dealt with. Thus, one of the keys to unlocking the dysfunction of an agency is understanding our dysfunctional coping mechanisms on an individual basis, perhaps by participating in a series of seminars or workshops that help us to: (1) see our relationships with co-workers as possibly similar to those existing in our own families, (2) gain an insight into the workings of our dysfunctional coping mechanisms, (3) be encouraged to examine our coping mechanisms and allow them to fade away in an atmosphere of mutual revelation and exploration, (4) have opportunities to learn about ourselves and one another, (5) learn that it is okay to have and express intense feelings we may have repressed, (6) be encouraged to join with our co-workers to recreate our past so we can become liberated from the restricting influences of our early childhood which we have brought with us into the agency, and (7) experience the support of others in discovering the universality of our struggles, which will help us to become our own persons.

Another key to unlocking the dysfunction of an agency is to understand the dynamics of job descriptions and how they are used to control individuals. A dysfunctional agency seeks to control its employees, because commodity production to satisfy the industrial/political bosses has become more important than the process of cementing human relationships. Yet we must ask ourselves: Is not human dignity more important than commercial products? What is an agency if it is not people? How can we rectify the problem?

We can rectify the problem by having the courage to take responsibility for our own thoughts and actions within our job descriptions. Rather than being functionaries for the corporate/political machine, we can question the validity of what we do for the good of society beyond a few special interest groups.

We are our family, and we are the agency. We joined the agency to serve people through our professional expertise. Human dignity must therefore be the foundation and primary product of an agency, especially an agency of public service. As it now stands, however, in order to get any kind of "real" attention in a dysfunctional agency, as in a dysfunctional family, a person must somehow threaten the homeostasis.

There are two reasons for this. One is that we no longer know for whom or for what we are working, or so we are told. In the Environmental Protection Agency, for example, I was told by a high-ranking supervisor that I worked for the Bush Administration and therefore had to comply with their wishes. "No," I said, "I don't work for the Bush Administration. I work for the public; in fact, I work for the children." The other reason for this sad state of affairs is that we lack leadership and are overrun with managerialship.

Leadership is of the heart and deals intimately with human values and with human dignity, because one must lead by example, as noted by Francis Bacon when he said, "He that gives good advice, builds with one hand; he that gives good counsel and example, builds with both." A leader knows and does what is right with a moral conviction, usually expressed as justified enthusiasm, which causes people to want to follow with action. Essentially, a leader is one who gives people value and can motivate people by sensitive negotiation so that a perceived need is raised to strong desire. The irony is that such leadership is perceived as a danger to a dysfunctional agency.

Managerialship, on the other hand, is of the intellect and pays minute attention to detail, to the letter of the law, to doing the thing "right" even if it is not the "right" thing to do. A manager relies on the external, intellectual promise of new techniques to solve problems and is concerned that all the pieces of the machine's procedure are properly accounted for, hence the epithet "bean counter."

Good managers are thus placed at a disadvantage when put in positions of leadership, because all such people can do is rise to their levels of incompetence and then remain there. Similarly, a leader placed in the position of managerialship is equally likely to be a dismal failure because the two positions require vastly different skills.

We need excellent leaders who have excellent managers to support them in a team effort, and both are in exceedingly short supply. What we therefore have is primarily agency-oriented, careerist managers, who often are in over their heads in leadership positions—and very few leaders with any authority to lead.

In my experience at least, a dysfunctional agency carefully, and for the most part unconsciously, screens all candidates and selects careerist managers who lack vision and will do as they are told with little interference from their feelings of right and wrong. If, however, a misjudgment occurs and the person chosen by the machine does, in fact, have vision and the courage to follow it, the person is soon shunned and thereby rendered powerless, as only the machine can do.

Another point to consider in breaking the cycle of a dysfunctional agency is that we must ask new, morally right, future-oriented questions, which means that public agencies must restate their visions and missions for the future:

> ...a strong, independent Forest Service can be the best trustee for these lands [national forests]. The Forest Service has always been at its best when it has used the sciences to listen to the commands from the ground, and when it has been the champion of the future, not the captive of the past.
>
> We should begin an earnest and far-ranging debate on whether public use really is the concept of the future. If it is, then the Forest Service should anticipate and embrace public use [rather than clinging to commercial use for timber, or livestock, or mining, etc.]. In so doing, it would take a giant step toward fulfilling its high calling of serving as the trustee for these lands that have come to embody many of the best dreams of the people of the modern American West.[33]

To accomplish this, we must clearly define the professional boundaries of expertise in a functionally integrated way—and adhere to them—if we are to have a sustainable planet for human society present and future.

Although it can be and often is argued that new people entering an agency bring new thoughts and ways with them as older people leaving the agency take once pioneering thoughts and ways with them, this influx of "new blood" does not absolve us as individuals from the responsibility of so changing ourselves that we can and will honor the dignity of each and every person with whom we work, whether we agree with them or not. In the end, it is you and I who must learn to value one another and to live by one simple rule: to accept responsibility for our own feelings, thoughts, and actions. We must learn to be kind and respectful to ourselves and one another without forfeiting our principles. Each agency must have a clearly stated vision and mission, as well as clearly stated goals and objectives, that honor both the intent and the heart of the highest laws of the land if we are to replace our dysfunctional agencies with functional ones.

Before we can replace our dysfunctional agencies with functional ones, however, we must understand the nature of conflict, so we can move beyond it to resolution and build bridges over which to cross into the future.

4

CONFLICT IS A CHOICE

I was taught that conflict between people is simply a mindless condition of life, a necessity of survival. Finally, in my mid-forties, I began to understand that conflict comes about because frightened people perceive the need to defend themselves from the potential loss of what they think and feel they must have to survive: control of their own lives as they perceive it. Control in this sense is synonymous to each person's "right of survival," however that is defined. This perceived security of our right of survival is weighed against the number of choices we think are available to us as individuals.

These choices are in turn affected by the supply and demand for natural resources—the world's source and supply of energy, which all life requires in one form or another. The greater the supply of a particular resource, say clean water, the greater the freedom of choices an individual has with respect to that resource. Conversely, the smaller the supply, the narrower is the range of choices. The variety of available choices thus dictates the amount of control we feel we have, which consequently affects our sense of security about our survival. Thus, the seed or the germ of all conflict comes from a perceived loss of choice, which we interpret as a threat to our survival.

We relieve our fear of being out of control by creating enemies onto whom we can project blame for our fears and thereby justify them. But what or who is the enemy? An enemy is one seeking to injure, overthrow, or confound an opponent; something harmful or deadly. We are not the enemy because we are convinced that *our* position, *our* values, are the *right* ones, and "the enemy" is wrong. This is what we are taught. This is the eternal verity around which conflict rallies.

The problem is that when both or all sides feel this way, there is little understanding that an enemy is anyone or anything that is perceived to threaten our sense of survival. Herein lies the great irony: conflict in one way or another is the spawn of misunderstandings, miscommunications, and misperceptions.

Conflict is thus a mistake or a misjudgment of appearances that is avoidable because it is only a choice of responses to a given circumstance.

Once one or the other side perceives a threat to its survival, the most important single precipitating factor in conflict is in fact misperception, which manifests itself in a leader's image of himself or herself, in a leader's view of his or her adversary's character, in a leader's view of his or her adversary's intentions toward himself or herself, and in a leader's view of his or her adversary's capabilities and power. Once misperception is in play, miscommunication closes in and joins hands with misjudgment to foster a distorted view of the adversary's character, which helps to precipitate a conflict.

If a leader on the brink of war believes that his or her adversary will attack, the chances of war are fairly high. If both leaders share this perception about each other's intent, war becomes a virtual certainty.

Now, however, at the very moment humankind has the power to destroy the Earth, human beings have also begun to perceive the planet as a whole. Similarly, environmental problems and the equity of resource allocation will be surmounted on a global basis or not at all. In both cases, the brute logic of the insensate machine is dictating a modicum of world order: the terror of nuclear fire and the prospect of choking in our own waste. Out of this terror springs a recognition of the need for flexibility and change, because the nuclear fire must not become the earth, nor must the earth become a sewer.[34]

The lesson war has to teach is that conflict of any kind is a cycle of attack and defense based on the *misjudgment of appearances*. Appearance is an outward aspect of something that comes into view, and judgment is the process of forming an opinion or evaluation by discerning and comparing something believed or asserted. Therefore, those whom we define as enemies are those onto whom we affix blame for our perceived sense of insecurity, our perceived threat to our survival.

Our judgments are necessarily incorrect, however, because nothing is as it appears since appearance is external. If we could understand the inner motive of our "enemy," we would find a mirror reflection of our own fears for our survival. In that reflection we would find that we had made a mistake about our enemy, which means to make an incorrect judgment of character or ability based on inadequate knowledge.

If we are not one other's enemies, what is the enemy? What we are really afraid of is change—loss of something we value through circumstances we cannot control, circumstances that we perceive as a threat to our sense of survival.

Control, often used as a synonym for power, is an interesting phenomenon. We pay dearly for control, but regardless of the price, there are limitations. I cannot, for example, control the wind, but I can trim my sails. The wind is the circumstance beyond my control, but by trimming my sails I can choose how I respond to the wind, and in my response, I am in control of myself, which de facto controls the circumstance.

That we cannot control circumstances is a given, although we continually try, which results in either inner or outer conflict of some magnitude. However, we can control how we react to circumstances, and therein lies both the problem and its resolution.

The inability to control circumstances in any meaningful way translates into fear of change because every circumstance causes change in some way, whether relatively minor or catastrophic. Change is thus perceived as a loss of control that threatens survival. We therefore want to control circumstances whenever we can so that other people—our perceived enemies—will have to risk change, but not us.

When, however, we focus our attention on human enemies, we are really focusing incorrectly. The other person is not the enemy; the enemy is fear. Conflict is an attempt to move away from that fear, away from some unwanted circumstance. Conflict is a choice of behavior that we resort to because that is what we have been taught to do in order to cope with circumstances that we perceive as threatening to our survival.

It is necessary to understand that every circumstance we encounter in some way evokes an unanticipated change in our participation with life. In turn, each change we are obliged to make is a compromise in our sense of control, which is frightening to most people in an increasingly complex world. In the United States, our ultimate attempt to control another person or persons is through use of the court system.

THE ENEMY IN THE COURTROOM

Although each court case, however small, in our struggle to redefine our land ethics is a stitch in the changing tapestry of an evolving world, it is of diminished value if integrity and dignity are lost in the legal process. To date, all I have seen the court used for is an arena for a power struggle that is fueled by short-term economics and resistance to change.

What happens to a society whose land ethics must be decided in court? What happens to a society that obeys the letter of the law and violates the heart of the law? First, we perceive one another as enemies. Second, we too often destroy the land's sustainable biological capacity in the ensuing struggle for power and control. Third, we foreclose the options for future generations. Who is it, however, that owns the land—public or private—over which we have the arrogance to fight? Are not all of us only custodians for those who follow?

We go to court with any and every legal standing we can find, which is usually procedure and seldom substance, and the attitude that the ends justify the means. But what is the environmental and social cost of a contest that does not think about or question the validity, the social responsibility or ramifications, or even the truth of the issue at hand? Because human values can neither be legislated nor legalized, do the ends really justify the means when my "victory" legally forces you to change your behavior (termed "compliance") so that I will not have to change mine?

Who is "right" when we are all right from our own individual points of view? If everyone is right, then who is wrong? Right or wrong is always a human judgment, and judgments can only deal with appearances, not with reality. Everyone loses when land issues are "settled" by judgments of right or wrong because, again, judgments can deal only with appearances. In addition, we do battle in court over procedural matters, and the land loses in the end, which means all of humanity has lost. If, on the other hand, the court could assess and would rule on substance, such as the long-term ecological health of the land, then the land and humanity could win.

The arguments I have seen always dealt with short-term economics: What is the maximum return I can get in products from the land for as little capital outlay as possible? There is no need for a court to intervene if both sides, industrialists and conservationists, play by the same rules and neither side cheats. The problem is that both sides cheat and justify it in the name of being right.

Let's look at a typical case dealing with a patch of old-growth forest. The attorneys for both the government and the industrialists argue either that it is "just one sale of old growth" of which there is plenty (but they do not know how much there is or where) or they argue about lost jobs (which is also an untenable argument when mills are being automated specifically to eliminate jobs and save money). In either instance, the argument is about special cases and is out of relation to the whole, the common denominator: the relation of the stand to be cut to other immediate old-growth stands, to the watershed, and to the landscape on the one hand and to the nurturing of loggers, log-truck drivers, and mill workers on the other.

With respect to timber, both the U.S. government attorneys and those for industry use generalizations about theoretical lack of harm to the land by liqui-dating old-growth forests—with no data—and demand specific data to "prove" that old growth is needed for ecosystem health. Conservationists must "prove" the impacts of cumulative effects based on scientific data. Even the scientists themselves must prove, and defend, their own data before the court, while industrialists can legalize disregard for the data based on procedure. A final government–industrialist argument is that because the land has already been damaged, a little more will not hurt.

I had a different experience in January 1991, however, when I testified before the Board of the Administrative Tribunal for the "Class Environmental Assessment of Timber Management" in Toronto, Canada, which was a contest of values as they relate to the use of Ontario's boreal forests. Unlike my experiences in courts of the United States, which have all been exceedingly violent and adversarial, my experience in Canada was more that of a classroom, where genuine learning was possible, should people chose to learn.

One of the things that made the experience so wonderful was the absence of a lawyer or a judge presiding over the hearings. Instead, two "ordinary" people (including the vice-chair in charge of the hearing) were sincerely interested in learning. They were two of the most open-minded people I have ever had the privilege to serve.

The second thing that enabled the experience of learning was knowing that I could say whatever I thought needed to be said and that I had nothing to defend. I therefore chose to be as defenseless as possible. To this end, I submitted five premises not only on how I would testify but also on how I would answer the "interrogatories," which were questions put to me by the people of the Ontario timber industry and the Ontario Ministry of Natural Resources in seeking clarification of points made in my written witness statement.

The premises that I submitted and to which I adhered were as follows:

1. I am here to share my understanding of data after more than 20 years as a research scientist. This I will do. I am *not* here to defend either myself or my understanding and interpretation of the data.

2. I am here to make observations, based on my understanding of the temperate coniferous-forest ecosystem. Those observations will encompass what I think are ecologically "wise or unwise" decisions in terms of the biological sustainability and ecological resiliency of the forests for future generations. I am *not* here to make value judgments about "right or wrong."

3. I am here to point out the commonalties of ecological processes and principles and the most probable ecological consequences of various types of management within and among different portions of the coniferous-forest ecosystem. This I will do. I am *not* here to defend or to quibble about a given piece of research, a management practice, or an idea because it was or was not done, used, or thought of within a particular political boundary. Such arguments have nothing to do with the biological sustainability of the forest in question. (Both sides surely recognize that political boundaries have nothing whatsoever to do with ecological processes and principles, consequences of management practices, or ideas, and they use available data accordingly.)

4. I am here to speak the truth as I understand it. My *only* interest is that the best set of ecologically sustainable options are passed as a legacy to the

generations of the future. Thus, I am *neither for nor against* either side in this hearing.

5.   To me, the burden of proof lies with whomever would foreclose an option for the future by liquidating a resource for today. The burden of proof *does not* lie with those who would save some of the resources as a legacy for the generations yet to come.

When it came time for the lawyer for the timber industry (the "opposition") to cross-examine me, there was little she could do, because it takes two to make a fight, and I chose to be defenseless, a decision and its subsequent behavior over which she had no control.

The part of society's legal system that applies to use of the land, as I have experienced it, is adversarial, aimed at winning a point of view rather than discerning truth. Peace and truth are afforded no place that I have found in our American legal system with respect to use of the land, because American land-use law is of the intellect and its judgment and is all too often played out through legal procedure.

That the lawyer was not trained to deal with truth became obvious at one point in my cross-examination when the she asked me in exasperation if I, in fact, would continue to refuse to defend my data. When I answered "Yes," she asked, "Do you really expect the Board to accept what you say on *faith*?" "Yes," I said, "because in the end, all facts, all data are only a matter of faith. The Board either has faith that my data are correct or that someone else's data are correct, but they never *know*. Therefore, I'm interested only in expressing my truth, that which I know in my heart. I'm not interested in facts, because they're no more than an expression of the faith I or someone else has in the intellectualization of data that may be proven incorrect next year or the year after. Facts can only be disproven but never known; truth, on the other hand, can only be known but never proven. Facts, therefore, are merely knowledge that is translated into points of view, which one may feel obliged to defend, but truth is the knowing beyond knowledge and needs no defense, because it is truth."

When my cross-examination was over, I asked the vice-chair for a few moments to make a brief closing statement. I repeat my statement here, because I think it is also relevant to our legal system in the United States as we struggle to redefine our sense of values and to negotiate a new social ethic for habitation on and use of the planet:

> What you've undertaken with these hearings is the beginning of a great and wonderful adventure—a changing view of the art of forestry. It will be a difficult adventure, because it's a journey into the unknown. As such, it's fraught with fears. There are those who would forge blindly ahead, and there are those who will resist change at any cost. The path lies somewhere in between.

Change, which is the soul of the Creative Process, is the master of the world, not us—much as we might wish it otherwise. Ours is but a tiny voice in the process of change. We can accept change or reject it; we can work with it or be dragged kicking and screaming by it. But we can no more stop change than we can halt the tides of the sea.

I was asked to come here to speak about science. I have done so to the best of my ability. And I leave you with a caution. Science, like economics, is a socially crafted, narrow way of thinking that is governed by the same stringent rules of immediate, social acceptability and conformity that are common to most, if not all, other social institutions of the day.

Thus science and scientists are no freer of social constructs, insecurities, and prejudices than are any other professionals in any other field of endeavor. Science is a veil that we draw before our eyes through which we see but dimly, because, as yet, science is a discipline that is isolated and imprisoned in the intellect and therefore hides us from the truth.

As science is of the intellect, so truth is of the heart. Thus the "hardness" of the intellectual, scientific data without the softening humility of the truth of the heart is at best dubious, at worst dangerous. This being so, neither I nor anyone else can tell you what is right or who is right. I, myself, do not know who is right. I only know what is right for me, and what I must do. Ultimately, after having reviewed the scientific data, I must follow the truth of my heart.

When we have learned to think with our hearts as well as we think with our minds, such hearings as this will no longer be necessary, because life is not a matter of economics, or technology, or science, or human law, or even sight. Life in its sum total is a matter of faith and of truth.

Although I think the courts will be with us for some time, I hope the time will soon be upon us when they will be used only as the last resort in settling disputes over land-use values. Resolving conflicts and negotiating settlements out of court, wherein all sides can emerge with their dignity intact, is the only way society can win, because in today's world, society wins or loses as a whole. However, if going to court becomes inevitable, we must recognize that we have a choice in how we use the court. By this I mean that we can go to court to punish (as we largely do today) or to teach and to learn.

If we go to court to punish and "win," what have we won? We have won the legal right to remain stuck within the rigid limits of our thinking. We have won the legal right to retain our fear of change and argue for our limitations at the expense of our potential. We have won the legal right to humiliate our opponent, because the court has awarded us our opponent's dignity as the legal trophy of "conquest." If, on the other hand, we go to court to teach, we all win, because teaching is a process of experiencing ourselves and one another as growing human beings with the courage to examine the issue, to allow and to help one another to change, and to help one another experience each other as we grow toward a new relationship.

Although we have not yet reached this peaceful use of the courtroom as a hall of learning, I know that there must be a better way to treat one another and the Earth. The burden placed on a court judge—without a thorough scientific background—to make ecological decisions that will have cumulative effects on the world forever is incomprehensible. In effect, a person who serves the people by working as an employee of the government of the United States, whether as a federal judge, a politician, or a scientist, must pass the tests described in the eulogy that Senator William Pitt Fessenden of Maine delivered on the death of Senator Foot of Vermont in 1866:

> When, Mr. President, a man becomes a member of this body he cannot even dream of the ordeal to which he cannot fail to be exposed;
>
> of how much courage he must possess to resist the temptations which daily beset him;
>
> of that sensitive shrinking from undeserved censure which he must learn to control;
>
> of the ever-recurring contest between a natural desire for public approbation and a sense of public duty;
>
> of the load of injustice he must be content to bear, even from those who should be his friends; the imputations of his motives; the sneers and sarcasms of ignorance and malice; all the manifold injuries which partisan or private malignity, disappointed of its objects, may shower upon his unprotected head.
>
> All this, Mr. President, if he would retain his integrity, he must learn to bear unmoved, and walk steadily onward in the path of duty, sustained only by the reflection that time may do him justice, or if not, that after all his individual hopes and aspirations, and even his name among men, should be of little account to him when weighed in the balance against the welfare of a people of whose destiny he is a constituted guardian and defender.[35]

Two years after Senator Fessenden delivered this eulogy, his vote to acquit Andrew Johnson brought about the fulfillment of his own prophecy.

Who is the enemy in the courtroom? Is it the U.S. attorney, the industrialist, the conservationist, the judge, the witness? There is no human enemy. The enemy in the courtroom is fear, an enemy it takes the utmost courage to face, for each of us must face our own fear in the depths of our soul. There are few arenas that test a person's courage like facing change, especially change over which one has no control.

5

CHANGE:
THE UNIVERSAL CONSTANT

The only constant in Nature is change—the continual flow of unknowable and unpredictable relationships, which lead to choices to be made, which lead to actions, which become events. Thus, everything is in a constant process of becoming something else, and the only question one can ask is: What if one choice had been made instead of another?

It is through actively observing and consciously participating in life that one of Nature's great lessons is revealed: knowledge represents the historical surety of the past, change flows as the ongoing current of the active present, and unpredictability is the womb of uncertainty and future possibility.

We cannot relive history and know in advance what is going to happen; nor can we look into the future and foresee impending actions, because what happens depends on choices made. For example, a spider builds its web along the edge of a forest bordering a farmer's field. A fly encounters the web and becomes entangled. The spider eats the fly. A passing sparrow eats the spider. A hunting hawk chases the sparrow, which flies toward the farmyard for safety. The farmer, thinking all raptors are after his prize chickens, shoots the hawk. Walking back into his barn to put away his shotgun, he trips and falls, driving a small, dirty splinter into his hand. Not even noticing the splinter, he continues his daily routine. Two days later he is slightly ill. In four days he is confined to bed, and he dies by week's end.

What if the spider had constructed its web elsewhere? Would the fly have been caught? What if the fly had avoided the web or had become entangled at a different time? Would the spider's activity have attracted the sparrow, which attracted the hawk that chased it? Had the chase occurred a few minutes earlier or later, the farmer might not have seen the hawk, or the farmer could have decided not to shoot it. Then, with no reason to get his gun, he would not have tripped and gotten the splinter in his hand that became infected and brought death by week's end.

All we have is the present, the here and now, with which we *must* interact. How we interact with the myriad stresses of continual change and why we behave as we do depends on whether we understand change as a fluid process to be embraced or as a terrifying condition to be avoided.

We tend to choose the latter. Our insatiable need for factual information is a desperate attempt to predict and control circumstances and to avoid the inevitable cultural changes they bring. Thus, one great obstacle to change is the illusion of knowledge, which economic imagination draws with certainty and bold strokes, instantly serving hopes and fears, while scientific knowledge advances by slow, uncertain increments and contradiction.

Fearing the unknowable future, people twist the best and most current information (informed denial) to justify avoiding change, which serves the purpose of not having to acknowledge the perceived truth of a situation, of rendering it invisible, or at least of describing it as purely subjective in a society that demands the lie of objectivity. In this way, as previously stated, threatening voices that prophesy the necessity of change are silenced, and the whole truth of available knowledge remains hidden.

Must we face disaster before we are willing to seriously consider changing our thinking and our behavior? Must we, as a society, experience an environmental disaster of such unimaginable magnitude that we are in the maw of social—or even biological—extinction before we are willing to concede that we must change *now*? It seems we must have such a disaster, because as long as a person thinks he or she can win agreement with his or her point of view, fundamental change is not even considered an option.

The deadlock between timber interests and conservationists over liquidating the remaining old-growth forests is an excellent example. At no point has the national leadership unequivocally told one side or the other that its point of view (to cut or not to cut the remaining old-growth forest) is no longer viable and therefore unacceptable. Consequently, both sides, still hoping to win agreement with their respective entrenched positions, avoid considering fundamental change as an option.

If there is a political compromise, timber interests will have won another year's worth of old-growth forests to cut, which at best will prolong the agony of inevitable job loss as commercially available old-growth timber runs out. Conservationists, on the other hand, will feel they have once again lost to a politically important special interest. And the battle will continue.

Timber interests will fight for just one more year's worth of old-growth timber to avoid making fundamental changes in their ways of thinking and acting. If it worked last year, why not this year and the next? Conservationists will be even more adamant that with 90 percent or more of the original old-growth forests already cut, there is nothing left with which to compromise. Lines are reinforced and the battle continues.

Has any change occurred? Yes, and it will continue to exert its influence despite both sides being fortified in their polarized foxholes.

THE DYNAMICS OF CHANGE

Change is to cause to be different or to alter. Change is definable mainly in terms of its opposite: constancy. A constant is something that is invariable or unchanging. If everything were constant, change would not exist.

We are comfortable with that which appears constant, because it lulls us into thinking that we know what to expect, that we are in control. We therefore do our best to avoid change in ways of which we are not even aware. Yet in reality it is not possible to avoid change.

Change is the continual flow of cause-and-effect relationships, which fit precisely into one another in time and space. Each cause creates an effect that becomes the cause of another effect *ad infinitum.* Thus change comes on many levels and in many dimensions, such as that which we can control and that which we cannot control, large and small, gradual and sudden. Change is forever changing, because all things have within them the seeds of becoming something else.

All dimensions of change are fluid and dynamic, flowing together as rivulets that flow together as streams that flow together as rivers that flow together into the sea, where all waters merge and become dimensionless, only to form again in the great cycle of raindrops and ice crystals and snowflakes. Change by its very nature is the creative process and is a constant in the Universe. Change, therefore, is the Universe.

We think of change, however, as a 16-mm film loop. We have the notion that change comes in discrete packages, much like an individual frame in the film loop, and that we can simply cut out a frame, a snapshot in time, and stop the continuance of change. If, however, we cut out a single frame, we no longer have a film loop, because the very act of cutting out the frame becomes the cause of an effect—a film strip—and we have unthinkingly created change.

The film loop, which was a continuous, unbroken cycle, is now a film strip, which is a self-limiting, straight line—like linear thinking. The film strip ceases forever to cycle unless it is again connected to form a film loop, albeit a different one because a frame is missing, the one that was cut out and removed from its relationship to the cycle.

With the first snip of the scissors, the frame was disconnected from its neighbor, and it ceased to exist in its old relation to the original film loop; the film loop was simultaneously converted into a strip. With the second snip of the scissors, the frame was disconnected from its other neighbor, which is still attached to the film strip. Now the frame is alone, in a different relationship with both the nonexistent film loop and the film strip. By removing that one frame, the frame, the film loop, and the film strip have forever been changed.

Understanding change is a matter of consciousness of the effect caused by a thought and subsequent action; the more conscious we are, the more flexible is our thinking. Unconsciousness, in contrast, is the lack of understanding of the relationship between a cause and its effect and the lack of discipline to achieve that understanding. The less conscious we are, the more rigid is our thinking.

CHANGES WE CAN CONTROL

Have you ever thought about the power you have to change the world? As an example, consider what happens when you put your finger into a glass filled almost to the brim with water. If you put your finger in far enough, you may cause the water to overflow because your finger takes up space and displaces the water. What would happen if you put your finger into an ocean? Would you cause it to overflow? Probably not, but your finger still takes up space and displaces water, and, theoretically at least, you have the power to raise the level of the ocean.

As a further example, it you pick up a small object and put it down again somewhere else, you have changed the face of the whole earth, and it can never be the same again. Each person is changing the world all the time and is in control of that change: what it is, where and how it is made, how much is made, and why and when it is made. Each person who plants a flower or litters a roadside is creating a "new face" for the world. It is the consciousness with which we act that makes the difference. Each choice either raises or lowers the consciousness of humanity as a whole, despite the fact that there are some things we cannot control.

CHANGES WE CANNOT CONTROL

At a secluded beach along the Oregon Coast, where I went as a lad, much driftwood collected and two huge, rugged rocks stood above the sand and the normal tide. At high tide the water swirled around and splashed over the outer rock, in whose tide pools I could find crabs, small fish, starfish, and sea anemones. The beach seemed to remain the same, except for the reshuffling of the driftwood with each winter's storms. One year, however, everything changed. A terrific storm cleaned out all the driftwood, filled the tide pools, and all but covered both of the huge rocks with sand.

In a stream called Flamingo Wash, near Las Vegas, Nevada, incredible changes took place after each thunderstorm. Sometimes, for example, the whole area just downstream from the bridge was a foot or two deep in rocks five to six inches in diameter, only to wash away at some later date and to be replaced by a much thinner layer of pebble-sized stones and sand. Most of the time, the water was relatively quiet and clear; at other times it was chocolate-colored as it carried the soil of the land suspended in its roiling current. The most fascinating part was that the thunderstorms often occurred so far away that I did not even know about them until the stream changed.

The above stories are about Nature and about time, which is often likened to a river of passing events. Something is no sooner brought into sight than it is swept by and something else takes its place, and this too will be swept away. So it is in the lives of people, who every day face changes, which they cannot control.

There are many kinds of change over which we have no control. Some are uncomfortable, while others can be exceedingly beautiful. Nevertheless, each change provides a new experience and an opportunity to plumb the depths of our being and to redefine our relationship to the Universe and all it contains.

CHANGE IN HUMAN TERMS

Because change is creation and creation is change, the variety of universal changes is infinite, but what does this mean in human terms? How do we deal with change? There are at least six ways in which we attempt to confront change. They are by no means discrete, but instead grade one into another at different stages in our lives.

Change by Exception

In essence, this means that an exception proves the rule. Old belief systems remain intact when we allow for change by considering an anomaly to be the exception that proves the rule. In reality, however, the exception disproves the rule. In this sense, "prove" actually means "test," and the exception shows that the rule fails the test.

Incremental Change

Incremental change occurs so gradually that it is difficult if not impossible to perceive. For example, someone involved in a situation on a daily basis may not detect change. On the other hand, someone who has been removed from the situation for some time may see that it has changed considerably since his or her last encounter. The person in daily contact is simply too close to the situation to detect the incremental changes constantly taking place.

Pendulum Swing

Change by a swing of the pendulum is the abandonment of one established, closed, and certain belief system in sudden favor of another. In so doing, all of one's own prior experience is rejected. A pendulum swing fails to recognize and integrate what is good from the old and fails to discriminate what of value may or may not be hidden in the overstatements of the new.

Paradigm Shift

Each new paradigm or system of belief is built on a shift of insight, a quantum leap of intuition with only a modicum of hard data, scientific or otherwise. Those who cling to the old paradigm demand irrefutable proof that change is needed or even desirable. Such irrefutable proof is not forthcoming in an ever-changing Universe.

The irony is that the old paradigm began as the new and was also challenged to prove that change was necessary or even desirable. Time and human effort proved the old paradigm to have been more correct than its predecessor, but still only partially correct. So it is with the new paradigm; it too is more correct than the old and eventually will be proven only partially correct and in need of change.

Any paradigm that has become comfortable has also become self-limiting. New data will not and cannot fit into the old paradigm, because each paradigm is a carefully constructed, impervious membrane of tradition in which we become encrusted. Thus, like the shell of a crab, which hardens with age, a paradigm must be broken periodically if a new thought form is to grow and a new vision is to move society forward. This is incredibly difficult for those whose total belief systems and personal identities are invested in the old paradigm.

In turn, those who subscribe to the new paradigm must understand that it can survive only because it is supported on the shoulders of preceding paradigms, all of which were at one time new and daring. In addition, we must understand and accept that there have been no failures on the part of those who adhere to the old paradigm, only changes that may leave them behind.

Coercion

Coercion is the act of forcing someone to behave in a given manner, to forcibly control, and it comes in many guises. Coercion, after all, is the *art of control,* forcibly changing someone's behavior through the practice of humiliation as a means of control. This form of change is external only in that it forces an alteration in behavior but not necessarily in thinking.

Inner Dictate

"The more faithfully you listen to the voice within you," said United Nations Secretary-General Dag Hammarskjöld, "the better you will hear what is sounding outside. And only he who listens can speak." Change by inner dictate, the still small voice within, is the most profound change of all. It may come gradually or in a sudden, blinding flash, but in either case, it just happens.

Those who listen to and live by their inner dictates may do so without question, but that does not mean there is no struggle involved. It only means the struggle to comply with the inner direction is a willing one.

When profound change comes, it often shifts our entire perceptual foundation of the world and its relation to the Universe. The world has not "changed" in a physical sense, but our perception of the world has changed, and so our relationship to the world has changed in a "real" sense. After all, the physical processes function perfectly; it is our perception of how they function that is imperfect. Therefore, each time we comply with an inner dictate, our perception of the world is given an adjustment toward perfection and our understanding of the ongoing process of creation. As we change, so the world changes, and that is the power of each person.

CAN WE STOP CHANGE?

We cannot stop change. We can only respond to it, and by our response we may be able to alter to some extent its trajectory, speed, and outcome.

The wave of uncertainty called the future is coming. We can accept it, flow with it, and seek its opportunities, or we can resist it and fight it, but we cannot stop it.

The Universe is always in creation and never created. Notice that the word "created" is past tense, and whatever is past tense cannot exist, because it is history, that which has already happened, not that which is now happening. Thus, we can only interact with and in the present.

Strangely, however, it is our perception of the future that often becomes our truth about the present, a truth that is rooted in lessons from childhood, because that is how we have learned to cope with the change that determines the tenor of the rest of our lives.

I CANNOT CHANGE CIRCUMSTANCES, I CAN ONLY CHANGE MYSELF

Modern society depends on organizational systems for much of its information, particularly with respect to the assessment of such large-scale technological projects as management of forests, and organizations tend to distort information to meet organizational needs. Such distortions do not depend on dishonest behavior on the part of individuals. Rather, the tendency to distort information is a systemic property of the organizational system itself. As the power of modern technology grows, the consequences of distorted assessments will become more serious and potentially catastrophic.

Change is inevitable, however, and we can learn something about change from Buddhism, the whole philosophy of which is based on the *acceptance of change*. The Buddha taught the Four Noble Truths, two of which are pertinent to our discussion.

The First Noble Truth—Truth of Suffering—states that the outstanding characteristic of the human situation is suffering or frustration, which comes from our difficulty in accepting that everything around us is impermanent and transitory. "All things," said the Buddha, "arise and pass away." The root of Buddhism is that flow and change are the basic features of Nature, and suffering arises whenever we resist the flow of life, whenever we try to control circumstances and cling to fixed forms, such as things, events, people, or ideas.

The Second Noble Truth—Truth of the Cause of Suffering—deals with clinging or grasping. It is futile to grasp life from a wrong point of view, from ignorance. We divide the world we perceive into individual and separate things out of ignorance and thus attempt to confine fluid forms of reality in unchanging, mental boxes. As long as we do this, we are bound to experience one frustration after another.

Trying to create anything fixed or permanent in life and then trying to cling to its perceived permanence is a vicious circle, which is driven by karma, the never-ending chain of cause and effect. As stated by the Buddha, "It is the everlasting and unchanging rule of this world that everything is created by a series of causes and conditions and everything disappears by the same rule; everything changes, nothing remains constant."[19]

This idea, that everything is constantly changing, that nothing is permanent, can be viewed another way: *acceptance of what is*. What is, is. It cannot be otherwise. I cannot, for example, control circumstances, but I can control how I respond to circumstances. If I simply accept the circumstance, I am in control of myself; if I fight the circumstance and try to control it, it controls me. What we resist persists.

One fascinating way in which people resist political and social change is to project their biases onto Nature. Professor Duncan Taylor cites a couple of interesting examples:

> ...in the seventeenth century...during the English Civil War[,] the bee-hive, with its queen, drones or "nobles," and its workers, was regularly employed by Stuart supporters to defend the concept of feudalism and social hierarchy. This tendency to project human values onto nature and then use such values to lend support to a particular world-view or social structure can again be witnessed throughout the nineteenth and twentieth centuries. Thus, for...William Bateson, the natural hierarchy of the biological world was seen to legitimize British class structure. Indeed, for a number of late nineteenth and early twentieth century thinkers, such concepts as biological hierarchy and homeostasis [a state of physiological equilibrium] were employed to validate and support those traditional values that were being eroded away in a rapidly expanding industrial world.[36]

Acceptance of a circumstance—that which is—is based on the notion that you cannot move away from a negative; you can only move toward a positive. To illustrate, you are near timberline on a mountain that is rich in patches of huckleberries. While peacefully picking berries, you suddenly come face to face with a large bear also eating berries. Without thinking, you start to run away from the bear. Because you are running away from the bear and looking at it over your shoulder to see how close it is, you will either run into the tree you wanted to climb or run past it. Your other choice is to run toward the tree, not away from the bear. In this case, you focus all your attention on the tree. You do not know where the bear is and you do not care, but you know exactly where the tree is and you care about that very much.

You can only accept what is if you are present in the here and now. People in Western culture spend an inordinate amount of time wanting things and/or circumstances to be different; frustration results from refusing to accept what is as it is now. We cannot control circumstances, be they how a forest functions or how the market for wood-fiber products acts over time. We can only accept what is and control how we react to it (Figure 53).

Because nothing is fixed or constant, no matter how much we insist on thinking it is, nothing is as it appears to be. As physicist Fritjof Capra wrote: "Whenever the Eastern mystics express their knowledge in words—be it with the

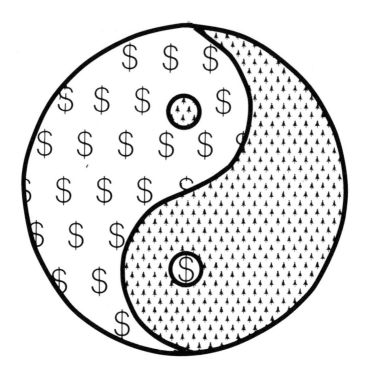

FIGURE 53 Everything is cyclic in Chinese thought; this notion is expressed in a symmetric arrangement of the dark *yin* and the bright *yang*. The rotational symmetry forcefully suggests a continuous cyclic movement. As the *yang* returns cyclically to its beginning, the *yin,* attaining its maximum, gives place to the *yang.* The two dots symbolize the idea that each time one of the forces, *yin* or *yang,* reaches its extreme, there already is contained within it the seed of its opposite. In this figure, biological sustainability and economic sustainability are in a dynamic cycle. The dots represent the idea that a healthy forest is the most economical, and a reinvestment of mineral and organic capital in the forest is required to ensure its health.

help of myths, symbols, poetic images or paradoxical statements—they are well aware of the limitations imposed by language and 'linear' thinking. Modern physics has come to take exactly the same attitude with regard to its verbal models and theories. They, too, are only approximate and necessarily inaccurate."[8] The same is true in managing forests; everything we do is *only approximate and necessarily inaccurate.* There are no absolutes.

So how do we deal with change? Once again, Professor Duncan Taylor made an astute observation in regard to this question. Throughout Western literature, says Taylor, our descriptions of the natural world reflect the values and biases of a given period in our history. Our perceptions of nature often tell us less about what is actually "out there" in the landscape and more about the kinds of mental topography and projections that we carry in our heads. It is natural, therefore, that as values change, so do our perceptions of Nature.

The form Western knowledge has taken is predicted on the objectification and control of other people as well as Nature. We are at a stage in history, however, when—if only for our very survival—it becomes increasingly necessary to realize that our ultimate security lies in a consciousness of wholeness and integration. "And so," concludes Taylor, "in order to step successfully into the future, we must find the courage to step first into the deepest recesses of ourselves.[36]

We cannot change history, and we cannot change one another; we can only change ourselves. As we change ourselves, our perception of one another and of everything else changes, and that is the bridge between conflict and vision.

Part III

PLANNING:
THE BRIDGE FROM
CONFLICT TO VISION

The dogmas of the quiet past are inadequate to the stormy present.
We must think anew and act anew.

Abraham Lincoln

I am but one person; what can I do? The answer is always the same: I can do *something*. Ours is not to question the size or value of our individual contributions. Our task in life is simply to give from the essence of who we are. Each gift is unique and valuable, and each adds a necessary piece to the whole.

To understand the value and power of each person in the context of collective thoughts and actions, pretend for a moment that we humans are snowflakes. We are part of the first snow of winter. Numbering in the millions, we fall one by one out of a quiet sky, touching our neighbors as we whirl and spin to earth. As we fall, we magnify one another until we blot out the sky.

The pioneers, the first flakes to fall, landed on warm soil and melted, disappearing without an apparent trace. But are they really lost? Have they really had no effect? Each flake that landed on the soil, only to melt and disappear, has given its coolness to the soil until, after enough flakes have landed and melted, the temperature of the soil drops.

Finally, because of the cumulative effect of all the flakes that have gone before, the soil has cooled enough for us—you and me—to survive as we land, and still the flakes fall one by one. It snows all night, and by morning, a glittering, transformed world greets the rising sun. As far as the eye can see is a world of winter white—one flake at a time—as we add our collective beauty to the wonder of the Universe.

Humanity is a multitude of people, each with a unique and priceless gift to give. British philosopher James Allen was speaking of these gifts when he wrote:

> The greatest achievement was at first and for a time a dream. The oak sleeps in the acorn; the bird waits in the egg; and in the highest vision of the soul a waking angel stirs. Dreams are the seedlings of realities.
>
> Your circumstances may be uncongenial, but they shall not long remain so if you but perceive an Ideal and strive to reach it. You cannot travel *within* and stand still *without*.[37]

Land-use planning is our opportunity to give the gift of our love, talent, and skill rather than to judge the effects of our giving. No one's gift is better, more splendid, or more important than anyone else's. They are only different. And each is necessary to the wholeness of the world. Like the flowers on a tree, if one falls, the tree is to that degree diminished of its beauty. Omit one person's gift and the potential for the world remains a fond imagining.

6

VISION:
THE FRONTIER
BEYOND CONFLICT

It is now the twilight of the 20th century, a century in which the abundance of natural resources has rapidly dwindled toward scarcity while the world's human population has grown at an exponential rate. We must now address a moral question for the next century: Do those living today owe anything to the future? If you answer no, then surely we are on the correct course, polluting the world and spending our children's inheritance from Nature as though there were no tomorrow.

If, on the other hand, you answer yes, then consider that we must now determine what and how much we owe, lest our present course continue unabated into the 21st century, with its rapid destruction of environmental options for all generations to come. Be forewarned, however, that meeting that obligation—whatever we determine it to be—will require a renewed sense of personal and social justice, one that causes us to act now for the simultaneous benefit of both the present and the future.

To fulfill our acknowledged obligation requires fundamental changes in our social consciousness, changes that will demand choices different from those heretofore made. We must therefore reach beyond conflict to environmental vision—the bridge planning provides, albeit an often treacherous one. Planning is about choices: those with which we imprison ourselves and those with which we set ourselves free.

To change anything, we must, through the choices we make, reach beyond where we are and where we feel safe. We must dare to move ahead, even if we do not fully understand where we are going or the price of getting there, because we will never have perfect knowledge. We must become students of processes and relinquish advocacy of positions and embattlements over narrow points of view.

No biological shortcuts, technological quick-fixes, or political hype can mend what is broken. Dramatic, fundamental change, which is both frightening and painful to most people, is necessary in order to improve the quality of life—even that of next year. The question is not *can* we change, but rather *will* we change. Change is a choice—the choice of individuals reflected in the collective of society.

True progress toward an ecologically sound and socially just environment will be expensive in both money and effort. The longer we wait, however, the more disastrous becomes the environmental condition and the more expensive and difficult become the necessary social changes.

The concept of environmental justice by its very nature asserts that we owe something to the future, but what exactly do we have to give? We owe the only things we can give: our love, our trust, and our respect embodied in every option and every choice we pass forward to the children of today and of tomorrow and beyond. And it is exactly because options embody all we have to give that today's land-use planning must be carried out within the context of ecological sustainability and environmental justice for all peoples and all generations.

Agencies and individuals responsible for the welfare of our nation's natural resources have functions that are prescribed by law but not necessarily specified by law. A wide range of administrative discretion is therefore permitted and allocated by legislative bodies. Although this is as it should be, the system lacks a guiding precept for "public service," one that in fact means serving the whole of the public with the impartiality of justice for the common good of all generations.

These agencies and individuals are often under political pressure from special-interest groups. The pressure they exert results in dedicated public servants being captured both by the history of their organizations and by the fear and political weaknesses of their superiors. This means that these same public servants are subjected to conflicting demands and receive no assurance or ethical governance from the public or any overseeing body. The result is that our system of caring for the nation's natural resources has neither an ethical standard or "ethos" nor a sense of social justice within society itself or toward the environment that nurtures and sustains society.

Ethos, a Greek word meaning character or tone, is best thought of as a set of guiding beliefs, which are all too often absent in the land-management agencies of the U.S. government. In phrasing this guiding direction, a distinction needs to be made between "ethos" and "policy."

Policy can be written in explicit terms and can be in the form of an order—the letter of the law. Ethos, on the other hand, is implicit and includes a guiding set of human values—the heart of the law—that is understood but cannot as such be written. Yet ethos can be translated into policy.

Instead of a clearly articulated ethos, however, we are, as Leopold wrote in 1933, confronted by a contradiction: "To build a better motor we tap the uttermost powers of the human brain; to built a better countryside we throw dice." The dice we throw are those of arrogance and greed.

Arrogance, although often attributed to ignorance, is born out of an uncritical assumption of knowledge. Greed, on the other hand, is born out of the fear of loss, the fear of never having enough in the material sense. And both are justified in the economic theory underlying our current capitalistic way of doing business, which is but a reflection of our social psyche out of which arise the environmental conflicts.

Here, then, is the question: Can environmental conflicts by resolved? The answer is emphatically, yes! Conflicts are created by the choices made and can be resolved by electing different choices with resolution firmly in mind and a collective vision of the future toward which to build. By working together in land-use planning, we can reap the following benefits:

1. Planning defines a course of action and helps ensure that the selected course has a good chance of success.

2. The planning process serves as the foundation on which all our organizational activities are based and, as such, must answer what, where, when, why, and how actions are to be taken and who will conduct the actions for whom. If these questions are answered in a manner satisfactory to all, the chance of conflict is greatly reduced; if perchance conflict arises, it can usually be transcended within the context of the plan.

3. A well-conceived plan helps determine who is responsible for implementing which tasks and provides people with the opportunity to gain a clear insight with respect to their specific tasks in relationship to the function of the organization as a whole.

4. Although planning is a field with its own rules and jargon, it is basically a process of rational logic that works well when performed well. Therefore, a good planning process helps a particular group of people communicate to others that it is thoughtful in what it is doing and stands a good chance of accomplishing its stated purpose.

5. A good planning process will aid in monitoring and evaluating an organization and its achievements.

6. Periodic evaluation of a group's progress toward meeting its vision, goals, and objectives identified in the plan is critical for evaluating its success and whether or not it is providing the promised services to its customers and supporters. This step is essential for the health and growth of any organization.

7. The planning process is a vehicle through which the collective long-range vision of the people involved with an organization can be realized. Planning helps people focus their energy on their vision, goals, and objectives and thereby helps an organization achieve maximum utilization of its human talents and financial resources.

8. Finally, strategic planning helps us to influence the future to the benefit of those who will follow. All this, however, is dependent on a shared vision of the future.

COLLECTIVE VISION: A LESSON FROM INSECTS

Society is composed of individual human beings much as the compound eye of an insect is composed of individual facets. Each facet has its own light-sensitive element and its own refractive system, and each forms but a portion of the image. Just as there are as many points of view in the compound eye of an insect as there are facets, there are as many points of view in a society as there are people, and although everyone is right from her or his point of view, no one person has the complete image.

Reality is what reality is, and in its vastness none of us sees it all. If we could in fact see reality, we would realize that "wrong" does not exist, because while we may all *look* at the same thing, we look at it from different physical and psychological perspectives. We thus each *see* it in a unique way, from a unique point of view.

It is precisely because each of us has a point of view, established after considering the data and reaching a conclusion, that I cannot convince you of anything. If I am to convince you that my point of view is "right," then I must simultaneously convince you that your point of view is "wrong." You will resist, of course, first because I have stolen your dignity and second because your point of view is as correct from your interpretation of "your data" as mine is from my interpretation of "my data," even though the initial database may be the same.

Although I cannot convince you that you are wrong without stripping you of your dignity, I can give you new data that allows you to reach a new conclusion while maintaining your dignity. In doing so, I have *raised the value* of your making a new decision, one based on new information. This allows you to change your mind in privacy, to move as it were to a new position so you can see more.

For example, suppose that a ship is sinking. The entire crew and captain can see an island, but before they decide to attempt getting there, they have to decide that is where they want to go. Once that decision is made, the captain looks at the island and sees no way to reach it because a reef intervenes. The captain then asks the crew what they see, because each person is peering out of a porthole. Eight crew members say they see no way to reach the island, but the cook says, "I see a way to get there."

Each member of the crew is correct from his or her point of view, his or her line of sight. Each sees something they agree is an island, but each sees something different because each sees it from a different point of view. The only way the captain can see what the cook sees is to move to where the cook is standing and look out the cook's porthole to see from the cook's point of view. The captain must have the courage to risk moving from a known, comfortable position to an unknown, uncomfortable position in order to see more.

To save our forests, indeed to save our planet and society as we know it, we must be willing to risk moving in order to see more, to validate one another's points of view. The world can be seen in totality only when it is seen simultaneously from all points of view. Such vision demands an open mind in a collaborative process of intellectual and emotional exploration of what is and what might be, the results of which are committed to paper.

Few people, however, know what a statement of vision is, what a mission is, what a goal is, or what an objective is or know how to state them. A statement of *vision* is a general statement that describes what a particular person, group of people, agency, or nation is striving for. Gifford Pinchot, the U.S. Forest Service's first chief, for example, had a vision of protected forests that were productive of commodities for people in perpetuity. In them he saw the "greatest good for the greatest number in the long run." Pinchot's vision was converted into a statement of *mission,* which is a simple, generalized statement of the Forest Service's perceived responsibility: "To protect the land and serve the people."

A *goal* is a general timeless statement of intent that will remain until it is achieved, the need for it disappears, or the direction changes. A goal is a statement of direction; it may be vague and is not necessarily expected to be accomplished. It does, however, serve to further clarify the statements of both vision and mission. Without a goal, where you end up is left to chance, or "If you don't know where you're going, any path will take you there." This was Alice's dilemma when she met the Cheshire-Cat in author Lewis Carroll's story of *Alice's Adventures in Wonderland.* Alice asked the Cheshire-Cat:

"Would you tell me, please, which way I ought to go from here?"

"That depends a good deal on where you want to get to," said the Cat.

"I don't much care where—" said Alice.

"Then it doesn't matter which way you go," said the Cat.

"—so long as I get somewhere," Alice added as an explanation.

"Oh, you're sure to do that," said the Cat, "if you only walk long enough."[38]

An *objective,* on the other hand, is a specific statement of intended accomplishment, which is measurable and observable, has a reference to time, has an associated cost, and is attainable. The following are additional attributes of an objective: (1) it starts with an action verb, (2) it specifies a single outcome or result to be accomplished, (3) it specifies a date by which the accomplishment is to be completed, (4) it is framed in positive terms, (5) it is as specific and quantitative as possible and thus lends itself to evaluation, (6) it specifies only the "what, where," and "when" and avoids mentioning the "why" and the "how," and (7) it is product oriented.

Consider the following goal: "My goal is to see Timbuktu." It can be restated into an objective: "I will see Timbuktu on my 21st birthday," which clearly specifies a date. Now the stated objective is action oriented: "I will see." It has a single outcome: seeing Timbuktu. It specifies a date and is framed in positive terms. It lends itself to evaluation of whether or not the stated intent was achieved, and it clearly states "what, where," and "when." Finally, it is product oriented: to see a specific place.

As I strive to achieve this objective, I must accept and remember that my objective is fixed, but the plan to achieve it must remain flexible and changeable. A common human tendency, however, is to change the objective (devalue it) if it cannot be reached in the chosen way or by the chosen time. It is much easier, it seems, to devalue an objective than it is to change an elaborate plan that has shown it will not achieve the objective as originally perceived.

It is important to understand what is meant by vision, mission, goal, and objective because collectively they tell us where we are going, the value of getting there, and the probability of success. Too often, however, we decide where we are going based on the cost of getting there.

First, we must determine what we want by the perceived value and purpose of the outcome. Second, we must make the commitment to pay the price, whatever it is. Third, we must determine the price of achieving the outcome. Fourth, we must figure out how to fulfill our commitment (how to pay the price) and make the commitment to do so. Finally, we must do it.

Once the statements of vision, mission, goals, and objectives have been completed, it is possible to answer the following questions concisely: (1) what do we want, (2) why do we want it, (3) where do we want it, (4) when do we want it, (5) how much (or many) do we want, and (6) for how long do we want it (or them)? If a component is missing, the result may be achieved by default, but not by design.

Only when all of these questions have been concisely answered can planning begin. We know where we want to go, have some idea of the value of going there, and can calculate the probability of arrival. Next we must determine the cost, make the commitment to pay it, and then commit ourselves to keeping our commitment.

Although we can define our vision, mission, goals, and objectives, it must be accepted first and foremost that it is the land that limits our options, and we must keep these limitations firmly in mind. At the same time, we must recognize that they can be viewed either as obstacles in our preferred path or as solid ground on which to build new paths. As previously stated, nature tends to deal in short- and long-term trends and not in absolutes. Second, habitat (food, cover, space, and water) is a common denominator among species, and we can use this knowledge to our benefit. Third, sustainable forests require sound, long-term ecological goals and objectives before short-term economic goals and objectives can be considered.

One of the main reasons industrialists have succeeded in liquidating the majority of the old-growth forests in the Pacific Northwest is because they know exactly what they want (old-growth timber), why they want it (to make windfall profits on a free, finite resource, in which they have no investment, which they do not own but which they can acquire for relatively little capital outlay), where they want it (at foreign markets of the Pacific Rim or, barring that, at the mill), when they want it (now), how much they want (all of it), and how long they want it (as long as it will last, to the last available tree).

Let's look at a short example. A 1300-acre parcel of land containing Douglas-fir trees was logged. The trees were estimated to have been between 800 and 900 years old, making them the oldest known stand of trees in the state of Oregon:

> In 1984, the U.S. Forest Service sold about 200 acres in the heart of the stand to Willamette Industries. Nearly 150 acres of the sale have been clearcut. The remaining 56 acres were scheduled for cutting this summer. When the Oregon Natural Resources Council (ONRC) brought suit earlier this year to halt all cutting in the area, however, Willamette Industries apparently decided to force the issue.
>
> Last week, the company selectively cut the largest and oldest trees in the stand, effectively nullifying the ONRC suit. The trees are to be turned into plywood and fiber board. Apparently the company has broken no laws, and plans to complete the clearcut ahead of schedule will undoubtedly go forward.[39]

Officials of Willamette Industries, right or wrong, were the only ones who had clearly defined their goals and objectives, and they carried them out boldly, within their legal bounds. These episodes, all too common with finite, natural resources, smack of the courtroom where procedure and the letter of the law prevail over substance and the heart of the law.

The sad part of this case is that only about 50 percent of each old-growth tree at the logged site had been used because they were so rotten. The other 50 percent was on firewood piles or burn piles. Furthermore, the stand was not old growth in the strict ecological sense; it was scattered old-growth trees surrounded by mid-aged trees. The latter could have been logged while leaving the old-growth trees standing.

Such confrontations gain little but enmity. Instead of combat, we must build together a vision of a sustainable future in which the core values and expertise of each individual are acknowledged. To accomplish this, we must exercise the good sense and humility to ask our children what they think and how they feel about their future, because each generation must be the conscious keeper of the generation to come—not its judge. It is therefore incumbent on us to prepare the way for those who must follow, which will entail, among other things, wise and prudent planning.

RATIONAL THOUGHT: A REQUIREMENT FOR RATIONAL PLANNING

The ultimate test of human beings is to share the labors and fruits of society. To share the best society has to offer, we must offer the best we have to society by learning to work together as equals, as one another's keepers, and as one another's learning partners. Only then can we achieve a collective vision for the future, which is inclusive and responsible, yet simultaneously allows and protects individual autonomy; only then can we plan wisely for a sustainable environment.

To learn how to accomplish this, we need to return to the ancient forest of what is now southeastern Canada and the northeastern United States and relive the story of the Peacemaker, who established a confederacy of peace that has lasted over 400 years.[40] The story begins around the year 1450 among the Iroquoian-speaking people who lived on the north shore of Lake Ontario, in what is now eastern Ontario, Canada. According to tradition, a baby boy was born under circumstances so mysterious that to this day some Iroquois will only pronounce his name during ceremonial recitations of his story. Otherwise, they simply call him "the Peacemaker."

As a young man, Peacemaker is rejected by his own people and leaves his village. This is a time of bloody feuds in which each killing sparks a downward spiral of vengeance and reprisal. Into this world of unrelenting bloodshed, Peacemaker paddles his canoe across Lake Ontario to the land of a tribe later known as the Mohawks. He seeks out the most ferocious of the surviving warriors of this undeclared war to begin his humble mission of spreading the political philosophy of peace.

His words require considerable thought. Peacemaker's premise is that although all human beings possess the potential for rational thought (rational logic), they must first work their way through the clouds of existing irrational thought (emotions, which give rise to irrational logic).

Peacemaker's message is simple. Because the Universal Laws are rational, Nature, which obeys these laws, is rational, and human beings as a part of Nature must therefore also possess rational thought. However, while Nature is rational, human beings also possess irrational thought and thus imperfectly understand Nature's rationality. It is therefore wise to act with humility in one's relationship with Nature.

He goes on to point out that unless we believe that every human being possesses rational thought, we are powerless to bring about peace short of total human annihilation, because we cannot negotiate with one another so long as our thinking is irrational. This being the case, the first principle of negotiation and the power to act with confidence is the belief that *all people are rational*; only then is it possible to reach accord among people.

Peacemaker holds forth the promise of hope in the future, a future in which people can gather together and use their minds to create peace—not war. He raises the idea of rational thinking to the status of a political principle of survival. He promotes clear thinking as the highest human level, saying that people have the potential to use their minds to create a better life through peace, the power of the individual, and clear thinking.

In addition to the importance of rational thought itself, he teaches that with judicious application of rational thought, a society can be created in which people can form governments dedicated to the proposition that no person shall abuse another, a society in which all people defend the rights of each person and each person defends the rights of all people. This is possible, because Peacemaker knows that any person who uses rational thought will inevitably seek peace and that the belief in these principles will surely lead to the organized enactment of the vision.

Peacemaker points out that the test of these principles lies in their converse. If you do not believe in the rational nature of another person, then you believe that you cannot negotiate with him or her. If you do not believe that rational people ultimately desire peace, then you cannot negotiate confidently toward peace with your opponent. If you cannot negotiate with your opponent, you are powerless to create peace, and if you cannot organize around these beliefs, then the principles cannot move from the minds of people into the actions of society.

Peacemaker is eventually joined in his mission by an Onondaga exile named Ayawentha, or Hiawatha. (Hiawatha is an Iroquois name mistakenly used by Henry Wadsworth Longfellow for the Ojibwa hero in his *Song of Hiawatha*.) Both Peacemaker and Hiawatha understand that people eventually reach a place of such psychological pain, fear, and rage that a sense of hopelessness prevails. They recognize that it is difficult to reach clear thinking at such times and therefore direct much of their attention to dealing with these feelings and their resulting sense of hopelessness. Together, they bring the message of hope, dignity, and peace to the warring people by persuading them to join in a "Great Peace" based on a "great binding law."

With all fifty chiefs of the first Grand Council assembled on the shore of Lake Onondaga (which is now surrounded by Syracuse, New York), Peacemaker plants the original Tree of Peace, a magnificent eastern white pine, beneath which the warring nations bury their weapons of war. Four long roots, the "white roots of peace," stretch from the tree in the four sacred directions, and Peacemaker proclaims: "If any man or any nation outside of the Five Nations shall show a desire to obey the laws of the Great Peace...they may trace the roots to their source...and they shall be welcome to take shelter beneath the Tree...." Thus began the Great Law of Peace, based on strength through union and embodying the Iroquois notions of free expression and representative government with checks and balances.

As Peacemaker taught in the spiritual language of his time, it is helpful to look at his message in the language of our time, which necessitates understanding the meaning of and the relationship between two words: emotion and logic. Emotion is a disturbance, excitement, or a state of feeling, such as anger or fear. Conversely, logic is a science that deals with the canons and criteria of validity of inference and demonstration. Emotion and logic appear to be mutually exclusive.

Emotion, which produces an irrational logic, is the energy that gives us values and feelings and drives us. Logic, when rational, is the steering wheel that allows us to negotiate the values contained in our emotions. Both aspects of thought are valid, because they are different and not substitutable. For example, emotions, whether negative (such as pain, fear, and despair) or positive (such as love and compassion), must be validated and transcended before logic can be accepted. Thus, negative emotions can only be brought to logic when all parties are open and honest, where love, trust, and respect can eventually prevail.

When our irrational, emotional views have been validated, they can be transcended and expressed rationally; only then can we begin to understand one another and to accept one another's set of values. Then, and only then, can rational thought and negotiation proceed to a satisfactory conclusion built on hope, dignity, and peace, when "a gentle answer turns away wrath."[41]

A major problem in the world today is the apparently irreconcilable split between irrational logic, which is false logic obtained in the face of repressed emotions, and rational logic, which is reached when emotions are validated, understood, accepted, incorporated, and transcended. We shall remain in darkness until we transcend our irrational thinking and bring it to a rational wholeness, and through wholeness gain the light of understanding and mutual respect. Only through rational logic can we achieve the insight (which means to see within) necessary to *see* the world and plan wisely and responsibly in the future for the present.

RATIONAL PLANNING REQUIRES
OUR TOTAL PRESENCE

We cannot be present if we are thinking either about the past or the future. *We are mentally where we think we are.* The same is true in managing our forests.

The present is seldom quite right or seldom seems to be good enough, particularly when dealing with natural resources, especially renewable ones. The past is thought of as "the good old days," with unlimited resources and few or no regulations controlling their use. The future—the next planning cycle, the next stand of young trees—will be better. The point is that the forest industry, forest economists, and forest managers take past values, both in dollars per unit of raw materials and quantity of "renewable" raw materials produced per acre, and project them into the future—skipping the present. The conservation-oriented public, on the other hand, wants results now.

The present is all we have. Yesterday is gone. We cannot change it. Tomorrow is not here, and we have no idea what it will bring. In reality, "tomorrow" is something that is always coming but never arrives.

To be *present,* in the sense of being *mentally here, now,* is a difficult concept to define because there is no word that means mental presence as opposed to simple physical presence. Let's examine what it means not to be present. Have you ever driven from point A to point B and not realized how you got there? Perhaps you are thinking about last year's vacation (remembering the past). It was your first trip to Hawaii, and you had a very rough flight, although you had a marvelous time once there. You are flying to Hawaii next month for your long-anticipated vacation and you begin to think about how much fun you had last year. Then you vividly remember last year's flight and become afraid that next month's flight might be the same. You are either in the past or the future, but suddenly you are jerked into the present as your car sputters and you coast to the side of the road out of gas.

Being present—here and now—is absolutely necessary for rational planning because everyone involved in the planning process must in some way change her or his thinking and relinquish the rigidity of old belief systems.

MODIFYING OUR BELIEF SYSTEMS

Change is the death of an accepted belief system that has become synonymous with our identity and therefore our security. We die daily to our ideas and belief systems, and in so doing, we go through the five stages of dying,[42] which really are five stages of grieving. These stages are necessary as a process that prepares the way for change, a dying of the old thoughts to make way for the birth of the new:

1. Denial (refusing to admit reality, trying to invalidate logic, resistance to change) is the first stage of a dying belief system in which we isolate ourselves because we do not accept change. We become defensive, fearful, and increasingly rigid in our thinking; we harden our attitudes and close our minds. If I become defensive, if I start to form a rebuttal before someone is finished speaking, if I filter what is said to hear only what I want to hear, I am in denial.

2. Anger is the violent projection of uncontrollable fear. A person is emotionally out of control because she or he can no longer control circumstances perceived as threatening.

3. Bargaining is trying to find a way out of having to deal with what is. It is looking for a way to alter circumstances based on "acceptable" conditions. In forestry, it could be called fertilization, which is an impatience with Nature's timetable that results in looking for an "acceptable" shortcut. We bargain with Nature.

4. Depression is when we become resigned to our inability to control or change the "system," whatever that is, to suit our desires. We feel helpless and deliberately give up trying to alter circumstances. We feel that we become victims of external forces, and our defense is to become cynical and distrustful. A cynic is a critic who stresses faults and raises objections but assumes no responsibility. A cynic sees the situation as hopeless and therefore espouses self-fulfilling prophecies of failure regardless of the effort invested in success.

5. Acceptance, the final stage of grief, is creative and positive. With acceptance returns trust and faith in the goodness and justice of the outcome. Acceptance of what is (for example, an unplanned change) allows us to define the problem and to transcend it, but acceptance must come before a resolution is possible.

Why do we fear change so much? We resist change because we are committed to protecting our existing belief system. Even if it is no longer valid, it represents the safety of past knowledge in which there are no unwelcome surprises. We try to take our safe past and project it into an unknown future by skipping the present, which represents change and holds both uncertainty and accountability. Thus, when confronted with change, we try to control the thoughts of others by accepting what to us are "approved" thoughts and rejecting "unapproved" thoughts. Such control is a defense against change, which, after all, is in the mind. As author George Bernard Shaw said, "My own education operated by a succession of eye-openers each involving the repudiation of some previously held belief."

Becoming too comfortable with our belief systems is comparable to a turtle. A turtle has only two choices: it can pull its head into its shell where in safety it starves to death or it can stick its neck out and risk finding something to eat and live.

In this sense, organizations often tend to be too self-protective and systematically distort information in self-serving ways. Such distortions do not depend on deliberate falsifications by individuals. Instead, competent, hard-working, and honest people can sustain systematic distortions by merely carrying out their organizational roles in an uncritical (and therefore personally safe) manner. Unchecked by outside influences or the undeniable realities of catastrophic failures, organizational systems can sustain self-serving distortions, even though the potential for catastrophic consequences is significant.

A technological culture such as ours faces two choices: it can wait until catastrophic failures expose systemic deficiencies, distortions, and self-deceptions (the turtle with its head sucked into its shell) or it can provide social checks and balances to correct for systemic distortions prior to catastrophic failures (the turtle with its head outside its shell, risking a view of the world). The latter is a more desirable alternative, but it requires the active involvement of independent people who take the initiative to ask "unfavorable" questions and pursue "unfavorable" inquiries. Without such initiative, checks and balances are undermined and catastrophic possibilities are likely to increase as the scope and power of organizational technology expands.

As we move forward in land-use planning, remember the turtle. The crisis is in the will and the imagination and not in the possibilities.

decisions, Decisions, DECISIONS

How we think determines the decisions we make. We make hundreds of decisions every day. Each decision is a fork in the road of life; each fork is an option, an alternative, a choice. The direction of our lives is a result of many little decisions, most of which we do not even remember. We usually remember the big decisions, but we seldom realize that a single big decision is merely an expression of little decisions along the way.

The life cycle of a salmon epitomizes the choice of destination. A reddish orange egg is deposited in a redd (the gravelly stream bottom that serves as a nursery for salmon) in the headwaters of a Pacific Coast stream. There the egg lies for a time as the salmon develops inside. In time, it hatches and struggles out of the gravel into the open water of protected places in the stream. There it grows until it is time to leave the stream and venture into life. It can go only one way—downstream to larger streams and rivers until at last it reaches the ocean.

After some years at sea, the inner urge of its species drives the adult salmon along the Pacific Coast to find the precise river it had descended years earlier. It must make a critical decision. If it selects an incorrect river, it will not reach its destination, regardless of all the other choices it makes. If it swims into the same river it had descended, it is on the correct course, until it comes to the first fork and must choose again.

Each time the salmon comes to a fork in its journey, it must make a choice and must accept what the chosen fork has to offer and forego the possibilities of the fork not taken. It can only return to the redd where it was deposited as a fertilized egg if it knows where it is going and when it has arrived. Its objective is to reach a particular place in a particular stream within a particular time to deposit either its eggs or sperm. The salmon will die, but some of its offspring will live to run the same gauntlet of decisions when their time to spawn arrives.

Our lives have a common thread with that of the salmon, because every decision we make determines where we are, where we are going, and where we will end up. Our stream in life is the collective thinking of parental, peer, and social pressure. Like the salmon, which goes downstream with the current to the ocean, we accept the route of least resistance of collective thinking.

Although most of the salmon die and become part of the sea, a few survive and begin swimming against the current to fulfill their life's purpose. As we mature, most of us will drown in the ocean of mass thinking, going with the current and seeking our sense of value outside ourselves through the acceptance of others who are also drowning in mass thinking.

A few, however, will chart their course against the current, driven by an inner need to find their life's fulfillment in the excellence of achievement. Having dared to risk the unknown (change), they will leave behind the seeds for even greater achievements by the next generation.

No decision is easier or more difficult to make than another. All decisions are the same and are easy to make. The difficult part is getting ready to make the decision, which is a process of making many little, often unconscious, decisions to assess risk and benefit.

We simply cannot get away from decisions. To avoid a decision is still to make a decision, usually an unwise one. Nevertheless, we are not victims of life; we are products of our decisions, and our willingness to risk change dictates the boldness of our decisions.

We always make the best decision we can at a particular time, under a particular circumstance, with the data on hand. This does not mean that, given similar circumstances, we would make the same decision today or in the future. It only means that it was the best decision we could make at that time.

It does not mean that others will necessarily agree with our decisions or we with theirs. It only calls attention to the fact that I must accept your decision as your best because I cannot judge. I do not know why you did what you did; I only know what you did and how that appeared to me.

A gentleman in the U.S. Forest Service taught me much about judgment. I was giving a speech in Spokane, Washington, about fire in forested landscapes and explaining new data and new points of view. When I was finished, he came up to me and, with a quivering chin and misty eyes, said, "I've been with the Forest Service 29 $^1/_2$ years and I'm going to retire in six months. Do you mean to tell me I've been wrong my whole career?" "No sir," I said, "I'm not telling you that at all. You did the best you could with the data you had on hand. Now, however, with much new data, we can make some different choices, different decisions than you could during your career."

Looking at him, listening to his faltering question, I realized with searing insight how incorrect we are when we presume to judge and that we are doubly incorrect when we presume to judge from hindsight. Everyone does his or her best within his or her level of understanding. If, therefore, we feel the need to voice another point of view during the planning process, let's be gentle with one another and treat one another with mercy because we are each doing the level best we can at any given moment.

Besides, whether we realize it or not, we need one another. Consider, for example, an ancient forest, particularly large, old trees. Each signifies primeval majesty, but only together do they represent an ancient forest. Yet we do not even see the forest for the trees. If we could see belowground, we would find gossamer threads of mycorrhizal-forming fungi stretching for billions of miles through the soil. These fungi grow as symbionts on and in the feeder roots of the ancient trees. Not only do they acquire food in the form of plant sugars through the roots of the ancient trees, but also they provide nutrients, vitamins, and water from the soil to the trees and produce growth regulators that benefit the trees. These symbiotic fungus–root structures (mycorrhizae) are the termini of the threads that form a complex fungal net under the entire ancient forest and, as evidence suggests, connect all trees one to another.

Like the ancient trees, we are separate individuals, and like the ancient forest united by its belowground fungi, we are united by our humanity—our need for love, trust, respect, and unconditional acceptance of one another. In determining whose judgment in a decision is right or wrong, we must therefore remember that everyone is right from his or her individual point of view.

Part IV

FUNDAMENTAL ISSUES: THE SILENT DILEMMA

The survival and the well-being of the next seven generations
depend on every thought we think and every action we take in the
now....We are all responsible for all of the tomorrows to come.

Jamie Sams

Rational planning requires dealing with the fundamental issues, instead of addressing only symptoms. Consider the following psychological model. Stress exerts an influence on a person. An individual's internal personality characteristics, shaped and influenced primarily by his or her family of origin and secondarily by his or her environment during the formative years, determine how he or she deals with stress.

In a functional person, stress may accumulate to the point that it causes a symptom to manifest itself. Recognizing that something is amiss, this person seeks help in understanding the problem at its internal core or root level and deals with it at the source. By connecting the stress with the underlying cause, the symptom disappears, because the person was willing to accept change as a process of growth and opportunity. This person becomes progressively freer and more functional each time she or he is willing to accept the inner work necessary to change on a fundamental personality level.

A dysfunctional person is unlikely to seek help and do the inner work necessary to understand and break the stress–symptom relationship. The stress continues to mount and recycles its destructive energy in a self-reinforcing feedback loop that continually strengthens the symptom's manifestation, because the person perceives change as a condition to be avoided at any cost.

Such a person may turn to alcohol or drugs to alleviate the symptom through denial, displacement, and avoidance, which does nothing to break the stress–

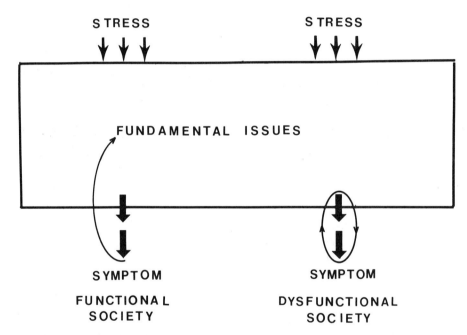

FIGURE 54   This figure represents a forest. Stress (socially induced stress) exerts an influence on the forest. The personality characteristics of a society, shaped by the collective traits of its people, determine how the society deals with the stress it is causing. In a reasonably *functional society* (left-hand side), stress may accumulate to the point that it causes a symptom (the fundamental issue) to begin manifesting itself. The society, recognizing that something is amiss, seeks help in understanding the problem at its fundamental core level and deals with it. A *dysfunctional society* (right-hand side) is unlikely to seek help and do the inner work necessary to break the stress–symptom relationship. In this case, the stress continues to mount and recycles its energy in a self-reinforcing feedback loop (continual deforestation with all its ecological and economic consequences) that continually strengthens manifestation of the symptoms because the society perceives change as a condition to be avoided at any cost.

symptom relationship. This person becomes more rigid and dysfunctional each time she or he resists the fear and discomfort of change.

To extrapolate, Figure 54 represents a forest. Some type of stress (socially induced) is exerting an influence on the forest. Society's personality characteristics, shaped by the collective traits of its people, determine how the society deals with the stress it is causing. In a reasonably functional society (left side), stress may accumulate to the point that it causes a symptom (the fundamental issue) to begin manifesting itself, for example, a well-defined, long-term population decline in northern spotted owls. Society, recognizing that something is amiss, seeks help in understanding the problem at its fundamental core level (in this case habitat loss) and deals with it.

Understanding the relationship between the stress and its symptom (overhauling forestry, including forest economics, and its practices) causes the symptom to disappear because the society is willing to accept changing itself as a process of growth and opportunity. This society becomes progressively more flexible and functional each time it is willing to accept change as a process.

A dysfunctional society (right side) is unlikely to seek help and do the necessary work to break the stress–symptom relationship. The stress therefore continues to mount and recycles its energy in a self-reinforcing feedback loop (continual deforestation with all its ecological and economic consequences) that continually strengthens the manifestation of the symptom (the owl's population declines unto extinction) because the society perceives change as a condition to be avoided at any cost.

Such a society may turn to an argument ("owls vs. jobs") to alleviate or displace the symptom (a growing timber shortage), which does nothing to break the stress–symptom relationship, because the cause (liquidation of old growth through short-term, economically driven clearcutting) is not accepted and dealt with, despite mounting evidence. This society becomes more rigid and dysfunctional each time it resists change.

We as individuals are largely dysfunctional, and we carry our dysfunction into our societal consciousness and behavior. This is manifested in continuing deforestation and the purposeful extinction of species, despite mounting evidence of the illegality of such acts and of the negative long-term ecological/economic cost to society.

7

TECHNOLOGY, SCIENCE, AND UNCERTAINTY

Broadly speaking, technology is the body of knowledge available to a civilization that is of use in fashioning implements, practicing manual arts and skills, and extracting or collecting materials. Technology has no sensitivity, makes no judgments, and has no conscience. It is a human tool and is as constructive or destructive, as conservative or exploitive as its user.

The fact that science-based technologies function to the degree they do when applied outside of the laboratory cannot be marshalled as evidence that our base of knowledge is about Nature. Rather, it is evidence that the controlled conditions of the laboratory can be sufficiently reproduced in the field to achieve comparable results.

As an example to illustrate this dynamic, consider a man who has for some years been trying to teach ranchers a different way of operating their ranches. Two ranchers do exactly what the teacher says. Even though they both do the same thing and they both do it correctly, it works for one but not the other. The difference is not in the technology that was employed, but rather in the properties of the ranches on which it is carried out.

Science-based technology is thus the seed of human exploitation of the world's resources, including people. As such, it is here appropriate to examine what the word "resource" means, because the term is sadly misused. Resource originally meant a reciprocal gift between humans and the Earth, but today it is defined as the collective wealth of a country or its means of producing wealth; any property that can be converted into money.

First we covet what Nature has produced. We then exert "ownership" over that which we covet, and finally we convert Nature into money. The technological efficiency with which we convert Nature into money has even become the measure of social success and stature. We thus transform spirited and lively mutual gifts, including the human resource, into lifeless commodities.

Let's take another look at the term "resource." Resource = *re* and *source*. *Re* means to put back, to restore, and *source* means the original supply, the point of something's origin. Interpreting the word in this way can be the inspiration for the rebirth of its original meaning. How would this change in meaning affect our sense of forestry and the world?

We did not always know we were ignorant about forestry; too often we still do not. In the beginning of forestry, people used crosscut saws, axes, oxen, and horses to harvest an apparently unending supply of giant trees. Because we knew nothing about forests or forestry, we were unaware of "making mistakes." As the west was settled, technological advances moved logging equipment and techniques from flumes, splash dams, and log rafts to log trucks, chain saws, balloons, and helicopters. Exploitation of the forests was and still is limited only by technological capabilities; with each advance, therefore, we have the opportunity to expedite cutting the world's forests. And each technological advance seems to bring with it a false sense of security that whatever problems exist in the forest have been solved.

Although on the surface this seems innocuous, each technological advance has one or more hidden trade-offs in the invisible present, such as soil compaction, soil erosion, or the pollution of both soil and water, which are not immediately apparent. These hidden trade-offs, or cumulative effects, are really hidden costs in terms of the health and productivity of future forests.

Cumulative effects are not well understood. We only learn about them when we see the outcome of our actions—when we have crossed the threshold of no return. Cumulative effects impair the forest's ability to function until it has healed.

In an age of rapidly increasing technology, we too often rely on the promise of new techniques to solve problems, rather than questioning the ecological validity of profit-oriented goals. To minimize economic trade-offs within an ecological system, we forge ahead in developing new technology, and then we spend inordinate amounts of time and money trying to predict the outcome of trade-offs on profit margins. We become obsessed with predictability and predictions.

We face an interesting dilemma in our desire to avoid uncomfortable surprises by predicting the future. In this sense, the only thing we might predict is our own behavior. As Sherlock Holmes pointed out to Dr. Watson, "While the individual man is an unsolvable puzzle, in the aggregate, he becomes a mathematical certainty." He goes on to explain that it is impossible to foretell what any individual person will do, but an average number of people are always predictable because, although individuals vary, percentages remain constant.[43]

While human behavior in some aggregate is predictable, the problem in trying to predict the behavior of the forest is that it rests not on human behavior but on ever-changing relationships of both living and nonliving components of itself as part of the ecosystem. Ecological understanding is an inexact, nonstatistical subject. Cumulative effects cannot therefore be rendered statistical, because ecological relationships are far more complex and far less predictable than statistical models lead us to believe. Although we cannot foresee the moment when cumulative effects become irreversible, we can visit history.

LESSONS FROM HISTORY

"History is but a glimpse into the eternal cycle of Creation, a perceived reflection of what is, a ghost of what might have been, a dream of what might yet be. Creation is that which has, is, and will inexorably draw humanity and the ancient forest into the crucible of Cosmic Interrelatedness where the forest will mirror for humanity the consciousness of its own evolving self."[44]

The forest has often mirrored humanity to itself, and that image has been both positive and negative, as we shall see. First we will visit the Anasazi of Chaco Canyon, Arizona; then we will visit the Easter Islanders,[45] and finally we will spend some time with the Mayans.[46]

Anasazi

Today, Chaco Canyon is mostly saltbush, nothing higher than one's knee, and no trees in sight; 1800 years ago, however, it was a woodland of pinyon pine and juniper, according to the remains of vegetation in the fossilized middens of the bushy-tailed woodrat. The middens, which are the remains of the woodrat's food, range in age from 10,000 years ago to the present and have yielded 10,000-year-old remains of spruce and Douglas-fir, as well as a record of pinyon pine and juniper up to 1200 years ago. Middens are missing, however, from the period of 1200 years ago to 500 years ago, which likely means that the Anasazi hunted the woodrats for food. When the middens reappeared, none showed traces of pinyon pine. What happened to change the landscape so dramatically? Mounting evidence strongly suggests the hand of the Anasazi culture itself.

Over 1000 years ago, the Anasazi Indians moved into the area and established one of the most advanced civilizations that pre-industrial North America ever knew. The Anasazi (Navajo for "ancient strangers") built huge pueblos and developed sophisticated networks of politics and trading. Then, in the 12th century, they mysteriously disappeared. The reason why, which has long been sought, may now have an answer.

First, however, it is important to understand that indigenous peoples did not always live in perfect harmony with their environment. In fact, a number of ancient cultures (such as the Anasazi) were actually quite adept at ecological destruction. Although it may be impossible to prove direct cause-and-effect relationships, it is interesting to note that many cultures collapsed just after major human-caused ecological changes took place.

The peak of the Anasazi civilization occurred between 1075 and 1100, at which time the two Anasazi pueblos, named Chetro Ketl and Pueblo Bonito, were the largest buildings north of Mexico. They were five stories tall and contained more than 500 rooms each. Their three-foot-thick walls were made of flat, inch-high stones mortared with the red mud of the canyon, and the roofs were supported with massive timbers of spruce and fir, neither of which have existed in the canyon since the Pleistocene, more than 10,000 years ago. This means that the Anasazi had to import the huge timbers from distances of 50 miles or more.

Because the Anasazi brought the massive timbers for their pueblos from 50 miles or more away, they could also have imported firewood from similar distances. The decline of the culture must therefore have been caused by more than simple deforestation. It may have been caused by the concordant collapse of the elaborate network of politics and trading, which in turn might have been accelerated by a rapid increase in the population of the region even as its resources decreased. Such an increase in population would place a tremendous strain on the resources, from wood to water to arable land. If the land could not support the population, the culture would collapse, along with the habitat on which the culture was based.

Easter Islanders

Easter Island is a tiny, 43-square-mile piece of land in the South Pacific 2400 miles off the coast of South America. The oldest pollen dates go back some 30,000 years, long before the first people, wandering Polynesians, arrived. At that time, based on the pollen record, the island was forested.

The Polynesians settled on the island in about 400 A.D. They began to gradually clear the land for agriculture, and they cut trees to build canoes. The land was relatively fertile, the sea teemed with fish, and the people flourished. The population rose to about 15,000, and the culture became increasingly sophisticated. Eventually trees were cut to provide logs for transporting and erecting hundreds of statues, or *moai,* some of which are about 32 feet high and weigh as much as 85 tons.

Unfortunately, when the trees were cut, they did not grow back, as shown in the pollen record. Deforestation began about 1200 years ago (a few hundred years after the first people arrived) and was almost complete by 800 years ago. The people also exploited many of the island's other resources, such as its abundance of birds' eggs. The result was ecological disaster.

So much of the forest was cleared that there were no trees to build canoes for fishing. Deforestation led to soil erosion and reduced crop yields, and the eggs of the sooty tern were probably exploited to the point that the bird no longer nested on the island.

The downward spiral had begun. Fewer fish, eggs, and crops led to a shortage of food. Hunger, in turn, brought warfare, even cannibalism, and the entire civilization was pushed to the brink of collapse. By the time European explorers arrived in the 1700s, only 4000 people remained on the island. Today, all that remains of the culture of Easter Island are the statues that once stood erect on specially built platforms, others that lie abandoned between the volcanic quarries of their origin and their planned destinations, and still others that remain unfinished in the quarries.

Mayans

The Mayans inhabited the Petén region of northern Guatemala, an area that is sparsely populated today, but which was an enclave for the Mayans between 1000 B.C. and 1000 A.D. The population of the area grew as much as tenfold over hundreds of years, and the forest was cut down for planting, building, and fuel. Because the tropical forest held most of its nutrients in the plants and little in the soil, most of the available nutrients were lost as the trees were cut. As the habitat disappeared, the animals that provided a major source of protein also vanished. It is as if the Mayan civilization choked on its own success.

There is an interesting twist to the story, however. As the Mayan culture was collapsing, the forest was rejuvenating itself. What is sobering about this scenario is that the Mayans were very ingenious, knowledgeable conservationists. They knew a great deal about their environment, and their systems of land management were very sophisticated, as demonstrated by the Mayan farmers who fed their huge population in the tropical forest of the Yucatán peninsula. Rather than cutting down the forest to practice the destructive slash and burn agriculture of today, they managed the tropical rain forests with ecological acumen and cultural harmony long before the Spanish conquistadors set foot in the New World.

The Mayans practiced sustainable agriculture by constructing *pet kotoob* (plural of *pet kot,* which is Mayan for round wall of stone), which are rock walls two to three feet high that enclose a small area about the size of a backyard garden. Within these *pet kotoob,* the Mayans grew ramón (a tree whose seeds could be ground as a substitute for maize), fruit-bearing sopodilla, mamey, sapote, and citrus trees, as well as cocoa plants, mahogany trees for wood, and other herbs and shrubs not native to the region.

The Mayans did not move in, raze the land, and move on. The changes in the landscape were so subtle (the invisible present) that they took place over a period of 2000 years and were probably imperceptible to the people at any given time.

Yet the hidden environmental damage of centuries undoubtedly played a large role in the eventual collapse of the Mayan civilization.

What Will History Record About Us?

Today, even more so than in the past, the transition from wild forests to culturalized ones is fraught with many changes and uncertainties, because today we have proportionately greater technology available and proportionately less wisdom with which to use it. Our experiment is vast, unplanned, and unwitting in its change of the spatial and temporal structure of the landscape. Although it seems clear that the mechanisms stabilizing Nature's landscape are largely ecological in nature, such stability emerges from a complex of factors and the manner in which biological communities are arrayed in time and space.

Disturbance has been a common historical feature throughout the forests of the west. Much of the Cascade Mountains of Oregon and Washington was burned in massive forest fires during the 1300s, and numerous smaller fires have burned since then. Most fires in the Coast Range of Oregon date from a large fire during the middle of the 19th century. East of the crest of the Cascade Mountains, extending into the interior forests of the northern Rocky Mountains, both major fires and periodic outbreaks of insects appear to be part of the normal historic pattern. These disturbances were part of the ecological pattern to which the organisms that comprise the forests were adapted.

On the other hand, foreign disturbances (those that in one way or another defeat the adaptive mechanisms that confer stability on the ecosystem) are potentially destabilizing. Such foreign disturbances as those introduced by humans all too often initiate the invisible present, which sooner or later confronts a people with an irreversible threshold of severe environmental alteration and its accompanying collapse of their culture, such as the Anasazi, the Easter Islanders, the Mayans, and many other civilizations in Eurasia.

What about today? Have we learned from the past? Deforestation in the Amazon, as well as in the coniferous forests of the Pacific Northwest and boreal forests across Canada, shows that we have learned little. Imagine, for example, the forests of the Cascade Mountains in western Oregon and Washington, the Klamath and Siskiyou Mountains in southwestern Oregon, the Coast Range in western Oregon and Washington, and the Olympic Mountains in northwestern Washington all burning in one year. According to Boone Kauffman, a fire ecologist at Oregon State University, that is "roughly equivalent to the amount of land that burned in only one year in the Brazilian Amazon." Each year, an area that is 80 percent the size of the state of Oregon burns in the Brazilian Amazon alone.[47]

The major cause of deforestation, which leads to such extensive burning, is the conversion of tropical forests to pastures for cattle. Simple harvesting of timber also causes problems, however, because once the canopy of the forest is opened, the understory environment changes drastically and the forest can no longer sustain itself. Never in the history of humanity has so much of the world's tropical forests been disturbed in such a foreign and catastrophic way on such a large scale as during the last 30 years. The significance of this statement lies in the fact that tropical rain forests—one of the world's oldest ecosystems—occupy only 7 percent of the Earth's surface but are home to more than 50 percent of all species. What does this mean in terms of the Amazonian tropical forest?

An intact rain forest creates its own internal and external climate in which about half of all the rainfall originates from moisture given off by the forest itself. When large areas are deforested, local and regional climatic patterns change. Once the forest is gone, drought is likely to occur, which increases the probability of fire and decreases the probability that the forest will ever return.

The environment in the deforested areas of the Amazon has been altered to such an extent that the ecological processes that once maintained the tropical forest are unraveling in irreversible change. Once the forest has been even partially cleared or logged, the environmental conditions change swiftly and dramatically. Removal of the trees not only alters the internal microclimate of the forest by exposing its heretofore protected, moist, shaded interior to the sun but also leaves behind large accumulations of woody material that are exposed to the sun's drying heat. Daily temperatures soar in the deforested areas by 10 to 15 degrees, which causes the woody fuels to dry and become extremely flammable.

Now it is not a matter of *if* the area will burn, but instead *when* it will burn. The ultimate result is a quick, dramatic change from a dense, closed-canopy forest virtually immune to fire to a weedy, flammable pasture in which fires are common and often occur repeatedly—to the exclusion of a new forest.

In the preceding stories, decisions about profit margins played an integral part in converting today's renewable abundance into tomorrow's finite scarcity. These decisions, which determine the direction of the invisible present, all too often manifest themselves as a future threshold of no return.

Today's danger lies in the technology for harvesting timber and utilizing the wood, which advances much faster than our scientific understanding of how a sustainable forest functions (Figure 55). Added to this equation is the elusive, invisible present of hidden, cumulative effects and the only real absolute—a growing *uncertainty* about the outcome of our decisions, and the more uncertain we become, the more we try to predict and to force results that are within our intellectual comfort zone. None of our efforts, however, can make the forest conform to our simplistic desires for predictable, short-term absolutes as opposed to Nature's variable-term trends.

According to Marilyn Jesmain, an archaeologist working for the U.S. Forest Service, our challenge, which has been the challenge of humanity from the beginning of its history, is to somehow transcend the conflict between develop-

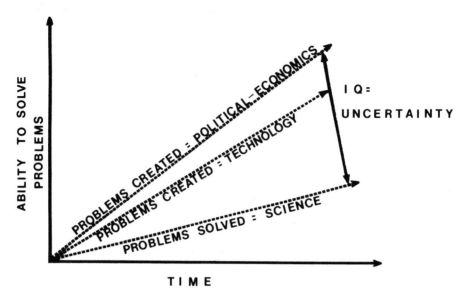

FIGURE 55 Technology, driven by short-term political economics, advances much faster than scientific knowledge. The widening gap between the two (IQ or Ignorance Quotient) causes a great deal of uncertainty in decisions. Uncertainty is the difference between what is currently known about how to solve a problem and what is ultimately needed to actually solve it.

ment and the need to understand and conserve our ecological relationship with our environment within the technological context of the future. The problem that now confronts us is that we have to know the questions to ask before we can find the answers, but the questions are affected by and often hidden in the biases of our culture and our history. This is an important concept, because "every society, every culture bases its existence on an awareness of a common past."

"Our ideas about the past," she states, "have been largely determined by regional perspectives and Western values. What we have been calling 'world history' has actually been no more than regional history in a global setting." Although our view of the world is biased by the cultural environment in which we live, we must learn to view history and its cumulative effect on the world as a whole.

History is common to all human beings, all levels of culture, and all societies. It is our social mirror, because history explains our motives and therefore the consequences of our activities. When the past is ignored, our brief history inevitably appears far too important and out of context with the world. We must learn to read history, she asserts, not in the way it has been perceived, but in how it is carved into the globe itself. We have stripped the forests, polluted the waters, and ignored the lessons of the past. Some countries are even ignoring the current lessons from across their borders.

The Canadian government, for example, is allowing large timber companies to destroy the forests of Canada in the same way that large timber companies were allowed to destroy most of the forests in the United States. As Canadian author Jamie Swift points out, Canadians had not learned from their mistakes in 1983.[48] They still have not, nor have they learned from mistakes made in the United States.

In the coming years, much of today's culture will disappear with the alteration of landscapes in the name of progress. We can, if we so choose, consciously guide the impending changes. Yet here we sit, stuck in the present while dreaming of past abundance and resisting the need to face the future. The future, however, is suddenly upon us, and unless we have the courage to face it with vision and firm resolve, remembering always the importance of learning from the past, our forests may well go the way of those of the Anasazi, the Easter Islanders, and the Mayans.

8

SHORT-TERM
ECONOMIC EXPEDIENCY

Although short-term economic expediency, which privatizes profits and commonizes costs, has been a mainstay of human society, its environmental consequences probably first became evident in the forests of the world, where they have persisted through the centuries. In his book entitled *A Forest Journey*,[49] John Perlin presents a comprehensive history of global deforestation and its consequences.

The book begins in ancient Mesopotamia 4700 years ago, where civilization first emerged and people started to exploit forests extensively, and closes in the late 1800s in the United States, the last Western nation to leave the wood age. By that date, the United States had already lost most of the eastern American forest, one of the greatest forests ever to grow on earth. Consequently, reliance on wood as the primary fuel and building material gradually gave way to coal and iron.

Perlin ends his book with a quote from an 1882 issue of *Harper's Monthly*; the article was written by N. Egleston, one of America's leading authorities in forestry:

> We are...following...the course of nations which have gone before us. The nations of Europe and Asia have been as reckless in their destruction of the forests as we have been, and by that recklessness have brought themselves unmeasurable evils, and upon the land itself barrenness and desolation. The face of the earth in many instances had been changed as the result of the destruction of the forests, from a condition of fertility and abundance to that of a desert....The masses of the people...should have set before them the warnings of history.

In 1908, President Theodore Roosevelt convened the first meeting of all the governors of the states to address the topic of the environment. His opening address to the conference is as pertinent today as it was then. He began:

> I welcome you to this Conference at the White House. You have come hither at my request, so that we may join together to consider the question of the conservation and use of the great fundamental sources of wealth of this Nation.
>
> So vital is this question, that for the first time in our history the chief executive officers of the States separately, and of the States together forming the Nation, have met to consider it....
>
> This conference on the conservation of natural resources is in effect a meeting of the representatives of all the people of the United States called to consider the weightiest problem now before the Nation; and the occasion for the meeting lies in the fact that the natural resources of our country are in danger of exhaustion if we permit the old wasteful methods of exploiting them longer to continue.

Later in his speech he said, "Just let me interject one word as to a particular type of folly of which it ought not to be necessary to speak. We stop wasteful cutting of timber; that of course makes a slight shortage at the moment. To avoid that slight shortage at the moment, there are certain people so foolish that they will incur absolute shortage in the future, and they are willing to stop all attempts to conserve the forests, because of course by wastefully using them at the moment we can for a year or two provide against any lack of wood." He went on to say that "Any right-thinking father earnestly desires and strives to leave his son both an untarnished name and a reasonable equipment for the struggle of life. So this Nation as a whole should earnestly desire and strive to leave the next generation the national honor unstained and the national resources unexhausted...."[50]

Gifford Pinchot, the first chief of the U.S. Forest Service, also recognized this point when he wrote in his 1914 book *The Training of a Forester:*

> The forest is a national necessity. Without the material, the protection, and the assistance it supplies, no nation can long succeed. Many regions of the old world, such as Palestine, Greece, Northern Africa, and Central India, offer in themselves the most impressive object lessons of the effect upon national prosperity and national character of the neglect of the forest and its consequent destruction.[51]

Before we travel back in time to look at some additional ancient and not so ancient examples of what human society has done and is doing to the forests of the world, it is important to understand that no human action takes place without first the thought to act.

GREECE

The effects of deforestation were apparent even in early Greece. In the time of Homer, for example, there still were "deep, endless, shadowy" forests, but Greek civilization had already begun taking its toll through wood cutting for shipbuilding, construction of houses and fortifications, and illumination and heat. The forests were further affected by livestock grazing, windstorms, and wildfires in the 8th and 7th centuries B.C. Because the forests around Athens were finally decimated from years of exploitation for warships, which resulted in severe soil erosion, the Athenian Empire founded the colony of Amphipolis in 436 B.C. When war erupted between Sparta and Athens, Sparta blocked the timber supply by besieging Amphipolis, which caused the capitulation of Athens in 404. Plato, who lived from 427(?) to 347 B.C., was one of the first to observe, in an altogether brilliant way, the relationships of deforestation, water supply, and soil erosion in the mountainous region of Attica, the hinterland of Athens.[52]

MEDITERRANEAN

We turn next in our recital of ruined lands to the Mediterranean Basin, where the main agents of deforestation were cultivation, exploitation for timber and fuel (much of it for shipbuilding), industry, wars, population pressures, human-caused fires, and grazing. Archaeological evidence indicates that, in addition to deforestation, the abandonment of wide areas of land came about because of economic deterioration, political instability, nomadic incursions, and decline of urban activity.

Commentators on the effects of past deforestation have sometimes been led astray because they lacked appreciation of the renewable nature of the forest. The record clearly shows continuing and often extensive exploitation, *but given a chance,* it is evident from recurrent references to the same forested regions that many of these forests were able to return to a productive status despite periodic depletion. Undoubtedly, rehabilitation of these forests was achieved more by general happenstance than by intent, but it did occur.

In addition, there were considerable clearances of the forest for agriculture, and it is probable that such clearings led to the loss of land to cultivation and grazing, at least for a period. Apart from the early loss of the thin soils of Greece, however, in most parts of the Roman world an intensive crop husbandry maintained soil fertility even on cleared slopes, while the remaining forests were able to meet timber requirements.

During the post-classical period, people continued to collect fuel and use wood for construction. With the revival of civilization and ordered life, requirements for wood increased. Despite the demands of ever-increasing architectural standards, the greatest drain on the forests was again shipbuilding. This demand continued as long as the world's commerce was carried and wars were fought in wooden hulls. The Crusades gave a major impetus to the development of the shipping of the European Mediterranean states and to the development of their eastern maritime commerce. The great medieval fleets of the Mediterranean powers (the Byzantines, the Turks, and the Italian maritime states of Venice, Genoa, and other cities) and the pirate flotillas were launched at the expense of the Mediterranean forests.

The loss, as distinct from depletion, of the Mediterranean forests cannot be ascribed solely to exploitation for timber, fuel, or even cultivation; yet it is clear that depletion has been catastrophic. The bare mountains of the readily accessible ocean shore are witness that excessive exploitation of the forests initiated the cycle of destruction.[53]

MIDDLE EAST

In the Levant, or Middle East, we find the strongest evidence that the deterioration and desiccation of the Mediterranean environment is the combination of a vulnerable environment and a long history of society's effect on natural resources. Here the effect of society was as destructive as in other places, even if it was gradual and in ignorance of cumulative effects. In few other areas has it been as convincingly shown that we have the ability to both destroy our habitat and to reclaim it, given the will and socio-political motivation. However, such hard-won rehabilitation, resulting from many years of dedicated effort, may be set at naught by short-term political greed.[53]

FOREST DECLINE

There is now growing evidence of decline in productivity over large areas of intensively managed forests in central Europe, China, and North America. There is also a sizeable body of information from integrative research at the ecosystem level that clearly shows important connections among the wide variety of processes that operate within a forest. Further, there is solid evidence that diversity of plant and animal species (species richness) is a critical factor in maintaining these processes.

Where there is a problem, there is a cause. The Society of American Foresters recognized this: "In the face of mounting evidence that forest productivity is declining worldwide, researchers [and managers] know too little about complex ecosystems and the cumulative effect of subtle stresses from atmospheric pollutants."[54] The Congress of the United States stated: "...regulations, under the principles of the Multiple-use Sustained Yield Act of 1960...shall include...guidelines which...provide for diversity of plant and animal communities...." As stated elsewhere in the act, "...the Forest Service has both a responsibility and an opportunity to be a leader in assuring that the Nation maintains a natural resource conservation posture that will meet the requirements of our people in perpetuity...."[55]

It is worth noting that decades of scientific research have concentrated on every possible cause of forest decline *except* that it might be the direct result of intensive plantation management based on ignorance of forest processes. Plantation management is the profession that embraces the science, art, and business of growing a plantation of trees as an agricultural crop to reap the greatest economic return for the least economic investment in the shortest possible time.

The forests of central Europe are now dying; in fact, West Germany issued a postage stamp in 1986: "save our forests in the eleventh hour." Yet the effects of a century or more of intensive management based on short-term economic expediency are seldom discussed. Forestry professors Schütt and Cowling, for example, list a variety of reasons for the so-called Waldsterben (the dying forest); not one of them is directly connected to intensive management. They state that, "The stress factors inducing the Waldsterben syndrome are not known, but it is widely assumed (and we believe correctly so) that atmospheric deposition of toxic, nutrient, acidifying, and/or growth-altering substances is involved."[56]

I agree with Schütt and Cowling; however, the six general hypotheses put forth to explain one portion or another of the syndrome do not consider the cumulative effects of intensive management. The six hypotheses are acidification/aluminum toxicity, effects of ozone, deficiency of magnesium, general disturbance of physiological function, excess atmospheric deposition of nutrients (especially nitrogen), and air transport of growth-altering organic substances.

In fact, the cause or causes of Waldsterben in central Europe are generally seen as occurring from outside the forest, and much is said about acid rain. "One German scientist even believes a 'mysterious virus imported from Czechoslovakia' triggered Waldsterben."[57]

Schütt and Cowling conclude their article by stating that there is much concern among scientists and the public at large that the forests of central Europe "may not be sustainable." They reiterate that atmospheric deposition of chemicals may be involved and ask for help in understanding "...one of the most remarkable forest disease problems of this century." Again, I agree that atmospheric pollutants may well be playing a role in Waldsterben. Yet I cannot help but wonder how the cumulative effects of a century or more of deforestation and intensive plantation management and use may have strained the forests of

Central Europe and thus predisposed them to the "Waldsterben syndrome" seen today.

Are the forests of Central Europe really dying or are they only showing signs of stress? According to one study, Waldsterben is a threat for the future, not a historical fact, because despite air pollution, "forest resources" have increased in Europe between 1971 and 1990. What the authors refer to as "forest resources" is predominantly an increase in the biomass of plantation trees, much of which is a result of planting additional trees as opposed to solely increased growth in existing trees.[58]

Nevertheless, as noted in the study, there is no doubt that acidification of the soil has been increasing for a long time through air pollution. It correctly points out that pollutants, like other environmental factors, can have both negative and positive effects on wood production, depending on conditions, and the fertilization effects of pollutants appear to be overriding the adverse effects—at least for the moment. Thus, while the increase in the growth of trees will change little over the next five to ten years, severe changes in the climate could alter the picture.

Although a plentiful supply of wood is predicted over the next 10 to 20 years, which according to the study will allow business as usual, it warns that beyond those years both plantations and forests will be at risk because of drastic human-induced alterations in the quality of the soil.

Only a few studies have examined other aspects of central European forests, such as the correlation of climatic factors and damage to central European forests since 1851, the cumulative effects of intensive management above ground, and the cumulative effects of silvicultural practices on soil invertebrates in both central European and North American forests. There are numerous groups of soil animals that may have a combined greater number than aboveground animals, and they may be disproportionately important with respect to cycling and release of plant nutrients.

In the past, lime was applied to the forest soils in Germany in an attempt to reduce their acidity, and changes in the population structure among tiny soil plants and small soil animals could still be detected 50 years after an application of lime. The soil animals, such as mites, are dependent on the oil plants, such as bacteria and algae, for energy, and these plants are in turn affected by the numbers of animals that eat them; therefore, a change in one effects a change in the other.

In addition, the population density of mites was only half that of the mite population in control plots 26 years after the combined application of fertilizer and lime. That fertilizers usually increase the numbers of soil animals is often interpreted as beneficial, but we are only now learning that this is not necessarily so.

One seldom-discussed aspect of intensive forestry in central Europe is biological simplification. Figure 56A represents an unmanaged, mixed deciduous-

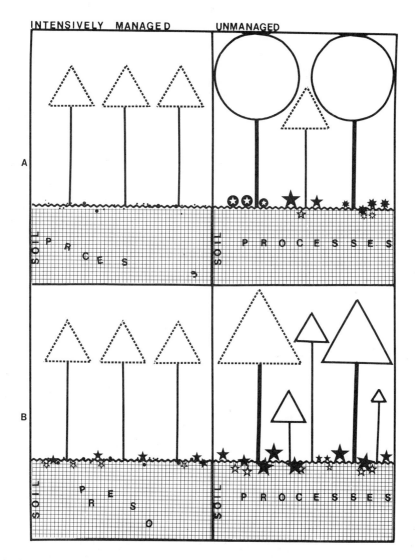

FIGURE 56   (A) The forests of southern Germany. The intensively managed Norway spruce forest exists today, as opposed to the original mixed conifer–hardwood or deciduous hardwood forests of European beech, oak, hornbeam, linden, and other species.[108] The symbols on and in the soil of the unmanaged forest represent woody debris, from large fallen trees to branches one inch or greater in diameter. These are missing from the soil of intensively managed stands. (B) The forests of the Pacific Northwest. The intensively managed forest of 70- to 80-year-old Douglas-fir depicts the future, when there will be little woody debris on or in the soil, unless it is left intentionally. The original old-growth forests often consisted of mixed-species stands and much woody debris, ranging from large, fallen trees to small branches.

FIGURE 57   The people of Bavaria, Germany, keep the floor of their plantations cleared of woody material.

coniferous European forest; Figure 56B represents a coniferous forest of mixed age classes. Note the mix of structural shapes of the trees (which include different species and ages) and the "woody material" on and in the soil. The belowground processes are fully functional in the unmanaged forest compared to the grossly simplified, intensively managed stand with only one even-aged species of tree and no woody debris left in the system. Such simplification is a product of hundreds of years in which all timberland has been overcut, overgrazed, and overraked for litter (Figures 57 and 58).

What it took Europeans centuries to do to their forests is taking us less than 200 years to accomplish—one third the life of a 600-year-old Douglas-fir. The only difference in the speed of modern deforestation is that our technology was unavailable 100 years ago. Deforestation is used here because in the German forests, which are our model, it took about a century for the negative biological consequences of intensive forestry to appear. Many of the pure stands of Norway spruce grew excellently in the first generation, but already showed amazing signs of stress in the second.

FIGURE 58   Woody material, even twigs, gleaned from the forest floor in Bavaria, Germany, ends up in the fireplace.

Although the cause or causes for the decline in forest productivity may not be readily apparent, some possibilities come to mind. The acronym for short-term economic expediency, SEE, is unusually apt here. Short-term economic expediency is the common denominator behind intensive, exploitive use of the world's resources, and the acronym *SEE* reminds us that in renewable resources we *"manage"* only what we *look* at and what occurs *above* ground. The health of the soil is therefore ignored.

Why does our form of capitalism blind us to the realization that healthy soils are the key to healthy forests? The answer lies in the observations made by Professor Donald Worster, who defined the capitalist approach to the land in jarringly stark terms, inviting us to look beyond the forest into our cultural ethic for the roots of the growing ecological dilemma:

> The land in this culture, as in any other, is perceived and used in certain, approved ways; there are, in other words, ecological values taught by the capitalist ethos. We may sum them up in three maxims.

1. **Nature must be seen as capital.** It is a set of economic assets that can become a source of profit or advantage, a means to make more wealth. Trees, wildlife, minerals, water, and the soil are all commodities that can either be developed or carried as they are to the marketplace. A business culture attaches no other values to nature than this; the nonhuman world is desanctified and demystified as a consequence. Its functional interdependencies are also discounted in the economic calculus.

2. **Man has a right, even an obligation, to use this capital for constant self-advancement.** Capitalism is an intensely maximizing culture, always seeking to get more out of the natural resources of the world than it did yesterday. The highest economic rewards go to those who have done the most to extract from nature all it can yield. Private acquisitiveness and accumulation are unlimited ideals, impossible to satisfy once and for all.

3. **The social order should permit and encourage this continual increase of personal wealth.** It should free individuals (and corporations as collective individuals) from encumbrances on their aggressive use of nature, teach young people the proper behavior, and protect the successful from losing what they have gained. In pure capitalism, the self as an economic being is not only all-important, but autonomous and irresponsible. The community exists to help individuals get ahead and to absorb the environmental costs.[59]

Although one may not want to characterize all of capitalism in just this way, it is certainly possible to define much of the current environmental degradation as stemming from the misapplication of the spirit of capitalism. The environmental costs of the purely self-indulgent elements of the capitalistic system outlined by Worster are rapidly destroying the world's natural resources. This point can be illustrated primarily with two examples: intensively managed forests in southern Europe and intensively managed forests in the Pacific Northwest, with a brief highlight from Asia.

Viewing world forests from a position of short-term economic expedience dictates a narrow perspective, a view that includes the notion that soil fertility, quality and quantity of water, quality of air, and quality of sunlight—all of which continuously interact with the forest—are constant values. In addition, this view demands gross simplification of the forest for maximum immediate profit and leaves the future of the forest to the future, because any commodity that is not used for today's profit is seen as an economic loss. No provision is made to reinvest any biological capital in the maintenance of a healthy forest soil to safeguard the options for the future.

Europe

First we will examine a generalized view of central European forests. Mid- and low-elevation forests in central Europe were historically either deciduous hardwoods (Figure 59) or mixed deciduous hardwoods and coniferous softwoods (Figure 60). Prior to the advent of forestry, European beech and oak often occurred together in these forests, with common hornbeam and linden as the main species of trees. Birch was also found in many areas in the north. Oak grew especially well on the more nutrient-rich soils, establishing excellent stands of trees where climatic conditions were favorable. The beech, with its more modest requirements of site and climate, covered large areas. In addition, many different species of trees grew along the river valleys.

Under favorable conditions of climate and site, vigorous mixed stands of Norway spruce, silver fir, beech, and Scotch pine appeared in various percentages over vast areas of the mid-elevation slopes. In addition, Scotch pine and oak together formed a mixed type of stand characteristic of the southern German landscape. At the higher altitudes, such as the Highlands of the Harz Mountains, stands of pure Norway spruce were found. Moreover, in the mixed forest of the mountains, the spruce was predominant and, chiefly in the Alps, formed pure stands at something more than 4225 feet (1300 meters) elevation, which then extended over into the mountain pine regions.

When humans arrived in central Europe, changes in the forests were at first quantitative (cutting wood and clearing land for agriculture) and then qualitative (changing species composition and converting sites from hardwoods to softwoods). This scenario resulted in the better soils being put to agricultural uses and the poorer soils left for forests. By the end of the Roman era, about a quarter of the land in western Germany had been cleared for agriculture and attendant communities.

Clearing of land in connection with urban and monastic settlement policies began in the early medieval period and was considerably intensified during the 11th and 12th centuries, resulting in the present partitioning of fields and forests. The forests that had served as the most important source of energy and raw materials for a growing and increasingly efficient work-specialized society were reduced to approximately one third their original extent during this time.

Specialized uses of trees began increasingly to change the species composition of the forest. The natural resources of the forests had been exploited to the edge of economic collapse by the end of the 15th century, and the iron producers of the 16th century could not maintain production because they could not get wood. The "great wood crisis" became the limiting economic factor with the rise of manufacturing in the 18th century.

The critical shortage of wood by the end of the 18th century finally forced the change to a planned forest economy. Deciduous hardwoods were replaced with

FIGURE 59   Deciduous hardwood (European beech) forest in Bavaria, Germany.

conifers (mainly Norway spruce and Scotch pine) in an effort to stock devastated areas with productive stands as quickly as possible and without incurring financial risk (Figure 61).

FIGURE 60   Mixed deciduous hardwood and coniferous softwood forest in Switzerland.

This concept of incurring as little financial risk as possible is based on the soil-rent theory, which, as discussed earlier, calculates the species of tree with the highest monetary return and the financial rotation with the highest internal rate of return on a given site. Stated another way, by selecting the fastest growing species of tree and then holding everything intellectually as an economic constant, except the age at which the trees are harvested, it is possible to calculate the age of harvest or "rotation age" that will give the highest rate of return on the economic capital invested.

The soil-rent theory is used in maximizing profits as the general objective of economic activities (economically sustained yield), which since its adoption by foresters has become the overriding objective for forestry worldwide. Because of that view, the conservation (carrying into the future) of "renewable" natural resources is not considered to be economically sound.

The criteria used for making decisions based on the soil-rent theory are (1) easy establishment of a stand, (2) short rotations, (3) rapid growth, (4) better uniformity of the stands, (5) greater percentage of usable wood fiber, and (6) better technology for utilization of wood fiber. The possibilities for exploitation

FIGURE 61   Pure coniferous plantation (Norway spruce) in Bavaria, Germany.

and maintenance of the forests caused a virtual reversal of the original proportions of approximately 75 percent deciduous trees and 25 percent conifers. The forests of the old German Federal Republic today are about 70 percent conifers (mostly economic plantations) and 30 percent deciduous trees. Norway spruce covers 40 percent of the area and Scotch pine covers 26 percent.

The short-term economic success of this forest conversion was and is extraordinary. From 1860 to the present, there has been a clear improvement in supplying the domestic market with wood, and there has been a general stabilization in the wood-related economic situation. These circumstances caused financial criteria to be applied to forestry for the first time in striving for a "modest" interest on land capital. From this effort, during the 19th century, the theory of financial rotation or rotation age developed, in which the economic aspect of forestry was given a clear priority.

Despite the enormous economic success, grave reservations about the vast monocultures of Norway spruce were expressed early and have never been completely silenced. In the final analysis, the arguments are focused on whether the strong preference for Norway spruce, which constitutes the financial backbone of forestry as the "bread and butter tree," is consistent with the long-term productivity of the soil. In many cases, the cost of improved yield was an increased risk to the fertility of the soil.

For example, calculations made 150 years ago without modern knowledge and the help of yield tables, site maps, and so on proved to be exact in one regard: the economic superiority of the softwoods over the hardwoods became an indisputable fact.

The biological consequences, however, are less pleasant. It took about a century for them to show up clearly. Many of the monocultural plantations grew excellently in the first generation of planted trees but showed amazing biological problems in the second. The reasons are complex and only a simplified explanation can be given. A plantation of spruce will serve as an example.

Planted in the soil of a former hardwood forest, the spruce roots could follow the deep root channels of the former hardwoods in the first generation. In the second generation, however, the root systems became shallow because of progressive soil compaction and the collapse of the root channels. As a result, the available nutrient supply for the trees became smaller.

The spruce plantation could profit from mild humus accumulated in the first generation by the hardwoods, but it was not able to produce a mild humus itself. Spruce litter rots much more slowly than does broadleaf litter, is more acidic, and is much more difficult for the microplants and microanimals to decompose in the upper layer of the soil. Raw humus is thus developed in most cases. When humic acids started leaching the soil under the humid climate of southern Germany, it impoverished the species richness of the soil plants and animals, which then caused an even poorer decomposition and a faster development of raw humus.

The whole nutrient cycle then got out of order and eventually nearly stopped. The nutrient accumulation in raw humus was difficult to use. The soluble nutrients of the upper soil layer were washed down beyond where the spruce roots could reach, which in its extreme can lead to the formation of a bog.

Nevertheless, the drop of one or even two or more site classes during two or three generations of a monocultural plantation of spruce is a well-known and frequently observed fact. This represents a loss in wood fiber production of 20 to 30 percent. The reactions of the soil to pure pine plantations are similar to those of spruce in many ways.[26]

In addition, the demands placed on the plantation can change considerably within one rotation. This is valid not only with reference to the lumber market and its demand for a specific assortment of goods but also with a view to the changing societal desires from the forest. For example, at one time society might demand the maximum production of wood fiber and at another time want to emphasize recreation or ecological processes.

The economic principle of sustained yield postulates the idea of a plantation continuing to produce wood fiber at a specified level in perpetuity, which requires not only an accommodation of cutting to forest growth but also assumes the constant capacity of the soil to yield the desired growth. However, the ecological knowledge that every form of commodity production from the plantation has a considerable influence on the dynamics of the soil raises the question of whether intensive management of the aboveground portion of the plantation is sufficient unto itself.

Generally speaking, intensive management of conifers, especially Norway spruce and Scotch pine, presents many more problems than does any species of deciduous tree. Conifers, for example, are practically the only trees in central Europe to be seriously threatened by such forest "pests" as the pine looper, pine beauty, pine, and Nun moths, saw flies, bark beetles, pine needle cast fungus, pine blister rust, honey fungus, red rot, and a complex of other species of fungi in Norway spruce. In fact, Germany's forests are so strained that conifers must have their bark peeled off when cut to avoid outbreaks of bark beetles (Figures 62 and 63).

Commenting on German forestry, Richard Plochmann, a professor at the University of Munich and District Chief of the Bavarian Forest Service, said that: "Our forestry will be carried on even under bad economic situations. We could better the return if we would be willing to give up the high intensity now maintained or if we gave up the principle of sustained yield. We cannot do both and do not want to do either. The first seems imperative for the multiple uses of our forest and the second for the benefit of following generations...."[26]

In 1989, Plochmann, now a professor of forest policy at the University of

FIGURE 62 Logs from coniferous trees in Bavaria, Germany, are peeled on site to avoid attracting bark beetles, which attack stressed trees.

FIGURE 63   Because most of the intensively managed coniferous plantations in Bavaria, Germany, are so strained ecologically, logs left in the plantations must have their bark removed to avoid outbreaks of bark beetles that kill live trees.

Munich, spoke to the College of Forestry at Oregon State University. He said that if we ask today what the consequences of the changing view of forestry in central European plantations will be, we will have to capitulate to the expectations of different interest groups with respect to future plantations:

- **From the point of view of recreation (under central European conditions, recreation does not include hunting and fishing):**   The suitability of forests for recreation will improve with their increasing age, their diversity of composition and structure, their accessibility to the visitor, and their resistance to damage by recreational visitors.

- **From the point of view of water resources:**   While our knowledge is still insufficient concerning optimal refill conditions for groundwater, we do know that in order to protect groundwater resources from contamination, large clearcuttings, fertilization, the application of herbicides, and the use of heavy machinery should be avoided.

- **From the point of view of landscape protection:**   The capacity of forests to protect against erosion, avalanches, and floods depends mainly on high stand densities for an indefinite period. This protective capability also depends on the forest's resistance to change and its resilience to disturbances. Naturally composed forests under selection or group selection management will meet such expectations best.

- **From the point of view of nature preservation:** Under the assumption that only 3 percent to 5 percent of the forest area will be placed under strict preservation, the large majority of forests should be managed as "mature systems," which means managing for long life cycles, natural rehabilitation, a sufficient amount of remaining dead timber, and the lowest possible input of artificial energy.

- **From the point of view of forestry:** What needs to be achieved is not the highest volume, but the highest value per unit area per year. That means diameters at breast height not below 20 inches and clean boles [stems or trunks] of 15 to 50 feet over rotations of 80 to 250 years, depending on species and site.[26]

Plochmann went on to say that today's concept of forestry would include mixtures of two or three species of trees, as opposed to the historic, single-species monocultures. In such mixes, at least one species must be indigenous. The ages of rotation (the age at which trees are cut) will depend on their highest value: quality wood as opposed to the historic, inferior, fast-grown wood. This will require rotations of 120 to 140 years. Wherever possible, natural regeneration, as opposed to genetically selected or manipulated seedlings from nurseries, will be used; therefore, clearcutting will be replaced by shelter-woods in which some of the mature trees are left to reseed the area by group cutting or by selection harvest. Further, herbicides will no longer by used, and insecticides and fertilizers will be used only rarely. Finally, there will be no highly mechanized operations within the plantations as they are brought closer and closer to the physical structure and biological functions of a real forest.

Asia

Europe is not alone in its forestry problems. In China, the problem of deforestation was already noted and clearly described by Mencius, a Chinese philosopher of the 4th century B.C. In modern China, a plantation of pure Chinese fir cannot be initiated more than two or three times on the same site; after two or three 20- to 30-year rotations, the soil is no longer able to support the normal growth of Chinese fir.

In 1986 *The Oregonian* carried a story about forestry in Nepal. Each year thousands of people from around the world flock to Nepal to witness its spectacular mountain scenery. When they lower their gaze, however, the glory fades and yields to a darker sight. Nepal is being destroyed.

Ecologists warn that deforestation threatens to transform the Himalayan foothills and turn the green landscape of Nepal into a desert. Once isolated but now open to the world, Nepal is suffering environmental self-destruction because once the trees are gone, nothing will hold the topsoil on the steep hillsides. Every year the rains wash some three billion cubic feet of topsoil down to the Bay of Bengal. Soil has become the kingdom's number one "export."

Life was better in centuries past. Nepali villages managed their communal woods themselves, and local leaders enforced rotational grazing to prevent the stripping of pastures. With the nationalization of forests in 1957 and rising population pressures, however, the balance between people and Nature was destroyed.[60]

A point of interest here is that while working in Nepal in 1966–67, I visited both the logging camps and the sawmill (located in Trisuli) and found that the U.S. AID mission was responsible for Nepalese logging practices and philosophy. We taught them how to log and mill their forests based on short-term economic expediency, but we did not teach the Nepalese about conservation of soil or anything else.

Under these circumstances, it is necessary to take a brief look behind the notion that foreign aid from the United States really does what it is conceived to do. To this end, a few of Mohandas K. Gandhi's ideas on national economic development make impeccably good sense.[61]

Although Gandhi looked forward to economic development, he wanted to prevent the Indian villages from catching the "infection of industrialism," which inevitably leads to the unrestrained pursuit of material goods. He realized that the economy was an organic whole and that if economic growth was to take place, it must be in harmony with all aspects of the society. Rapid industrialization would result in negative consequences.

Gandhi saw that if the major problems of the rural economy were not addressed concomitantly with industrial development, industrialization would get too far ahead of agriculture—at the expense of agriculture, including forestry—and would grind to a halt. This has already happened to some degree in a number of developing countries. "In the first decade of independence," observed *To the Point International* magazine, "not a single Black African country gave priority to agricultural investment, and expenditure on this sector represented only a tiny fraction of the total government disbursement. When industrialization did not deliver the goods…there was no agricultural base to fall back on." This was my sense of the U.S. AID mission's direction in Nepal: to bring forestry into the modern era by U.S. standards, standards that Nepal as a nation was neither ready for nor which were appropriate for Nepalese forests. In reality, therefore, we helped the Nepalese to harvest their trees and destroy their forests.

Nation after Third World nation finds itself in this same trap. Having committed themselves to rapid industrialization, they squander their foreign exchange and their natural resources on schemes of development that are not suitable for the particular country or its culture. When such schemes fail, they are financially ruined and often left with a severely damaged environment.

At this point, two options are available. The first is to apply for additional assistance from Washington, D.C. or some other industrialized nation or an international funding institution. Learning the language of ecosystems, both native and managed, should be a high priority for such funding agencies as the World Bank. Unfortunately it is not, and the economic relief they proffer comes at a supremely high cost: the loss of self-direction and self-esteem of the indebted nation. The other option is to exercise patience and adopt Gandhi's program of building or rebuilding self-sufficiency and self-esteem from the bottom up on a bioregional and national basis.

For whatever reasons, almost all developing nations choose rapid industrialization and simply trade political imperialism for economic imperialism. In India, for example, Nehru made this mistake when he turned his back on Gandhi's program and embraced rapid industrialization. Today, India has little ability to distribute anything to its impoverished masses from its relatively small but highly advanced technical sector.

In Gandhi's view, if the individual, then the village, and then the nation are brought step by step to economic self-sufficiency, it is then possible to attain and retain true political freedom for the whole. Accelerating the process means economic dependence at the expense of liberty and freedom of choice. For example, in order to protect its investment, the World Bank has frequently dictated domestic environmental policy as a precondition for approving a loan and then has commanded a supervisory role in the recipient's economy.

A Third World nation is thus exposed to grave risks when it opts to enter the international economic system. Once it has become reasonably integrated into this system, it may find that it has unwittingly imported all its problems as well, problems such as inflation, the effect of foreign recessions, uncertainties in the price and supply of oil, sudden unemployment, or employment that is skewed to the desires of foreign markets to the detriment of its own economy. Now the leadership of the nation is subject to even greater influence by foreign nations, and its economy is even more dependent on the international system.

The problem is that today's world leaders want to build from the top down, without a solid foundation of human dignity and environmental sustainability. This top-down approach to industrial enterprise is propelled by a vision of the ability of human intelligence to transcend material limits and thus serve the powerful economic elite. Although it is possible that some people become so enthralled with the vision or so dependent on the enterprise that they see beauty in the belching smokestacks, dams in rivers, endless power lines, and dumps of toxic wastes that accompany it, for the masses of humanity the crude, destructive, inequitable aspects of industry remain glaringly obvious. They are accepted, however, as the price paid for material progress.

Gandhi's plan for world order, on the other hand, is predicated on the voluntary cooperation and coordination of friendly states reaching out to one another for mutual benefit in such a way that they can approach the same goal from opposite directions. However, if an external force (such as imperialism) or an internal force (such as premature industrialization) disrupts the balance, debilitating problems become the norm. Gandhi would thus choose a biologically sustainable forest for the lasting benefit of everyone over a quick glut of short-term profits for the immediate benefit of the tiny but powerful elite.

United States

Our forestry is patterned after German forestry; our philosophy and practices are also based on and driven by short-term economic expediency. This comparatively youthful country is already starting to reap some of the same consequences seen in the forests of central Europe and China. The growth of southeastern pine plantations (one of the nation's largest sources of wood) is declining after decades of increase.[62] In addition, forest researcher Alex Shigo finds that dying forests are nothing new, especially in the northeastern United States. He states that acid rain, insects, and fungi are real problems that can kill trees, but wounded forests and starving trees are also realities and part of the total picture. Forest decline, a many-sided problem, has as yet focused attention on only a few factors. The blame for diebacks and declines cannot be placed on well-publicized short-term agents. Knowing how a tree or forest dies is as important as knowing the cause of death.

Trees suffer more than mechanical wounds to trunks and roots. Injuries can also be caused by compaction of the soil, alteration of drainage patterns, disruption of niches for soil microorganisms, disruption of the species composition of nonwoody and woody plants, and the list goes on. Starving trees, for example, face limitations beyond water, oxygen, other chemical elements, and energy. Trees may starve because of a reduction in storage space within the tree. As space to store energy reserves in the tree decreases, so do reserves. Trees can thus starve in the midst of plenty if storage space is sufficiently reduced.[63]

Let's look at what we are doing to forests in the Pacific Northwest. Note the mixed stand of conifers in the unmanaged forest and remember that Nature designed Pacific Northwest forests to be unique in the world: 25 species of conifers, with 7 major ones, which are the longest lived and largest of their genera. These systems also have much woody debris.

By designing a forest based largely on a single-species short rotation that is intensively managed, we are grossly simplifying forest systems. We are speeding up early successional stages as much as possible and liquidating mature and old-growth stages. We are eliminating snags and large down woody material over time as we emphasize short-term economic expediency instead of sustainable forest diversity and stability. Intensively managed stands have little or no wood in the system.

Although U.S. forests may never look as "neatly groomed" as European forests because of their diverse, rugged topography, the economic philosophy is essentially the same. After all, Gifford Pinchot, the first U.S. forester, was trained in British colonial forestry in France and had a correspondingly utilitarian view of forests and forestry. This philosophy of short-term economic expediency is epitomized by John B. Crowell, Jr., past Assistant U.S. Secretary of Agriculture for Natural Resources and Environment:

> ...I do not comprehend how forest plans can propose reductions in harvest levels if the economic analysis required by the National Forest Management Act has been appropriately carried out and given full consideration in arriving at preferred alternatives.
>
> I am, quite frankly, shocked that the Forest Service, which represents itself as being a professional land-managing agency, can possibly even be considering plans so shirking both good fiscal management and good forest management.
>
> The forest plans must provide for adequate levels of timber harvest if the national interest in having plentiful supplies of wood products at reasonable prices is to be realized and if wood manufacturing mills here in the Pacific Northwest are to be kept operating.
>
> If the plans that are ultimately adopted by the Forest Service provide only for decreased levels of harvest, any attempt by Congress subsequently to fund annual timber-sale programs at harvest rates greater than the plans allow can be frustrated by environmentalist-initiated court actions.
>
> Congress, of course, could overrule the entire planning process. But the political possibilities of achieving such a correction to a planning process gone haywire obviously would be very difficult.[64]

J. Laurence Kulp, Vice President of Technology Strategy for Weyerhaeuser Co., was reported as saying that new management practices between now and the year 2020 will double the amount of wood harvested per acre. Tissue-culture technology and genetic improvements will produce faster growing, hardier trees that are resistant to insects and disease. To keep costs down, he added, it will be important to utilize "all above-ground biomass," such as tree stumps, for energy or new products.[65]

Again, keep in mind that those who say that we can have more of everything simultaneously by intensifying plantation management and the utilization of wood fiber because we have the technology *assume* that the growth of trees is the only variable with which they must deal. Contingent on that assumption is the further *economic assumption* that six fundamentals of forestry—soil, water, air, sunlight, biodiversity, and climate—are in fact constants.

As forests are simplified aboveground in an attempt to maximize profits through short-term economics, they are simultaneously simplified belowground through alteration of the soil, biodiversity of microorganisms, nutrient cycling, and nutrient uptake processes. Thus, Nature "balances the books" to the eventual impoverishment of both the forest and the profit margins. Forest researcher Alex Shigo summed it up when he wrote that the "predisposition" of a forest for ecological problems results from stress, which is a reversible condition of lowered energy reserves that sets the stage for "strain," which is an irreversible condition, such as may now exist in the forests of central Europe. Energy is required to fuel biological functions of the tree: to build cells, maintain living functions, reproduce, and defend the tree after injury and infection.

Survival of all living things, Shigo said, depends on energy, space to grow, concentrations of water and chemical elements, temperature, time, and the genetic capacity to resist stress and strain. Trees cannot move, however, and therefore these survival factors are all linked because trees either grow on suitable sites, adapt to unsuitable sites, or die. Because survival factors are linked, any disruption in one affects the others.

Finally, Valadimir Molozhnikov of the Baikal Ecology Museum in Irkutsk, in the former Soviet Union, provided a friendly warning:

> ...I am a forest ecologist with 30 years' experience in the forests of Siberia. Not long ago, I was able to visit the forests of Oregon. I went to different spots in the forest, and I saw them from the air. I was stunned by the scale of logging.
>
> Earlier, I would refer to the literature describing the way forests in the United States are managed. In these sources, a bright picture is painted. We in the Soviet Union were often taken by your approach. You were the example [as Germany was for us] of a progressive country using your forests intelligently. Your example was even used to cool the heads of our aggressive forest industrialists.
>
> Now I'm here to have a look at the way forestry should be done. But what I've seen in Oregon won't make it possible for me to use your forest techniques as an example: a multitude of bare, forestless cliffs, slopes rib-boned with roads, intensive erosion of soils, silting of rivers and reservoirs, loss of animal habitat, the disappearance of recreation areas.
>
> So what do we do? What can future generations expect after us? It's often said now that the Earth is our common home. But if it's our home, then let us, by our common efforts, put it in order.
>
> I'm not trying to lecture Americans. All I care to do is, in a friendly way, warn you: Don't repeat our mistakes. Tremendous natural-resource use and planned transformations of nature have led our system to the point of ecologic crisis, even in Siberia. And with this have come economic crises as well.
>
> In closing, a little advice: don't cut down the limb on which rests the well-being of the people, or your fall will be even more frightful than ours. After all, you still have something to lose.[66]

RECENT EXPERIENCES

Slovakia

In June 1992, I visited Cergov, in northern Slovakia, to evaluate the condition of the native forest, which is primarily European beech with an admixture of white fir. The native forest is being rapidly clearcut and replaced with plantations of such nonnative species as Norway spruce, larch, and pine. The biological errors of forestry made in Germany, the United States, and Canada are being repeated in the forests of Cergov and for the same reasons.

If we as a nation claim technological superiority, then we must invest in one another as human beings first and in the products of other lands and cultures second. This means sharing our technology with other nations but at the same time clearly and openly *informing* them of its environmental and cultural costs.

Among the most graphic examples of the cost of technology involved in clearcutting is the loss of topsoil. Within one hour after each thunderstorm in Cergov, all the streams and rivers fed by clearcut slopes went from clear water to the appearance of milk chocolate as the soil of the forest was washed away to the sea.

Prior to importing the technology of clearcutting, the forest of Cergov was logged with horses. In addition, horse logging had been biologically sustainable for centuries, as were the economies of the small mountain villages. Now in the villages located in the upper valleys near the edge of the forest, the jobs once sustained by horse logging are gone like the topsoil of the forest.

Today, because of the uncritical acceptance of Western technology as the panacea of short-term economic problems, the people who once made their living from the forest must commute to the cities to find work and the villages have lost part of their cultural heritage. These are but two of the costs of Western technology when it is used blindly for short-term economic gain.

With the preceding in mind, a slightly modified version of my report to the people of Slovakia (which was presented on June 23, 1992 at a conference on the environment in Presov, Slovakia) is provided here. It is important to note that the intent was to provide neither criticism nor advice. The purpose was to look at the forest of Cergov and to make recommendations for its biological health and sustainability, as well as to point out the ecological consequences of certain actions based on what I saw and my experience as a research scientist in forest ecology. To prevent anything I said from sounding like a directive, I told the people what I would do *if* I were suddenly made chief forester and told that my job was to maintain Cergov in a condition of biological health and sustainability.

Definition of Native Forest

The native forest is a forest created by Nature in a particular place at a particular time without human intervention in the form of either introduced foreign species or extinction of native species. Native forests are the oldest living beings on Earth, and as such, they deserve the appropriate respect.

Why the Native Forest Is Important

Native forests are important because a healthy, sustainable native forest is the best way to ensure a good supply of quality water and good water-storage capacity in the soil. Native forests hold undiscovered species of plants that may be beneficial to society. Native forests have the best potential for maintaining healthy, fertile soils. Native forests have the best chance of producing a sustainable supply of quality wood, and they harbor biological, genetic, and functional diversity that allows them to adapt to sudden changes in global climate.

Areas of Concentration in Management

To maintain healthy, biologically sustainable native forests, the following must be carefully tended:

I. **Soil:**   Soil is the membrane of exchange between the nonliving and the living components of the world. It is where the nonliving and living components come together and are mixed with each other. Soil is like a placenta in that it nourishes all life that grows out of it. Without soil there can be no life on land.

II. **Water:**   Water is necessary to all life. Every species requires water of sufficient quality and quantity to live.

III. **Biological diversity:**   Biological diversity is the variety of living species of plants and animals, from the smallest to the largest. Biological diversity of living beings is responsible for the biological processes and functions that nourish all ecosystems and that keep all ecosystems healthy and sustainable.

IV. **Genetic diversity:**   Genetic diversity is the hidden diversity within all species. It is the key to a species' ability to adapt during times of stress, such as environmental change.

V. **Landscape patterns:**   Landscape patterns are the arrangements of habitats across the landscape. These patterns may be naturally evolved, artificially created by humans, or a combination of the two. Regardless of how the patterns are created, they are constantly changing.

VI. **Cultural evolution:** Every culture, such as a forest, evolves through time. Such cultural evolution means that cultural requirements and desires from any forest will change over the decades. It is therefore necessary to manage a forest both for present cultural requirements and desires and for those of future generations.

Plan of Action: The Forest Plan

Working with a forest plan, it must be understood that the forest—not the plan—is sacred. The purpose of any forest plan must be to serve the forest by protecting its long-term biological health and sustainability while at the same time allowing humans to use the forest. Too often, however, the forest is redesigned to serve the plan, which is what appears to be happening in Cergov. To counter the observed trends in Cergov, I would, as chief forester, direct the following to be done:

I. **Soil**

A. Protect the soil by vegetative cover at all times. To accomplish this, the following is necessary:

1. Stop all clearcutting immediately and permanently.

2. All logging is to be done with selective cutting and with horses.

3. Select a minimal permanent system of roads from the existing roads and, wherever possible, redesign as necessary to meet sound ecological standards. Design and construct new roads as required. All other roads are to be permanently closed and reverted to native forest.

4. Pay particular attention to leaving adequate amounts of large dead wood on the floor of the forest as a biological reinvestment in the long-term health and productivity of the soil.

5. As a rule, no organic material is to be burned in the forest as a result of timber harvest.

B. Immediately revegetate soil in damaged areas with native plants.

II. **Water**

A. All sources of water are to be protected immediately. Such protection is necessary because, as it now stands, every stream that is not protected by healthy native forest becomes a flowing ribbon of chocolate within one hour after each thunderstorm as the forest's soil is washed away. The following steps would counter such devastating and unnecessary erosion:

1. Stop all clearcutting immediately and permanently.

2. Establish buffer strips at least 60 meters wide (the height of a large, full-grown tree) along all streams (permanent and intermittent) and around all springs, bogs, and wet meadows. (A buffer strip is a strip of land along a source of water in which all human activities that in any way disrupt the ecological integrity of the protected area are prohibited.) The purpose of a buffer strip is to keep the water cold and to act as a source of large wood for habitat diversity in the stream in order to keep the stream and its banks in a stable condition.

3. Where logging has already occurred into the bottoms of streams, plant native trees to grow into protective buffer strips.

B. Put large wood back into streams where it has been lost due to past logging; this can be done with minimal additional damage to a stream's channel by using horses.

C. Do not allow roads within 60 meters of any source of water, with the possible exception of necessary bridges. Gradually relocate all roads necessary to horse logging that are now within 60 meters of a source of water, and revert the existing road to native forest. Close all other roads and revert to native forest.

III. Biological diversity

A. To maintain native biological diversity, it is necessary to stop introducing nonnative species and to remove all existing nonnative species of trees, whether they occur singly or in plantations. The following is to be done to accomplish this:

1. Make it illegal to introduce *any* nonnative plants or animals into the forest.

2. Prohibit all clearcutting.

3. Pull out all young trees that have been artificially planted in native meadows within the forest.

4. Cut and sell for Christmas trees all suitable nonnative trees.

5. Cut down all remaining nonnative trees.

6. Replace all artificially planted trees and tree plantations with native forest.

B. To maintain maximum diversity of all native species, it is necessary to protect not only the species themselves but also the quality, quantity, and distribution of their habitats. To accomplish this, the following is to be done:

1. Inventory, classify, and map all habitats and give management priority to the rarest.

2. Inventory, to the maximum extent possible, all rare and endangered plants and animals and relate them to their specific habitats.

3. Protect rare plants, animals, and habitats from human transgressions.

IV. Genetic diversity

A. It is absolutely critical to maintain the maximum diversity of the native gene bank in order to protect the forest's ability to adapt to environmental changes. The following is to be done to accomplish this:

1. Require that logging practices be designed to maximize natural seeding.

2. If artificial seeding is needed, the source of the seeds is to come from the area into which the seedlings are going to be planted.

V. Landscape patterns

A. Every landscape evolves patterns of topography, climate, and vegetative cover that not only form habitats but also determine a species' ability to move within and among neighboring areas of habitat. It is vital to protect the connectivity of habitats in order to avoid creating artificial islands of isolated habitats that may disrupt movements of species among vital habitats and thereby cause local extinctions of native species in Cergov. With global climate change imminent, it is also imperative to protect and maintain the forest's evolved patterns across the landscape, because the forest is dynamic and must be able to move in elevation as the climate warms.

VI. Cultural evolution

A. To protect the cultural rights of future generations, it is necessary that the following be done:

1. Develop an official mandate that requires a citizen's committee of sufficient size and variety of backgrounds to participate freely and fully in the formulation of all forest plans as a guaranteed voice for the generations of the future.

2. Guarantee the legal right of any interested citizen to have free and full access to any and all information concerning the forest plans.

Juarantee the legal right of any interested citizen to participate freely and fully in the formation of and in monitoring the application of any and all forest plans.

B. Language of the forest plans:

1. Every effort is to be made to write the forest plans simply, so that they are understandable to the citizen's committee. It is, however, incumbent on the members of the citizen's committee to educate themselves as necessary to the forester's terminology.

2. Any terminology or concept that is not understood by the citizen's committee is to be explained until it is clear. If necessary, a glossary of terms is to be prepared and a free copy given to each member of the committee. Other copies are to be available at cost for interested citizens.

In Conclusion

My final comment is actually a caution. Once cultural desires, such as economics, become intertwined with ecological principles, it is easy to loose sight of a sound ecological perspective.

For example, I was told by a forester that it was inevitable that the old trees of Cergov would be clearcut and that plantations of spruce would replace them. I questioned this statement because there was nothing inevitable either about clearcutting the old trees or planting spruce in rows. It was simply someone's choice to maximize immediate economic gain.

The forester told me that clearcutting was necessary because the ground was too steep for logging with horses. I found this statement to be particularly interesting because the forest had been logged with horses for centuries before the chain saw and log truck became available, and all the clearcuts I saw would have been easy to log with horses.

As chief forester, my first obligation would be to protect the soil. Therefore, in order to give the forest's soil maximum protection, my general working ethic would be that any land too steep to log with horses is too steep to log under any circumstance because of soil erosion. My second obligation would be to protect the quality of the water used by people in the towns. My third obligation would be to protect the biodiversity of the forest to ensure a biologically healthy and sustainable forest for the generations of the future. My fourth obligation would be to maintain, as far as biologically possible, a sustainable supply of quality wood for industry's mills.

Whether or not to log an area and how to log it is only a choice. It is a choice of short-term economics balanced against long-term ecology, as well as a choice of the present generation balanced against those of the future.

Japan

I spent part of October 1992 in Japan looking at the forests from which the Shinto priests of the Grand Shrine of Ise get the old-growth Japanese cypress with which they rebuild the Grand Shrine every 20 years. The Grand Shrine is located in Ise City.

Shinto, in its broadest sense, refers to the entirety of native culture, which is established against a background of hydraulic rice agriculture, a form of agriculture uniquely suited to Japan's warm and humid climate. In short, Shinto refers to indigenous Japanese spiritual culture. When used in the narrow sense, it refers to the rites offered to deities or "kami," primarily those of heaven and earth listed in classical Japanese works of the ancient period. The physical facility used for the performance of this worship is called "jinju" or shrine.

That Nature and natural phenomena are revered as deities is a result of the Japanese view of Nature as a kind of parent, which nurtures life and provides limitless blessings. Shinto shrines all over Japan are surrounded by luxuriant groves of trees. Backed by the Shinto view of untouched natural scenery as itself sacred, the "forests" surrounding the shrines are themselves an important composite element of each shrine.

FIGURE 64   Torri or gate to the Grand Shrine of Ise at the Uji Bridge across the Isuzu River. Note the solid logs from old-growth trees. (Photograph courtesy of Jingu Shicho–Office of the Grand Ise Shrine.)

FIGURE 65 Ceremonial cutting of Japanese cypress to be used in the 61st regular replacement (called the regular removal) of the Grand Shrine of Ise. (Photograph courtesy of Jingu Shicho–Office of the Grand Ise Shrine.)

About 1300 hundred years ago, Emperor Tenmu ordained the practice of removing the old shrine and rebuilding a new, exact replica next to it every 20 years. It is not clearly known why Emperor Tenmu stipulated the rebuilding of the shrine at this interval, but it is likely that 20 years was considered to be the optimum period for carefully preserving the Grand Shrine, considering that it has a thatched roof, unpainted or otherwise preserved structures, and is erected on posts sunk into the ground with the benefit of foundation stones.

Twenty years is perhaps also the most logical interval in terms of passing from one generation to the next the technological expertise needed for the exacting task of duplicating the shrine. The cultural knowledge has thus been passed on for 1300 years without change and will continue so into the future (Figure 64).

Because, as I was told, about 10,000 logs are required each time the shrine is rebuilt, it is necessary to have a biologically sustainable supply of old-growth timber in order to accommodate the continual rebuilding, which means that the forest must be selectively logged to secure the appropriate trees (Figure 65). The problem faced by the priests today is that the main "forest" at Kiso Fukushima from which they get their trees is no longer a forest but instead a carefully manicured plantation, created through long-term selective cutting. It is missing such components of native forest structure as large standing declining and dead trees, large fallen trees rotting on the ground, large wind-thrown trees, and multiple layers of vegetation. While the plantation may appear at the moment to be in good shape, it is headed for trouble in the future because large wood is not being *reinvested* into the "bank" of organic material in the soil.

The addition of such organic material allows the chemical elements in the soil to become nutrients, and it creates and maintains the necessary soil infrastructure to make the nutrients available to the trees. In other words, more wood is being withdrawn from the soil in the form of logs than is being replaced in the form of dead standing trees and dead fallen trees. Withdrawals without the balance of additions can only draw down the organic material in the soil over time and thus continue to impoverish the long-term productive capacity of the soil.

If these plantations are to produce the size and quality of logs needed to replace the Grand Shrine of Ise and all of the associated shrines every 20 years, the amount of organic material that is withdrawn from the soil must be balanced with an equal amount of organic material that is allowed to return to the soil, some of which must be whole trees. This is not now happening and has not happened for at least a century.

In addition, after conversations with Dr. Murao, Chairman of the Resources Programme and World Forestry at the University of Ehime, Japan, it seems clear that fires were once a vital part of the forest's cycle at Kiso Fukushima. The existence of early fires would explain the wide spacing among the old-growth Japanese cypress that are so valued for the reconstruction of the Grand Shrine. It is therefore reasonable to assume that if the Grand Shrine is to have trees of comparable value in the future, fire must be reintroduced into the care of the plantation at Kiso Fukushima.

Fire does things to a forest or a plantation that management without fire never can do. The role of fire cannot be replaced by any other management technique, and the long-term removal of fire from a forest or plantation is ecologically devastating, as is now apparent in the western United States following 80 years of fire suppression.

To maintain a sustainable supply of logs, I was told that the Shinto priests have maintained some sort of forestry plan for at least 500 years and that their present plan extends 200 years into the future. Because I have not seen the plans, I have no way of knowing if they are ecologically sound and based on the best and latest knowledge available today. Implementing the plans is still another consideration.

In addition to forestry plans, the priests of olden time planted cryptomeria trees within the Grand Shrine. Today, these trees have a 500-year written history (Figure 66).

FIGURE 66   Cryptomeria trees in the Grand Shrine of Ise. The large trees are 500 or more years old. The bamboo around the bases of the trees is to protect them from tourists who pick off pieces of bark for souvenirs. (Photograph courtesy of Jingu Shicho–Office of the Grand Ise Shrine.)

The fact that the original forest of Kiso Fukushima is today a plantation does not in any way detract from the conscious efforts of the Shinto priests through the centuries to maintain a sustainable forest.

This entire section is but an example of the invisible present, which reveals itself in time lags between events and outcomes as the decades pass. Such time lags are generated because the chain of cause-and-effect events accumulates the lags from each link in the chain, resulting in cumulative effects. Our impatience with and lack of understanding of ecological events and processes in the time scale of decades has been and will continue to be costly to human society, especially in terms of the sustainability of our forests.

9

SUSTAINABLE FORESTS = SUSTAINABLE HARVEST

A biologically sustainable forest is a prerequisite for a biologically sustainable yield (harvest). A biologically sustainable yield is a prerequisite for an economically sustainable industry. An economically sustainable industry is a prerequisite for an economically sustainable economy, which, finally, is a prerequisite for an economically sustainable society.

Put another way, sound bio-economics (the economics of maintaining a healthy forest) must be practiced before sound industrio-economics (the economics of maintaining a healthy forest industry) can be practiced before sound socio-economics (the economics of maintaining a healthy society) can be practiced. And it all begins with a solid foundation—a biologically sustainable forest.

We are not now headed toward sustainable forestry because plantation managers rather than foresters are being trained. A forester manages a forest. Forests are being liquidated and replaced with short-rotation plantations. We will have foresters only when we have sustainable forests in which we manage not just trees, but the constantly changing processes.

Everything Nature has done in designing forests adds to the cyclic dynamics of diversity, complexity, and stability in space through time. We, on the other hand, decrease diversity, complexity, and stability in space and through time by redesigning cyclic forests into linear plantations.

A unique forest exists in the Pacific Northwest, yet European plantation management is being taught and practiced. This confirms what the poet Matthew Arnold observed almost a century and a half ago, that we live "wandering between two worlds, one dead, the other powerless to be born." Arnold recognized that the perception of Nature reflected in our basic institutions, including science, was and is inconsistent with the world around us. As yet, however, the new world is unable to be born, in part because we learn from experience and all of our experience is in the old world. We are thus left to confront the new world with concepts, methods, and institutions that remain tenaciously rooted in the old, which we are loath to review and change because of blind loyalty to linear, materialistic thinking.

We need to develop our own forestry in updated, revitalized university curricula that stress forest ecology rather than product harvest and that teach good writing, speaking, and people skills in addition to cruising timber and laying out timber sales. We need to learn to see the forest as the biological, ever-dynamic, ever-changing, living organism that produces raw materials, such as healthy soils to grow trees and filter water; pure water to drink, with which to irrigate crops, and for electricity; salmon and steelhead, deer and elk; and the countless other products and amenities derived from forests.

We need to learn about reinvesting biological capital into the processes of the forest so that mills will have a sustainable harvest of timber. We need to understand that Nature cannot be constrained to absolutes, that a sustainable yield is a trend within some limits and that even the timber industry must be flexible and continually change as the forest changes. And schools of forestry must become leaders in research, management, and human relations rather than the last bastions against inevitable change.

In addition, the timber industry, which resists change at the expense of the global environment, needs to be redesigned and restructured. To this end, we must also change our historic view of the timber industry.

FOREST-DEPENDENT INDUSTRIES

The timber industry as it is usually thought of goes only from the *forest to the mill,* but the United States, Canada, and in fact the world as a whole are founded largely on an interrelated, interdependent suite of *forest-dependent industries.* Before discussing forest-dependent industries, however, the concepts of a forest and forestry must be revisited.

A *forest* is the most complex, terrestrial, biotic portion of the ecosystem and is characterized by a predominance of trees. *Forestry* is the profession that embraces the science, art, and business of managing the forested portion of the ecosystem in a manner that assures the maintenance of biological, genetic, and functional diversity and productivity for the perpetual production of amenities, services, and goods for human use.

A *forest-dependent industry* is any industry that uses raw materials from the forested portion of the ecosystem, including amenities and services, such as oxygen, water, electricity, recreation, and migratory animals (anadromous fish). A forest-dependent industry also includes any industry that uses extractive goods, such as minerals, wood fiber, forage for livestock, resident fish and game animals, and pelts from fur-bearing mammals.

Forest-dependent industries based on amenities and services are not extractive if the products either enter and/or leave the forest under their own volition. Such industries includes both the sport and commercial fisher who catches migratory salmon and steelhead, the farmer who uses water to irrigate crops, the person who markets those crops, the electrical company that uses water converted to electricity, and the municipal water company itself.

Forest-dependent industries that are based on extractive products, on the other hand, physically remove raw materials from the forest and, for the most part, make them available for further refinement and use that goes beyond the initial extraction. Such industries include timber companies that cut trees, ranchers who graze their livestock in forested allotments, miners who extract ore, and hunters, fishers, and trappers who kill and remove forest-dwelling wildlife.

Forest-dependent industries that refine the extracted products include jewelers, carpenters, boat builders, artisan woodworkers, anyone who uses paper, meat cutters and packers, and furriers. Finally, these forest-dependent industries are all interwoven because each industry uses one or more of the other's products, such as water, electricity, wood fiber, red meat, vegetables, and so on.

Today, however, the driving economic force is the extractive timber industry, which the public perceives as *the* forest industry, but which actually goes only from the forest to the mill. Sooner or later the timber industry must change, because the public will eventually recognize that there really is a suite of forest-dependent industries that serve the economic benefit of society.

How has the timber industry become the only industry with a voice? It is allowed to speak for all forest-dependent industries, when in truth it is only looking out for its own self-interest by grossly altering the biological sustainability of the forest for short-term profits, by shipping most of its product (logs whenever possible) out of the area to other countries, and by automating its mills to eliminate jobs.

As the timber industry affects the biological sustainability of the forest for short-term profits, it has a dramatic effect on all forest-dependent industries, often to the qualitative and quantitative detriment of the long-term product base and products of those other industries. Oregon, the Pacific Northwest, and indeed the world must have a balance between short-term profits for the timber industry and the sustainability of all other forest-dependent industries that rely on and must continue to rely on a biologically sustainable forest. Many of the existing forest-dependent industries, however, do not understand that they are in fact dependent on the biological sustainability of the forest. The quick profits they seek today, obtained at the expense of ecologically sound forest management, may well spell their demise tomorrow, because the "resource," the old, high-volume trees, has run out.

Today, as never before, the sustainability of our forests rests in our hands and in the wisdom of our decisions, which for better or worse will determine our legacy for tomorrow, a legacy that is becoming increasingly irreversible. Consider, for example, the music that great composers have committed to paper, the translation of which has been through orchestras. An orchestra, in turn, is composed of musical instruments and musicians that together give voice to the mute beauty on paper. A musician's ability to play a musical instrument is dependent not only on human skill but also on the quality of the instrument. Over the last two centuries, the violins made by Antonio Stradivari (1644–1737) have given to the human ear some of the world's most exquisite melodies.

To build a Stradivarius one not only must be an expert violin maker but also must have available ancient, fine-grained wood. Thus the quality of a Stradivarius is of yesterday and of today, but what about the quality of the wood for the violins of tomorrow?

I was once asked this question at a conference where I was speaking on rain forests. A young man who made violins by hand asked, "What will I do for a living when the ancient forests are all gone and there's no more special, high-quality wood, such as that of an ancient, tight-ringed, clear-grained Sitka spruce, for me to work with?" "There are," he said, "very few people who work the way I do, and we're rapidly becoming fewer." There was silence in the audience as people tried to assimilate the profound ramifications of his question and the frightening direction in which society is blindly racing. The haunting reality of his question remains.

If humans continue to destroy the ancient forests of the world and replace them with nonsustainable plantations that produce inferior wood, the symphonic beauty of the centuries will become but a hollow echo of dull tones from dead instruments as they play the requiem of the old-growth forests.

We can have sustainable forests, but only if that is what we are committed to, only if we constantly question and re-evaluate what we think we know along the way, and only if we retain all of the pieces—including native forests and old growth (ancient forests)—from which to learn. We can have a sustainable suite of forest-dependent industries, including the timber industry, that produce products for people, but only if the timber industry is redesigned to operate within the sustainable limits set by the forest.

In both cases, we must learn humility, which means that we must learn to be taught by Nature. We must become students of processes and not advocates of positions. Both schools of forestry and the Congress of the United States must become leaders, rather than the anchored resistance to change they currently personify. We must work together for a common goal, with a common commitment: a sustainable forest for a sustainable suite of forest-dependent industries for a sustainable environment for a sustainable human society. The latter, however, must ultimately remain within the cultural capacity of the planet. (For a discussion of cultural capacity, which is a chosen quality of life that can be sustained without endangering the productive ability of the environment, see *Global Imperative*.)[11]

Before we can have a biologically sustainable forest, however, we must have the humility to sit in Nature's classroom—the native forest, including the old-growth successional stage. Only when we have the foresight to save what is left of the old-growth forest, in order to learn how to recreate it as needed, will we understand forests and their biological sustainability.

10

WHY OLD GROWTH?

THE VALUE OF OLD GROWTH

There are many valid reasons for saving old-growth forests from extinction, as many perhaps as there are for saving tropical forests. One reason is that such forests as those of the Pacific Northwest are beautiful and unique in the world. Another is that such old-growth trees as those of the Pacific Northwest forests inspire spiritual renewal in many people and are among the rapidly dwindling living monarchs of the world's forests (Figure 67). Old-growth forests are the oldest living beings on earth, and as such they form a tangible link with the past and provide a spiritual ground in the present. Consider that trees can and do live for centuries or millennia. Bristlecone pine (Figure 68), for example, is known to live for over 5000 years.

What would happen to our sense of continuity or our sense of spiritual ground in the dynamic, unpredictable Universe if all the remaining commercially available old-growth forests, which constitute most of the best old-growth forests left in the world, were liquidated for short-term profits? How would you feel if these centuries-old living beings where suddenly replaced with trees that were not allowed to grow much older than a person? How would you feel if you knew that the oldest living beings on earth had all been converted into money and you would never again see them, and your children would never have a chance to see them?

A third reason for saving old-growth forests is that they are unique, irreplaceable, and finite in number. They exist precisely once in the world because whatever is created in the future will be different—and centuries away. Large trees can perhaps be grown over two or three centuries, but such trees will not be Nature's trees in Nature's landscape; they will be humanity's trees in society's cultural landscape. Although they may be just as beautiful as those created by Nature, they will be different in the human mind. And even if we start growing

FIGURE 67 Old-growth island surrounded by forest plantations of various ages. (USDA Forest Service photograph by J.F. Franklin.)

them today, neither we nor our children nor our children's children for several generations will be here to see them.

A fourth reason for maintaining old-growth forests is that a number of organisms, such as the red-cockaded woodpecker, the spotted owl, the northern flying squirrel, and red tree vole (Figure 69), either find their optimum habitat in them or require the structures provided by the live old trees, the large declining trees, the large standing dead trees, or the large fallen trees. In other words, old-growth forests are tremendous storehouses of biological diversity. As such, they are genetic reservoirs, harboring plants of potential use in medicine, agriculture, and industry.

This has already proven to be the case in the forests of southern Mexico and Central American with their exceptionally rich biotic diversity. In 1978, for example, a wild variety of perennial maize (or corn) was discovered in a forest of southern Mexico. Through cross-breeding, this new strain could enable the corn-growing industry to avoid the seasonal costs of plowing and sowing. In addition, the wild germplasm offers resistance to several viruses that attack

FIGURE 68 Bristlecone pine over 3000 years old, Wheeler Peak National Park, Nevada.

commercial corn. The economic benefits of this discovery could eventually reach billions of dollars.

In North America, the Pacific yew, a shade-tolerant tree in the understory of old-growth forests of the Pacific Northwest, was once considered a weed. It became of medical importance in recent years because it was the sole source of one of the most promising new drugs (Taxol) for the experimental treatment of

FIGURE 69   Red tree vole. (Photograph by the author and Ron Altig.)

ovarian cancer and more recently lung and breast cancer in women. Dr. Robert Schweitzer, President of the American Cancer Society, in a letter dated September 19, 1990, to Manuel Lujan, then Secretary of the Department of the Interior, stated: "One behalf of the two and one half million volunteers of the American Cancer Society, I write to urge that you take any and all actions to protect the Pacific Yew tree as a 'threatened species' pursuant to…the Endangered Species Act….The American Cancer Society believes that you should designate the Pacific Yew…as a threatened species. This action will ensure that women diagnosed with ovarian cancer in the years ahead will have access to a promising drug treatment."

The Pacific yew was once widely distributed from the southern tip of southeastern Alaska throughout the Pacific Coastal regions of British Columbia and the Olympic Peninsula in Washington. Although relatively rare in the Coast Range south of the Olympic Peninsula and north of the Umpqua River in Oregon, its frequency was and is greater in the Coast Range in southwestern Oregon and northwestern California. Today, however, the Pacific yew is rapidly disappearing because of clearcutting the old-growth forests and because of uncontrolled, commercial harvesting of its bark for medicine.

These circumstances prompted not only Dr. Schweitzer's letter but also a 12-page letter, dated September 19, 1990, from the Environmental Defense Fund to

Manuel Lujan requesting that the Pacific yew be declared a "threatened species." To this end, four major points are made in the letter: (1) "because the Pacific yew occurs principally within ancient forests and those forests have declined substantially over the past century, the yew is today a depleted species," (2) "the Pacific yew is extremely vulnerable to logging operations and is being further reduced with continued liquidation of the ancient forests," (3) "it is imperative that the Pacific yew be afforded protection since it currently serves as the major source of the anti-cancer compound Taxol and it will enhance efforts to produce an alternative source of Taxol, " and (4) "addition of the Pacific yew to the list of threatened plant species will authorize federal agencies to protect the species from habitat destruction and commercial exploitation."

Even if the Pacific yew is ultimately determined by the U.S. Fish and Wildlife Service not to be threatened with extinction and therefore not afforded protection under the Rare and Endangered Species Act, it prompts questioning what other species of plants are growing in the old-growth forests of the world, the benefits of which will never be known if these remaining habitats are liquidated for short-term profits.

A fifth reason to maintain old-growth forests is that each old-growth tree is a "carbon sink," a storehouse of immobilized carbon, the storage of which reduces the potential carbon dioxide in the atmosphere and thereby has a positive influence on the greenhouse effect. In a comparison of total carbon stored in Douglas-fir trees, there is 2.2 to 2.3 times as much carbon stored in a stand of Nature's 450-year-old Douglas-fir as there is in a 60-year-old plantation. In fact, landscapes with plantations that are cut on rotations of 50, 75, and 100 years would be able to store at most 38, 44, and 51 percent, respectively, the amount of carbon stored in the old-growth stand. And these values are conservative, because the intense utilization of wood fiber in plantations removes even more of the soil's capacity to store carbon.

Although the reintroduction of trees into deforested regions will increase the storage of carbon in the living organisms of the area, conversion of old-growth forests to young stands and plantations under the current conditions of harvesting and use has added and will continue to add carbon dioxide to the atmosphere. These circumstances are likely to hold in most forests in which the age of harvest is less than the age required by the trees to reach the old-growth stage of succession.

The amount of carbon storage that will be lost by converting old-growth forests to young stands and plantations will vary among the types of forests. In addition, such storage will depend not only on the age at rotation in a managed forest or plantation but also on the difference between the capacity of the young stand to store carbon when it is cut and a forest's maximum capacity to store carbon in the old-growth state.

The timber industry has argued for some time that liquidating the remaining old-growth forests in favor of plantations will help to reduce the carbon dioxide in the atmosphere because it is absorbed in greater quantities by the fast-growing young trees than by the slow-growing old trees. In light of the preceding data, the timber industry's argument is clearly erroneous.

A sixth reason to maintain old-growth forests is that such forests are the only living laboratories through which we and future generations may be able to learn how to create sustainable forests, something no one has so far accomplished. Let's examine this latter reason in more detail.

OLD GROWTH AS A LIVING LABORATORY

As a living laboratory, old-growth forests serve many vital functions, including the following four. First, old-growth forests are native and as such form our link to the past, to the historical forest. The historical view tells us what the present is built on, and together the past and the present tell us on what the future can be projected. As stated by South African General Jan Christian Smuts, "The whole, if one may say so, takes long views, both into the future and into the past; and mere considerations of present utility do not weigh very heavily with it."[67] Thus, because the whole forest cannot be seen without taking long views both into the future and into the past, to lose the remaining, commercially available old-growth forests is to cast ourselves adrift in a sea of almost total uncertainty with respect to the biological sustainability of the forests and plantations of the future.

We must remember that knowledge is only in the past tense, learning is only in the present tense, and prediction is only in the future tense. To have sustainable forests, we need to be able to know, to learn, and to predict. Without old growth, we eliminate learning, limit our knowledge, and greatly diminish our ability to predict.

Second, because we did not design the forest, we do not have a blueprint, parts catalog, or maintenance manual with which to understand and repair it. Nor do we have a service department in which the necessary repairs can be made. Therefore, we cannot afford to liquidate the old growth, which acts as a blueprint, parts catalog, maintenance manual, and service station and is our only hope of understanding the sustainability of the redesigned forests and plantation forest.

Third, we are playing "genetic roulette" with forests of the future. What if our genetic engineering, genetic cloning, genetic streamlining, and genetic simplifications run amok, as they so often have around the world? Native forests, whether old or young, are thus imperative because only they contain the entire genetic code for living, healthy, adaptable forests.

Fourth, intact segments of the old-growth forest from which we can learn will allow us to make the necessary adjustments in both our thinking and our subsequent course of management to help assure the sustainability of the redesigned forest. If we choose not to deal with the heart of the old-growth issue, which is the biological sustainability of present and future forests, we will find that reality is more subtle than our understanding of it and that our "good intentions" will likely give bad results.

Although there are many valid reasons to save old-growth forests, there is only one apparent reason for liquidating them: short-term economics. Economics, however, is the common language of Western civilization; therefore, it is wise to carefully consider whether saving substantial amounts of well-distributed, old-growth forests is a necessary part of the equation for maintaining solvent forest-dependent industries.

Can we really afford to liquidate our remaining commercially available old-growth forests? It has often been said that we cannot afford to save old growth because it is too valuable and too many jobs are at stake. We must be careful here, because scarcity not only increases the economic value of the remaining commercially available old-growth forests but also increases their ecological value. In the face of this scarcity, cutting most of the remaining commercially available old-growth forests will serve only a small proportion of the immediate generation of humans, whereas protecting most or all of them will serve all individuals within all generations to come. We must therefore be exceedingly cautious lest economic judgment and greed isolate us from the evidence that indicates that without ecologically sustainable forests, we will not have economically sustainable forest-dependent industries, without which there will be human communities in which we cannot have a sustainable economy.

Thus, if we liquidate the remaining commercially available old-growth forests—our living laboratories—and our plantations fail (as plantations are failing over much of the world), there will be no timber industry, and forest-dependent industries will also suffer greatly. If we liquidate the remaining commercially available old-growth forests, the timber industry will be the bath water thrown out with the baby, and we will have further impoverished our souls and those of future generations through the myopic drive for short-term profits.

Unless our minds and our hearts are set on maintaining an ecologically sustainable forest, each succeeding generation will have less than the preceding one, and their choices for survival will be equally diminished. While this is our choice—and we are limited only by what we think we cannot do—the consequences belong to the generations of the future, which may prove to be a double burden in the face of global climate change.

11

GENETICS, ADAPTABILITY, AND CLIMATE CHANGE

GENETIC "IMPROVEMENT"

Genetic manipulation of forest trees was born in the concept of short-term economic expediency. By necessity, this process ignores long-term ecological ramifications to and within the forest as a whole, because tenable management practices for short-term profits must be based on predictable results.

Some people consider "genetic improvements" to be the panacea of forestry. It is not surprising, therefore, that large areas of native forest in tropical, subtropical, and temperate regions are being converted into industrial plantations. In the short term, this is seen as maximizing returns on investments and as homogenizing the base of raw materials to the benefit of the timber industry. Weyerhaeuser, "The Tree Growing Company," for example, placed an ad in *The Oregonian* (September 17, 1990) stating that "We've planted about 280 million trees in Oregon over the last 50 years. It's a good beginning." The ad goes on to say that "The cornerstone of our forestry programs is stewardship—we manage our lands so we can have trees here forever."

Weyerhaeuser does replace native forests with genetically manipulated trees in plantations to benefit Weyerhaeuser. Beyond the short-term economic benefit to Weyerhaeuser and other timber companies, however, it is imperative to address the immense importance of long-term biotic, genetic, and functional diversity in relation to global climate change. Current criteria for the economic evaluation of plantations do not adequately account for these and other crucial issues that accrue from the management of whole forests.

In reality, the only "panacea" in forests is an unlimited supply of virgin timber (such as old-growth Douglas-fir) to feed the insatiable appetite of the linearly programmed, expansionistic timber industry. That is not possible, however, because native old-growth forests are an inheritance that we are given only once on each acre of ground. We have spent our own inheritance and are spending that of the future generations.

Genetic manipulation, the perceived industrial panacea, contains hidden costs. Four hidden costs can be shown for genetic manipulation of forest trees.

The first is lack of predictability in our unique forests of the Pacific Northwest; we cannot predict any results, because no one has yet grown a "genetically improved" forest (actually a plantation) for even one rotation. We are playing genetic roulette with future forests, a dangerous game about which more will be discussed later.

The second hidden cost is that by manipulating the genetics of the trees, the function of ecological processes in the entire managed forest is being altered by changing how the individual trees function. For example, if they grow faster, they will have larger cells and more sapwood and less heartwood, which changes the way they recycle in the soil. In turn, all other connected biological functions are altered. And we do not even know what these functions are, let alone what difference they will make in the long-term health of the forest. We can, however, make some educated guesses.

For example, a central thrust of modern agricultural technology (including forestry under this umbrella) has been to: (1) isolate such individual organisms as a particular species of tree which possess desirable economic characteristics, (2) enhance their characteristics through breeding, and (3) replicate them on a massive scale. This type of plant is termed an "ideotype" for such an ideal plant model, which literally means a form denoting an idea.

Any management, but particularly that which focuses on individual ideotypes, unavoidably changes other properties of the ecosystem in addition to productivity, which is the target of the change. Genetic diversity and fundamental structural relationships, the architectural aspects of native plant communities, are altered as well. Although numerous ecologists and foresters have raised concerns about this for some time, these concerns have too often been viewed as grounded in unproven criticism, which has little or nothing to do with economic need.

Quite to the contrary, system-level properties are likely to play a seminal role in the health of the ecosystem, and healthy, sustainable forests provide values that are no less real, because they cannot be traded in the marketplace. Society will continue to demand wood and forests will provide it, but forests both produce wood and play a central role in the dynamics of global climate, in addition to which they harbor immense biotic diversity.

The third hidden cost is illustrated in Figure 70. An unmanaged forest (Figure 70A) is like a numbered Swiss bank account that has a complete denomination of its own particular currency. The currency in the forest (called stored genetic variability) is unseen, as is the currency in the bank account. One does not have to see the currency in order to get the correct change. If there is an unlimited

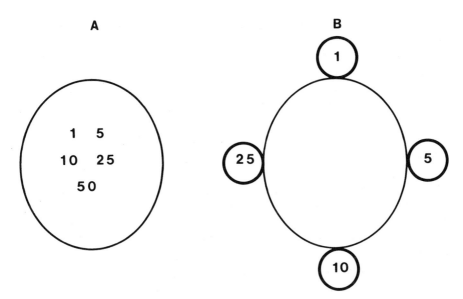

FIGURE 70 (A) An unmanaged forest has unlimited genetic variability, as represented by enough denominations of coin to make exact change from any denomination of currency. (B) The hidden trade-off in genetic specialization is the loss of genetic flexibility in that the forest is no longer as adaptable to changes in climate, pollution, etc. Gains in short-term, fast growth are at the expense of the long-term ability of the forest to adapt to constantly changing environmental circumstances. The long-term biological health of the forest is therefore being jeopardized for short-term economic gains.

amount of money in the bank account, as there is genetic variability in the unmanaged forest, then the exact change can be received for any denomination chosen, even from an automatic, mechanical teller.

This means that a forest can, within limits, adapt genetically to changes in climate from a human-caused greenhouse effect due to loss of atmospheric ozone, natural climatic changes, increasing air pollution, and so on. When genetic variability (currency) is withdrawn from the forest's genetic account (remove the 10 in Figure 70A), the ability of the forest to adapt to changing conditions—something it must continually do in order to survive—becomes artificially limited. Without the 10-cent coin, the exact change can no longer be received from the Swiss bank account; it becomes limited and loses flexibility in terms of the transactions that can be made.

Let's take this one step further. A forest is cut down across the landscape from northern Washington to southern Oregon and from the Pacific coast to the crest of the western Cascade mountains. "Genetically improved" Douglas-fir seedlings are then planted to grow quickly. In addition, seedlings planted in northern Washington are selected to withstand cold, those in the south to withstand heat, those in the west to withstand wet weather, and those in the east to withstand dry weather and a short growing season. In order to gain genetic selectivity, we have artificially adapted the trees to our set of values. In doing so, we give up flexibility, genetic plasticity—the tree's inherent ability to adapt to changing conditions.

As an analogy, suppose that you are traveling in Switzerland and that the central bank with your numbered account is in Geneva. Only the central bank has all denominations of currency (Figure 70A). As you travel in northern Switzerland, you find satellite, mechanical bank tellers from which you can make change, but they only have a denomination of 1 (Figure 70B). In the east you find a teller that has only 5s, in the south only 10s, and in the west only 25s.

The selected currency of the satellite tellers restricts your ability to adapt to unexpected necessities, and this loss of flexibility can cause considerable unforeseen hardship. It is the same with a forest. Nature allows for changes in climate and equips trees with the genetic ability to adapt and survive. What happens when a forest is converted to a plantation selected for warm and dry conditions and robbed of its "excess genes" when a long-term change in the climate results in wet and cold conditions?

Even worse than short-sighted genetic engineering for short-term profits is the concept of cloning. A clone is a group of genetically identical cells descended from a single common ancestor. Cloning, in turn, is the creation of a genetic duplicate of an individual organism, such as a tree with economically desirable characteristics, through artificial asexual reproduction.

Through cloning, genetic diversity—and therefore adaptability—is increasingly limited over time as genetically handicapped, identical trees are planted in an environment that changes freely in the invisible present. The more the environment changes, the more the cloned trees, which have been robbed of some portion of their adaptability, appear to remain the same, like a historic relict out of the past. Such relicts are holdovers that cannot adapt fast enough to keep pace with the changing conditions of their environment. Extinction, therefore, is always just around the corner.

The fourth hidden cost is what Thomas Ledig calls the "secret extinctions," which is the loss of locally adapted populations of species, such as trees, that have evolved over millennia, a loss that can be more or less permanent.[68] Thus, if locally adapted populations of a species are extirpated, they might never be replaced, because other populations of the species might lack the characteristics necessary to become reestablished in the habitat. They also might not be able to reach suitable habitat because of major environmental shifts due to the unprecedented speed of the changes brought about in the global climate.

In addition, hybridization between species, such as black spruce and red spruce, shows that modifying the genetic structure can, in some cases, lead to changes in the cycling of carbon, which must necessarily modify other ecological cycles. Bear in mind that these secret extinctions are part of the invisible present.

Altering the genetic adaptability of the forest is a concern because evidence indicates that global climate change is real. Global warming through the greenhouse effect is a phenomenon that most scientists agree is taking place now, but how is it happening?

CLIMATE

The average temperature of the surface of the globe is determined by a complex of factors, including the amount of energy received from the sun, the properties on the Earth's surface that absorb the sun's heat (such as the size and distribution of oceans, snowfields, forests, and deserts), and the absorptive properties of the atmosphere. A high proportion of the energy that is absorbed by the lower atmosphere and by the Earth's surface is emitted as heat. Some of the heat passes through the atmosphere into space, but the rest is absorbed into the atmosphere and passed back towards Earth.

At the root of the predicted changes in climate is the accumulation of an array of gases in the atmosphere that trap heat as it is radiated outward from the Earth's surface. Carbon dioxide has to date received the most attention, but other gases are involved as well, such as water vapor, ozone, nitrous oxide, chlorofluorocarbons, and methane.

Although there is uncertainty as to how the climate will respond, there is no doubt that greenhouse gases are accumulating. Atmospheric carbon dioxide began to rise in the latter part of the 19th century as a result of clearing forests and plowing prairies, both of which released into the atmosphere carbon dioxide that once had been stored in the living tissues of plants and in the soil. Burning of fossil fuels, such as coal, and accelerated deforestation since the 1940s have greatly exacerbated the problem.

Although public attention has focused on the problems caused by deforestation in the tropics, problems also exist in the northern hemisphere. On the one hand, burning of tropical forests converts about 50 percent of the badly needed nitrogen contained in the forests' biomass to a gas, which is lost from the forests to the atmosphere. It also causes vast amounts of smoke, which can so alter the internal physical structure of clouds that severe repercussions may be in store for the water cycle in the tropics. On the other hand, clearcutting old-growth Douglas-fir forests in the Pacific Northwest results in a net release of the carbon dioxide stored in the stems of the live trees into the atmosphere. It takes several centuries to recoup this loss, even when the cut forest is replaced with young, fast-growing trees.

In many areas of the world, including parts of Canada, the United States, and Central America, logged forests are not being successfully regenerated with young trees. In Central America, for example, deforestation has been rapidly accelerating during the past three decades. If it is not slowed, there will be little but scrub forest left by the end of this century. If important areas of forest, particularly those that serve the major rivers and watersheds of the region, are not protected soon, no amount of social or economic reform will be able to provide for the many basic needs of the increasing population of the region. This also means that the carbon dioxide lost to the atmosphere by cutting the old-growth forest is not being mitigated by the capacity of young trees to use carbon dioxide and thus remove it from the atmosphere.

With a doubling of the pre-industrial, atmospheric concentration of carbon dioxide (which will be reached some time during the next century), temperatures in the conterminous United States are predicted to increase in winter to the equivalent of a four- to six-degree shift southward in latitude. In summer, the temperature increase would be the equivalent of a five- to eleven-degree shift southward in latitude. This would be like shifting southern California to central Oregon, which is a difference of about ten degrees in latitude.

Note that a greater warming is predicted in summer than in winter. The magnitude of warming is also predicted to vary from east to west across the United States and Canada. The greatest warming will take place in the area of the Great Plains, where North Dakota may have a climate similar to that which presently exists in Texas. In addition, the models predict a warming in the mountains that will be equivalent to roughly a 2000- to 3000-foot decrease in elevation. This means that most, if not all, of the subalpine and alpine areas would probably disappear in the United States south of the Canadian border.

If current trends in the emission of greenhouse gases continue, increases in carbon dioxide and the other gases are predicted to result in "doubled carbon dioxide" temperatures in less than 50 years. Other greenhouse gases, although less abundant in the atmosphere than carbon dioxide, have a much greater warming effect per molecule. For example, one molecule of methane has 3.7 times the potential of one molecule of carbon dioxide for warming the atmosphere. A single molecule of nitrous oxide has 180 times and chlorofluorocarbon-12 has 10,000 times the potential capacity of one molecule of carbon dioxide to warm the atmosphere.

The models of climatic change are less consistent in predicting the greenhouse effect on precipitation, which is a crucial factor in how the ecosystem will respond to changes in climate. However, it is generally agreed that warming will be greater at high latitudes than at lower latitudes. The consequence of such warming will be the narrowing of the global temperature gradient, which in turn seems likely to alter global patterns of precipitation, but what these changes might be remains unclear. Nevertheless, a warming climate, even with no change in precipitation, should increase drought because of the greater evaporative demand brought about by generally higher temperatures.

Regardless of whether or not such a warming effect is already reflected in global weather patterns, the consensus among atmospheric scientists is that global warming, at some unknown intensity, is virtually imminent. In northwestern Ontario, Canada, for example, the climatic, hydrologic, and ecological records for the Experimental Lakes Area show that air and lake temperatures have risen by 3.6 degrees Fahrenheit and that the length of ice-free days has increased by 3 weeks over the last 20 years. (Hydrologic is the adjectival form of hydrology, which refers to the scientific study of the properties, distribution, and effects of water on the Earth's surface.)

Further, higher than "normal" evaporation and lower than average precipitation have occurred, resulting in a decrease in the rate of renewal of water in the lakes. In addition to other changes within the lakes themselves, the concentrations of most chemicals have increased in both the lakes and the streams because of the decreased renewal of the water and because of forest fires in the watersheds. These observations may provide a preview of the effects of increased greenhouse warming on boreal lakes.

Because some of the projected scenarios about global warming show increasing drought in various areas, such as the mid-continental "Great Plains," the greenhouse effect may also alter fire regimes and in some areas contribute to an increased frequency of forest fires. There may be some difficulty, however, in determining potential effects of global warming on the fire regimes, because most historical studies, termed chronosequences, are hampered by effects of unknown historical events, which can result in erroneous interpretations.

It is therefore particularly important to study the major ecological processes in an integrated fashion, because mechanisms are interdependent. From an ecological point of view, the variability in fire regimes is more likely to be important to plant communities than are the mean values computed from some arbitrary period of fire history.

For example, unusually long periods without fire may lead to the establishment of species of plants that are intolerant to fire. The simultaneous occurrence of such fire-free periods and wetter climatic conditions may also be extremely important to such species of plants as ponderosa pine that have episodic patterns of regeneration (specific, discrete episodes) as opposed to plants whose patterns of regeneration form a continual, yearly process.

Therefore, while statistical summaries of fire history over time are useful in understanding the general comparisons of fire regimes in different forests, the influence of fire on the ecosystem is a strongly historical process. Hence, forests of the southwestern United States may be more a product of relatively short-term and unusual periods of climate and fire frequencies than of average or cumulative periods of long-term histories of climate and fire frequencies.

How change in the global climate will affect forests (such as the boreal forests of northern Canada and Alaska, the Douglas-fir forests of the Pacific Northwest, or the ponderosa pine forests of the American Southwest) is not known exactly, because these conditions have never been faced before. Yet there is much that can be done, even as evidence is rapidly mounting that business as usual by the timber industry and wasteful consumerism are likely to be the undoing of forests as we know them. This is especially true in the face of a changing global climate that is predicted to take place with a speed unprecedented by anything known in geological history. Options and a different way of thinking will be discussed later.

For now, suffice it to say that armed with only short-term economic projections based on irrational, linear thinking, we are genetically "improving"—and biologically jeopardizing—our forests through substitution of genetically intact and adaptable forests with economically designed, genetically restrained, monocultural plantations wherever possible. And the allowable cut, now called the "allowable sale quantity" (not the amount of timber that may be cut but rather that which "shall" be cut—the "hard" target), is increased because of these projections.

This is not a new phenomenon, however, because modern forestry is touted as agriculture, which in some ways it is, as nicely put by Professor Donald Worster: "American agriculture has been powerfully persuasive in the world, even among those who profess to live by different principles. Its willingness to take risks for increased production has set a pace that other nations, such as the [former] Soviet Union, feel constrained to follow—just as less aggressive plains farmers have been led to emulate their more affluent entrepreneurial neighbors. There may be many reasons why people misuse their land. But the American Dust Bowl of the thirties suggests that a capitalist-based society has a greater resource hunger than others, greater eagerness to take risks, and less capacity for restraint."[59]

THE UNTESTED PRODUCT

Another point to bear in mind is that genetically "improved" trees represent a new, untested product (the first hidden cost), and American business has had much experience with new products. Consider the risk of new-product failure.

The timber industry, its customers, the American public, and society at large, especially future generations, have an important stake in the effective management of innovation and new-product process. Direct expenditure by the timber industry on research and development is large and must be judged by its ability to yield superior new products and processes. During development, the market success of a new product *is always in doubt.* In fact, a large proportion of all programs are terminated before achieving success in the market.

A number of industries have indicated that less than 15 percent of all development projects resulted in commercially successful products and that a full 37 percent of all products that reached the market proved to be commercial failures. An example is weak wood from fast-grown coniferous trees in economic plantations. The seven most convincing studies placed the failure rate of products introduced in the market at about 40 percent. It should be emphasized, however, that the types of products examined, the period of time covered, and the definitions used can all have a pronounced effect on the reported rate of failure or success.

Even though products discussed in the preceding paragraph were all conceived, studied, designed, studied, manufactured, studied, and marketed by people for people, there is no guarantee of success. When we "redesign" trees we also redesign the entire functional forest, even though we know little about the consequences. A similar caution is given by geneticists David Suzuki and P. Knudtson. In *Genetics, The Ethics of Engineering Life,*[69] they discuss human genetics, which is better understood than the genetics of trees and their symbiotic companions that make up a forest.

A new product on the consumer market succeeds or fails in months. The success or failure of a redesigned forest—the invisible present—will not become apparent for perhaps a century or more. If we fail, who will pay the price? It seems prudent, therefore, to experiment with small acreages of genetically manipulated trees rather than committing thousands of acres to potential disaster by stealing from the future and increasing the allowable cut based on short-sighted economic assumptions. With no data as collateral, there is no way to pay the ecological bill when it comes due and no way to be held accountable for our folly of corporate greed.

What about the vast majority of acres on which it would be imminently wise not to experiment with genetically "improved" trees? These acres could be the insurance policy that is set aside for the benefit of the future. They can be managed as "native forest" (that which has experienced no genetic manipulation), much the same as they are currently managed on public lands in the Rocky Mountains of Colorado and New Mexico. Here the forests are more often than not selectively cut or cut in groups that create small openings. In both cases, however, natural reseeding is allowed to take place, which maintains the native gene pool and thus the genetic adaptability of the forest.

This type of cutting means that clearcutting and genetically manipulated plantations are minimized, or even absent. It also means that while the managed, native forest may not, theoretically at least, produce the most desirable economic product in the short term, the forest has the best chance of survival over time and therefore the best chance of maintaining its biological sustainability and creating economic sustainability.

In our burgeoning, product-oriented society, however, one of the most insidious dangers to native forests, be they old or young, is the sadly mistaken perception that there is no value in maintaining or managing an area for its long-term *hidden potential* at the apparent cost of its short-term profits. Potential here refers to genetic adaptability, research value as an ecological blueprint of what a biologically sustainable forest is and how it functions, educational value, spiritual value, or any other value that does not turn an immediate, visible, economic profit.

This short-sightedness is understandable considering that: (1) whereas the Native North Americans viewed the land and all it contained as a "Thou," which is holy and is to be revered, and they therefore lived in relatively greater harmony with the land, Europeans viewed the same land and all it contained—including the indigenous peoples—as an "it," which is simply an object to be exploited; (2) Western civilization focuses predominantly on utilizable *products* from the ecosystem rather than on the *processes* that produce the products; (3) renewable "natural" products are largely manifested aboveground, whereas many of the processes that produce the products are invisible belowground; (4) we therefore think about and manage what is visible aboveground and tend to ignore the crucial biological processes below the surface of the soil; and (5) short-term economic gain is the driving force behind management of renewable natural resources and our society.

When these points are taken together, they form the foundation of Western economic culture. Reared with this historical background, most people find it difficult to really understand the risks to society's future that accompany the violation (particularly the genetic violation) of remaining native forests either in principle or in fact. Although this may seem a bold statement, consider that, in addition to representing a collection of native species of plants and animals with a given amount of genetic diversity, each area of native forest (particularly reserves, whether old or young) also represents a repository with a portion of the world's healthiest ecological processes and their attendant functions.

THE VALUE OF NATIVE FORESTS

The native, temperate coniferous forests of the Pacific Northwest, for example, are still relatively "healthy," whereas both the temperate coniferous forests and plantations of central Europe are dying. In addition, far more is known scientifically about forests of the Pacific Northwest and how they function than is known about European forests. Therefore, areas of managed and unmanaged native forest in the healthy Douglas-fir region of the Pacific Northwest are repositories for ecological processes that, although different in specifics, are similar in principles to those of the dying forests and the dying Norway spruce plantations of central Europe.

By analogy, rather than a historical transplant of a particular species to reintroduce it into an area from which it has been extirpated, we have the potential to perform *global process-information transplants* through ecological knowledge that can be gleaned from managed native forests and through unmanaged benchmark areas that represent Nature's blueprint.

Although benchmark areas of unmanaged native forests are likely to be the healthiest repositories of ecological processes, including reservoirs of "undomesticated" or "wild" soils, areas of managed native forests can also share in the positive legacy for the future. Yet today's economists and industrialists seem to see only "economic waste" in saving them and managing them for their "naturalness" and "nativeness."

Areas of native old-growth forest are more important now than ever before, because, as discussed earlier, we face a change in global climate, possibly a generalized global warming. Such a historically unprecedented warming would mean that forests must be adaptable. This includes both plants and animals, which constitute the interrelated, often symbiotic, biological processes of life.

The problem for human society is that no one knows which species or which individuals within a species will be able to adapt to such changes. We are only now trying to find out what is left in the genetic bank accounts of native forests and wilderness on which we have been drawing for generations. What is known, however, is that native species are much more likely to be able to adapt than exotics, even of the same species brought in from other areas. In this sense, adaptability equates to resilience in the face of sudden, dramatic, perhaps irreversible change.

Part of the process of maintaining ecological resilience in the face of a rapidly changing climate is to manage as much of our forests as possible in their native genetic state. This includes setting aside an ecologically adequate system of unmanaged natural areas of native forest, including old growth, as an unconditional gift of potential knowledge for the future. In so doing, present and future generations have a repository of both species (which more often than not are region-specific) and processes (which more often than not are worldwide in principle and application).

In addition to monitoring human-caused changes and maintaining habitat for particular species, it will be possible to learn from such repositories how to maintain, restore, and sustain biological processes in various portions of the ecosystem. Although a global framework already exists for identifying and maintaining some of these "repositories" in a program called "Man and the

Biosphere," the program must be expanded. In this sense, local and regional reserves of native forest, including old growth and to a lesser extent managed native forests, can augment the "Man and the Biosphere" program as parts catalogs and maintenance manuals for forests of the present as well as the future.

A QUESTION OF AFFORDABILITY

Can a single species, such as the northern spotted owl, really act as a symbol of the health of old-growth forests, or is the owl more a symbol of the battle of purposeful extinctions around the world for economic gain? (It is beyond the scope of this chapter, or even this book, to discuss the ecological significance of the extinction of species. For additional information, see *Global Imperative.*[11])

The northern spotted owl has become a surrogate for old-growth forests, a symbol in the struggle of conflicting values: short-term economics versus all other human values of old-growth forests in the Pacific Northwest. At issue here are values and the emotional interpretations they give rise to rather than the survival of the owl.

Recall that anger is not directed at a fact. It is an interpretation of a "fact" and its perceived effects that gives rise to negative emotions, regardless of their seeming justification or the intensity of the anger that is aroused. The intensity of anger may be only slight irritation, perhaps too mild to be even clearly recognized. Then again, it may take the form of intense rage, accompanied by thoughts of violence. All of these reactions are the same. They obscure the truth, which is never a matter of degree. Either truth is apparent or it is not. It cannot be partially recognized.

The foregoing brings us to the crux of the issue. The spotted owl is used as the symbol for the *survival* of the old-growth forest, but what does it really symbolize? The spotted owl is called an "indicator species" because its presence supposedly indicates a healthy, old-growth forest, but what does it really indicate?

The spotted owl may be seen as a symbol for the survival of old-growth forests, but in reality it is an indicator species for the *planned extinction* of old-growth forests. Although the spotted owl was selected as the symbol of the health and survival of the old-growth forest with good intentions, the results are bad if instead of focusing attention on the impending extinction of most of the old-growth forest, attention is deflected and focused on the owl or on jobs, because neither is the real issue. The real issue is the economics of extinction—the planned liquidation of old-growth forests for free, short-term, economic gains by the big timber companies.

If we really want spotted owls to survive, then we must also want old-growth forests to survive. Scientific evidence indicates beyond a reasonable doubt that northern spotted owls require the unique structural components of old-growth forests.

Old-growth set-asides on public lands (there are none that I know of on private timber lands), as now planned, will create self-destruct islands of time-limited old growth in a sea of young-growth plantations. Unless a portion of the existing mature forests is also set aside to replace the old growth as it falls apart with age or by unplanned catastrophe or is cut through immoral uses of policy on the part of land management agencies, the spotted owl is doomed. Then the only difference is time if a portion of the young-growth forest is not also committed to replace the mature forest as needed to maintain quality spotted owl habitat.

For years, we have been planning habitat set-asides for the spotted owl in terms of absolute minimums and the cut of old-growth forests in terms of flexible maximums. Then, in 1990, the spotted owl was officially listed as a "threatened species" by the U.S. Fish and Wildlife Service, based in large measure on the recommendations of a commission of federal scientists representing the U.S. Forest Service, the Bureau of Land Management, the U.S. Fish and Wildlife Service, and the National Park Service. After an exhaustive study, they published a report entitled "A Conservation Strategy for the Northern Spotted Owl."[70] The controversy over the study was immediate and intense, because the timber industry is determined to cut every last old-growth tree it can.

It must be clearly understood that I have for a number of years known five of the six people on the commission of federal scientists that wrote the report on the spotted owl, and I have worked with four of them, including Jack Ward Thomas. *Their honesty and their scientific integrity are impeccable.* Yet, to my knowledge, neither the Forest Service nor the Bureau of Land Management, both of which had professionals on the "spotted owl commission," has had the honesty, courage, or professional integrity to unequivocally accept and embrace the report.

The Bureau of Land Management, the timber industry at large, members of the U.S. Senate and House of Representatives from the Pacific Northwest, and the administration of President George Bush tried their best to deny the validity of the report and thereby negate the necessity of its implementation in saving the spotted owl from extinction, which de facto denied the necessity for change.

Yet the questions remain: Are we, as historian Arnold Toynbee determined for the 26 great civilizations that fell, unable or unwilling to change our way of thinking to meet the changing conditions of our world? Have we become so myopic in our economic view that we are willing to risk losing biologically sustainable forests because of the short-sighted, short-term, economic windfall to be had by a few people by cutting the remaining old-growth trees? Can society afford the continuing environmental costs of the economics of extinction? Can society afford the continual loss of species and their functions in preference to short-term economic profit? If you think so, I suggest you re-read "For want of a squirrel, a forest is lost" (in Chapter 1). The forest, after all, is not the trees. It is the sum total result of the interdependent functions of the species that comprise the forest as a whole, of which the trees are but one part.

12

A FOREST IS
A LIVING ORGANISM,
NOT A MACHINE

Everything—from automobiles to forests—runs on energy. If a car or a forest is otherwise in satisfactory working order, then it will continue to run until its supply of energy is used up. Consider the early model Volkswagen bug as an example. Because it lacked a fuel gauge, the driver had to poke a stick into the gas tank to measure the amount of available fuel. This was a critical exercise, because the small reserve tank might not be sufficient to reach the next gas station. Later model VW bugs, with their automatic fuel gauges, took much of the adventure out of travel.

Now consider yourself. Have you ever "run out of gas" when hungry and exhausted? When did you notice that you were getting low on energy? You know you are getting low on energy when hunger or fatigue (your fuel gauge) warns you that you have crossed an invisible threshold and need refueling.

Now consider a forest. Like you, it has an invisible threshold of energy, but it cannot be determined. And unlike the early VW bug, there is no way to gauge the amount of available fuel.

We spend much time gauging things. This is called measurement. We gauge our finances, how far we want to walk, how much to eat and drink, when to go to bed, when to get up, and on and on. We must have learned something of value from all of this gauging. Let's examine some of the differences and similarities between an automobile, a human body, and a forest:

1. An automobile has no living parts, whereas a human body and a forest do. It takes a certain amount of the available energy to keep the living parts in working order; this means that something less than 100 percent of the

energy is actually available to move the body or forest forward. Put another way, a living organism has three options: a surplus energy budget that allows growth and forward motion, a maintenance budget that simply keeps it alive and functioning, or a deficit budget with which it cannot maintain itself and begins to decline (Figure 71).

2. An automobile does not die if it runs out of gas. It has no threshold of life or death, as do human bodies and forests. We know where that threshold is for a human body, but not for a forest.

3. A gauge warns us when an automobile runs low on fuel. We know something is amiss when we get hungry, tired, or cranky; it takes only a few minutes, hours at most, for the symptoms to manifest themselves, and when they do, we know how to fix them. A forest, however, is a different story; it often takes decades, even a century or more, for the symptoms to show up, and when they do, we seldom know what they mean or what caused them.

4. An automobile has many gauges that appear to measure the functions of discrete parts of the engine, such as an oil pressure gauge, temperature gauge, etc. The only gauge in the human body, however, is to measure "how we feel," which we must learn to interpret. The forest, like the human body, also appears to have one gauge to which all others are joined: its energy-process maintenance threshold. This is the critical balance between the inputs and outputs of energy that keep the vital processes healthy and functioning. We can neither see nor feel this threshold (Figure 71).

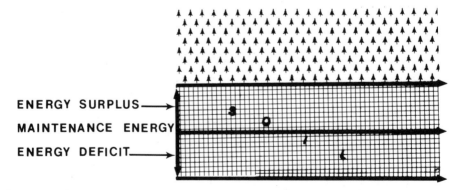

FIGURE 71   Where is the magic management line in the forest called the maintenance energy threshold, where inputs and outputs of energy—with intact functional processes—are balanced for survival of the forest?

5. How we use an automobile, our body, or the forest determines how fast the fuel is used up and how soon the processes are worn out. The more "conserving" we are, the longer each lasts; the more we abuse them, the faster they wear out. An automobile can be rebuilt indefinitely, but it will not be the same automobile. Parts of the human body can also be replaced, but it cannot be totally rebuilt and thus may be increasingly limited in function. A forest also cannot be "rebuilt" and remain the same forest.

A major difference between an automobile, the human body, and a forest is that because *we* designed and built automobiles, parts catalogs and maintenance manuals are available to manufacture and substitute various parts, which can be done by a service department. We also know a great deal about the human body. We have a parts catalog (anatomy texts), a maintenance manual (textbooks on disease and surgery), and a service department (the hospital). Some parts, such as blood, kidneys, etc., are somewhat interchangeable between humans, and some parts can even be manufactured, but less efficiently than an automobile, because we did not design and build the human body and still have a great deal to learn about it.

A forest, on the other hand, is more complex, has more pieces and moving parts, and has more hidden or unknown processes than either an automobile or a human body. We cannot ask it how it feels. Most importantly, we do not have a parts catalog or a maintenance manual, and we do not know which parts, if any, are substitutable.

There is no service department to rebuild a forest in any semblance of its original self, even if that could be defined. Further, the forest, like a human body, is a living entity and as such is not fixable in the mechanical sense. As discussed earlier, if you break your leg, the doctor can set it but cannot fix it. Being a living entity, it must be allowed to heal. So it is with the forest. Because we cannot repair a forest, the cost of its destruction is incalculable. And we may not even realize that we have crossed the final threshold until we have irreparably damaged the environment in which not just the forest but people as well must live.

It is easier to put a person on the moon than it is to manage one acre of forest. Putting a person on the moon is a black-and-white process—either a hit or miss. Managing an acre of forest, however, is an ever-changing kaleidoscope of gray. There are infinitely more pieces and total unknowns in an acre of ground than in a spaceship. A forest is a trillion-piece jigsaw puzzle, and most of the pieces appear to be the same indistinguishable shade with edges that shift and blur. No wonder we prefer to select and use the few pieces we recognize.

A FOREST IS CYCLIC, NOT LINEAR

The main reason we treat forests as commodities is that our thinking, as earlier discussed, has become linear and irrational. Essayist Wendell Berry poses an interesting point when he says that there are two fundamentally opposed views of the nature of human life and experience. One view holds that although natural processes may be cyclic, there is within Nature a human domain, the processes of which are linear. The other, much older view holds that human life is subject to the same cyclic patterns as is all other life. If the two are contradictory, Berry says that it is not so much because one is wrong and the other is right, but because one is only partial and the other is complete.[71]

The concept of linearity is the doctrine of progress that is represented by society's having moved across the oceans, continents, and into space on a course of no return, a course that is ultimately logical in the linear sense and one that is supposed to bring us into a human-made, material paradise. Within this concept, society moves through time in the same way, discarding old experience as the new is encountered. Thus, we never learn from history because in our minds, we never "repeat" the old ways, the old mistakes; in reality, however, we repeat them constantly.

We deny our repetition of historical mistakes in our blind drive for material progress. In this vision, there is but one definition of progress: onward and upward forever, an endless galactic voyage of discovery. To return is to come back to the used, whereas progress is to exploit the new and the innocent.

Characteristic of linear vision is the notion that anything is justifiable as long as it is immediately and obviously good *for* something else. Linear vision sees everything as a cause and requires it to proceed directly, immediately, and obviously to its effect. What, then, is it good for? And only if it proves to be immediately good *for* something are we ready to raise the question of its value: How much is it worth? This means how much money it is worth, because if it can only be good *for* something else, obviously it can only be *worth* something else. An excellent example of this type of thinking is presented in a paper entitled "Forests in the Long Sweep of American History," by industrialist Marion Clawson, who starts out talking about forests and ends up talking about timber and trees as products—a commodity.[72]

Consider that today's dictionaries define "resource" in a strictly linear sense as the collective wealth of a country or its means of producing wealth; any property that can be converted into money. Linearity therefore discounts intrinsic value in everything it touches. It is not surprising, therefore, that while there exist some local areas, philosophically speaking, where the intrinsic value of Nature is beginning to dawn, they are as yet few and far between.

The same can be said for the intrinsic value of human beings, when military capacity for the destruction of foreign human beings takes precedence over the domestic welfare and tranquility of human citizenry. Where does this kind of thinking lead as we consider ourselves and one another as "human resources?"

We can begin by looking at education. Education is stripped of its intrinsic value and becomes "training" as soon as we demand, in this lifeless, linear spirit, that it serve some immediate purpose and that it be worth a predetermined amount, such as the traditional training of a forester in the traditional school of forestry with which North America abounds. Once we accept so specific a notion of utility, all life becomes subservient to its use; its value is drained of everything except its use, and imagination is relegated to the scrap heap.

This type of reasoning also makes for difficult times for students and older people. They are living either before or after their time of greatest social utility and are thus robbed of their sense of purpose. This explains why so many species, both nonhuman and "primitive" humans, such as Indians in the Amazon Basin, are threatened with extinction. Any organism that is not contributing, obviously and directly, to the workings of the economy is perceived as having no value, which is a basic tenet of plantation management in the schools and practice of forestry. Many such "valueless" organisms are threatened with extinction, which means, as ecologists are showing, that human society is to the same extent endangered.

Endangered species are the result of linear vision that looks fixedly straight ahead with the notion that an economic endeavor must be ever-expanding in order to be healthy. It never looks back, because its premise is that there can be no return. Linearity is above all the doctrine of possession, which is not complemented by a doctrine of relinquishment, of sharing. It comes as no surprise, therefore, that our shallow, unthinking concept of use does not imply wise use, conservative use, or even good use. We simply trade quality in on quantity. Is it any wonder that we find ourselves wallowing in our own waste and disposability—"the end of living and the beginning of survival"—as Chief Seattle warned over a century ago.

Similarly, time and life are both squandered without respect for death. Through the lens of linear vision, death becomes accidental, the chance interruption of a process that might otherwise go on forever, and thus is an unacceptable surprise and always feared. After all, say the high priests of linearity, it is the length of life, the quantity, that counts; thus medical progress is to prolong life—regardless of its quality.

Such linear vision flourishes in fear, ignorance, and contempt for the processes on which it depends. We do not, for example, see the forest for the trees, because, as Berry says, in the face of these processes, our concepts and mechanisms of linearity are so unrealistic, so impractical, as to have the nature of fantasy.

Processes are invariably cyclic, however, rising and falling, giving and taking, living and dying in space on ever-expanding ripples of time. Yet linear vision places its emphasis only on the rising phase of the cycle: on production, expansion, possession, youth, and life. It does not provide for returns, idleness, contraction, giving, old age, and death.

Waste is thus a concept that can only be born from a vision of economic linearity. According to this notion, every human activity produces waste, because every human activity is linear. "This," says Berry, "implies a profound contempt for correct discipline; it proposes, in the giddy faith of prodigals, that there can be [everlasting] production without fertility, abundance without thrift. We take and do not give back, and this causes waste. It is a hideous concept, and it is making the world hideous...."[71]

The cyclic vision, on the other hand, ultimately sees life as a circular dance in which certain basic and necessary patterns are repeated endlessly in a cycle of use and renewal, of life and of death. This is the ethical basis of Native American religious thought, as exemplified by Black Elk:

> Everything the Power of the World does is done in a circle. The sky is round, and...the earth is round like a ball, and so are all the stars. The wind, in its greatest power, whirls. Birds make their nests in circles, for theirs is the same religion as ours. The sun comes forth and goes down again in a circle. The moon does the same, and both are round. Even the seasons form a great circle in their changing, and always come back again to where they were. The life of a man is a circle from childhood to childhood, and so it is in everything where power moves.[73]

The cyclic vision is accepting of the mysteries of the Universe and is humble in its perception. Humble in this sense means to be teachable by Nature, whose lessons are about intrinsic value and infinite relationship. It means to use something for its own sake and then to be the source of its renewal—to see it as a resource and to be its source of renewal.

This vision of the great circle sees everything as interdependent and nothing as independent. Thus, in Nature there exists no such thing as an "independent variable." Everything in the Universe is patterned by its interdependence on everything else, and it is the pattern of interdependence that is the only constant: change.

The cyclic vision, at once realistic and generous, recognizes the essential principle of return. What is here will leave and come again. What I have, I must someday give up. It sees death as an integral and indispensable part of live, for death is but another becoming.

Some cycles revolve frequently enough to be well known in a person's lifetime. Some are completed only in the memory of several generations—hence the notion of the invisible present. Still others are so vast that their motion can only be assumed. In reality, however, even they are not completely aloof because we are kept in touch by interrelatedness and interdependence. "It is only in the processes of the natural world, and in analogous and related processes of human culture, that the new may grow usefully old, and the old be made new...."[71]

CYCLIC FORESTS AND LINEAR MODELS DO NOT MATCH

A forest is a cycle of interdependent processes in relation to time, completing its cycle only in the memory of several human generations (Figure 72). We do not seem to understand this, however, or we ignore it, because all of our models—economic, managerial, and even ecological—are short term and linear. This is not only because we chose them to be so but also because we do not have the capability to construct them in any other way.

Thus while linear models can only predict in a straight line in the very short term, the cyclical nature of the forest touches that line for only the briefest moment in the millennial life of the soil, the womb from which the forest grows (Figure 73). Yet it is in this instant, with grossly incomplete, short-sighted knowledge and unquestioning faith in that knowledge that we base the sustained-yield prediction of all of our plantation management into the unforeseeable future. Thus, when we liquidate an old-growth forest, we do so thinking that we can forever have a rapidly growing plantation that has a magical sustained yield,

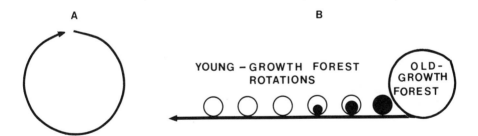

FIGURE 72   Two ways of viewing a forest: cyclic (A) or linear (B). In a cyclic view, there is no waste because, by definition, everything is recycled somewhere in the system. However, economic waste is built into the linear view of a forest. Short-term economics decrees that anything not used by humans is a waste. Short-term economic expediency is to liquidate old-growth forests and forever harvest even-aged, rapidly growing young forests with a sustained yield (open circles). Reality (solid circles) shows a decrease in productivity over time[25] because the forest's system is out of balance as a result of withdrawing capital without reinvestment, which impairs its ecological ability to function over time.

FIGURE 73 All of our models—economic, management, and ecological—are linear (A), while the forest is cyclic (B). Note that the only place where a linear model can reveal anything about a cyclic system, such as a forest, is where it touches the cycle (C), and then only for an instant in the life of the forest. Note also that the linear model projects beyond the cyclic forest based on the false assumption that soil, water, air, sunlight, and climate are constant values and that the only variable (the independent variable) is the volume of wood fiber in the form of growing trees.

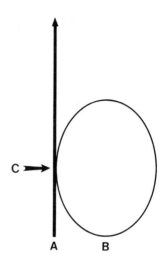

even as we ignore the foundation stones of forestry: soil, water, air, sunlight, climate, and diversity.

Consider a case in point. Like most industrial timberland owners, Weyerhaeuser Company, according to a newspaper story,[74] charts the growth, management, and harvest of its tree farms by computer, which can only handle a linear program. The company knows how much fertilizer will produce how much increased growth and when to apply it, which of course assumes that all processes in the soil are healthy and remain constant over time. The company knows when to send in the thinning crews to cut the spindly or defective trees. It knows how much damage animals cause and what it is worth to control it, again assuming all things remain constant and predictable, including animal populations and the factors that control them, such as the abundance and arrangement of food, water, and cover.

Trees are an investment with a long-term payoff, and Weyerhaeuser, according to the article, "wrote the book" on how to get the highest linear return. Within a few years, the article says, the company will achieve sustained yield on its Clemons Tree Farm. This means that it will be able to harvest 2 percent of its trees annually on a 50-year cutting cycle forever, which assumes that all things (soil, water, air, sunlight, climate, and diversity) will remain constant over the time of the investment.

What about the predicted changes in global climate? Will that not affect the growth and yield of trees on the company's tree farm? We are altering the quality of the sun's light as it affects the tree farm by altering the composition of the Earth's atmosphere, which in turn will alter annual temperatures and humidity, the hints of which have already been seen in the Experimental Lakes Area of Ontario, Canada (discussed earlier).

Further, changes in global climate will have profound effects on the processes and dynamics of the soil from which the tree farm grows. All of the changes in global climate are dynamic and will alter climatic cycles on Nature's scale of time and space, not humanity's. None of these alterations is quantifiably predictable—even in the short term, let alone the long term—and therefore they are moot in the linear sense. This leaves the company's linear computer predictions ecologically "deaf, dumb, and blind" over time.

Even apart from the above scenario, if we were to recalculate yield tables for a given acre after each rotation, we would probably find, as the German's learned, that it does not work the way we insist on thinking it does. Yields are not sustainable until we first learn how to sustain the basis for these yields: healthy and productive forests, including especially the soil.

BIOLOGICAL SUSTAINABILITY IS CYCLIC

Sustain is defined as to supply with sustenance, to keep up. Sustenance, in turn, means support or maintenance, nourishment, or something that gives support, endurance, or strength. How is sustained yield defined? Are three crops of corn, at the same number of bushels per acre, considered a sustained yield by a farmer? Three crops of corn take only three years to produce. Three 80-year rotations of trees, at the same volume of wood fiber per acre, take 240 years to produce. If three crops of corn are not considered to be a sustained-yield crop because the yield drops and the soil is finally exhausted, why are three crops of trees considered to be sustained-yield forestry? The only difference is the length of time it takes to grow three crops (3 years instead of 240 years), but it is still only three crops of corn and three crops of trees.

What does the concept of sustainability mean in terms of a forest? We cannot have an economically sustainable yield of any forest product (such as wood fiber, water, soil fertility, wildlife, or genetic diversity) until we first have an ecologically sustainable forest, one in which the biological divestments, investments, and *re*investments are balanced in such a way that the forest is self-maintaining in perpetuity.

Sustainability, as discussed earlier, is additive. We must have a biologically sustainable forest in order to have a biologically sustainable yield, a biologically sustainable yield in order to have an economically sustainable industry, an economically sustainable industry in order to have an economically sustainable community, and an economically sustainable community in order to have an economically sustainable society. When sustainability is put in purely economic terms, the additive economic relationship of the biological yield becomes clear.

We must first practice sound "bio-economics" (the economics of maintaining a healthy, biologically sustainable forest), before we can practice sound "industrio-economics" (the economics of maintaining a healthy, economically sustainable timber industry), before we can practice sound "socio-economics" (the economics of maintaining a healthy, culturally sustainable society). It all begins with a solid foundation, which in this case is a healthy, biologically sustainable forest.

Today's "forest practices" are counter to sustainable forestry. Instead of training foresters to manage forests, we train plantation managers to manage the short-rotation, *economic* plantations with which we are replacing our native forests. Forests have evolved through the cumulative addition of structural diversity that initiates and maintains process diversity, complexity, and stability through time. We are reversing the rich building process of that diversity, complexity, and stability by replacing native forests with plantations designed only with narrow, short-term, economic considerations.

Every acre on which a native forest is replaced with a plantation is an acre that is purposely stripped of its biological diversity and ecological sustainability and reduced to the lowest common denominator—simplistic economic theory. The simplistic economics of the soil-rent theory, whether in forestry or agriculture, has not proven to be ecologically sustainable anywhere in the world in the medium and long term. Thus, the concept of a "plantation," a strictly simplistic, economic concept, has nothing whatsoever to do with the biological sustainability of a forest. Under this concept, native forests are replaced with plantations of genetically manipulated trees accompanied by the corporate–political–academic promise that such plantations are better, healthier, and more viable than the native forests that evolved with the land over millennia.

"Sustainable," however, means producing industrio-economic outputs as the land gives us the ecological capability of the forest to do so in perpetuity. This in turn necessitates balancing socio-economic product withdrawals with bio-economic *re*investments in the health of the forest. It means maximizing first and foremost the health of the forest and harvesting all products and amenities within the forest's sustainable capacity.

To accomplish ecological sustainability, we must shift our historical paradigm from that of the exploitive colonial mentality (use it until it collapses, then someone else can deal with it) to the paradigm of trusteeship. Much as we might wish otherwise, humanity is not in control of nature. Therefore, if human society as we know it is to survive, we must become trustees of our natural resources. If we go back to the original sense of the word "re-source," we will find that the ecological sustainability of our forests is embodied in a word that we blithely use but do not fully understand.

"Sustainable-yield forestry" has not been practiced in the Pacific Northwest or Canada, because our "sustained yield" (which really equals sustained cut) has come from old growth we inherited from Nature and for which we can claim no credit. In fact, even the stated concept of sustained yield has been violated by continually increasing the cut of old-growth forests.

Consider, for example, that in 1949 the total cut from the entire national forest system in the United States was 2.6 billion board feet. In a 1989, however, U.S. District Judge William Dwyer wrote in his decision: "Statistics show that in the last ten years more than 30 billion board feet of raw logs...have been exported from Washington and Oregon to foreign markets, primarily Japan." Note that Judge Dwyer's statement does not take into account the volume of timber milled in the United States.[75]

Sustained yield, as we have practiced and propose to practice it, is short-term economic exploitation. We harvest the inherited principal of biological capital without reinvesting sufficient biological capital in the forest, either within or between rotations, to at least balance the account. Ecological principles of diversity, process interactions, and the forest's cycle through time are violated in order to practice exploitive forestry, and then the diminishing return is called "sustained yield."

Further, the proposed practice of "sustainable-yield forestry" excludes all other human values except production of fast-grown wood fiber. Young forests do not produce the highest quality water. They are not conducive to recreation. Spotted owls, marbled murrelets, and elk are not sustained by them through time. They have lost the attractiveness of diversity. In addition, genetically engineered, "improved" trees in fast-growing plantations produce problem lumber because of weak wood that tends to shrink, warp, and break under stress (Figure 74).

FASTER IS NOT NECESSARILY BETTER

The housing industry has documented over 700 cases of a new and unpredictable phenomenon called the rising truss. It is caused by fast tree growth, which increases the percentage of juvenile wood in timber harvested from plantations in the southern and western United States. Nevertheless, plantation foresters herald new, "improved," genetically engineered trees as an answer to future timber shortages. They may be right, but advocates of "bigger and faster is better" face a two-sided problem: maintaining the strength of the wood and its dimensional stability while simultaneously increasing growth in the diameter of the stem. The problem is at least three-sided when depletion of soil fertility and impairment of ecological processes are added.

The race to produce the maximum volume of timber per acre cannot ignore consumers' needs for forest products best produced by slow growth. Maximum volume, obtained by plantation culture or silvicultural manipulation, may be fine for pulp and paper, but not for building-quality lumber. An underlying concern

FIGURE 74 Lumber made from weak, rapidly grown plantation wood. Note the warping and bending (arrows). (Photograph by Brian Egan.)

for forest managers must be product quality. Fast-grown, short-rotation conifers lack the wood fiber needed for products that require structural strength.

The impetus for plantations came initially from the desire of the pulp and paper industry to maximize fiber output. The result was a successful research effort to select, genetically screen, and silviculturally manage particular species of trees in order to produce more volume per acre and shorten the time between rotations. This was a logical, short-term economic course of action for the paper industry, yet many pulp mills still prefer old-growth stems with narrower rings, higher density, higher pulp yields, and lower cost.

Maximum timber yield entails the additional costs of regular thinning cuts and short rotations. Wood processed into pulp does not return the profit per unit of wood used for other purposes. Inevitably, wood cultured and managed for pulp is being processed as unsafe lumber.

Ironically, the language of the Multiple Use Sustained Yield Act of 1960, although of good intent, is based on a linear economic assumption that is totally at odds with ecological reality. The assumption is that biological processes in forests remain constant while we humans strive to maximize whatever forest product or amenity seems desirable. The errors of linear thinking in central

Europe over the past several hundred years illustrate the results of ignoring the cyclic nature of ecological reality while attempting to maximize short-term profits from a system that ultimately is controlled by ecological laws.

Obviously we must change our thinking and our actions in managing forests for the long run. Forests are not the endless producers of commodities and amenities that we have heretofore assumed them to be.

The critical point is that before we can change our European, utilitarian paradigm that forces us to view the forest and all it contains simply as commodities to be endlessly exploited, we must devise a new paradigm. In this new paradigm, we must view the forest as a living organism with which we cooperate and through such cooperation are allowed to harvest products as the biological capability of the forest permits.

Finally, how would our thinking about our forests be affected if the Multiple Use Sustain*ed* Yield Act had been conceived of and written as the Multiple Use Sustain*able* Yield Act or even as the *Sustainable Forest* Act?

THE *INVISIBLE* MAKES THE DIFFERENCE

As noted earlier, the forest we see aboveground is in large measure a reflection of the ability of the soil to grow that stand of trees. Let's take another look at this relationship. The native old-growth forest has three prominent characteristics: large live trees, large standing dead trees or snags, and large fallen trees. The large snags and large fallen trees ultimately come to rest on the forest floor and become reinvested as biological capital into the soil, where myriad organisms and processes make the nutrients stored in the decomposing wood available to the forest and the live trees. In addition, the changing habitats of the decomposing wood allow nitrogen fixation by free-living bacteria to take place. (Nitrogen fixation is the conversion of atmospheric nitrogen to a form useable by such living organisms as trees.) These processes are possible through Nature's "rollover" accounting system, which includes such assets as large dead trees, genetic diversity, and biological diversity, all of which count as *re*investments of biological capital in the growing forest (Figures 75 and 76).

As previously discussed, the advent of intensive, short-term plantation management disallowed reinvestment of biological capital in the soil and therefore in the forests of the future, because such reinvestment has come to be erroneously seen as economic waste. We therefore plan the total exploitation of any part of the ecosystem for which we see a human use and the elimination of any part of the ecosystem for which we do not see a human use. With this myopic view, we have created the intellectual extinction of Nature's diversity through humanity's

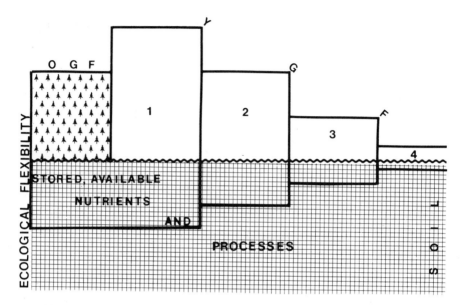

FIGURE 75 The original old-growth forest (OGF) created much large woody debris (fallen trees) which became incorporated into the forest floor and soil. These fallen trees, with all their attendant organisms and processes, are a vital storehouse of available nutrients and nutrient cycles on which the living forest depends. Note that the first rotation (1) in the young-growth forest (YGF) produces more wood fiber per unit area per unit time than exists in the old-growth stand. Rotations 2, 3, and 4, however, decrease in the amount of wood fiber produced per unit area per unit time. Note that the amount of woody debris in the soil also decreases and is not replaced with intensive forestry; this translates into ecological brittleness as opposed to ecological flexibility.

planning system, which inevitably leads to biological extinction of species and their functions within the ecosystem.

In May 1985, I examined soil pits three to five feet deep in the intensively managed plantations of southern Germany. There was no large woody debris in the soil of those plantations, nor had there been for well over a century, and productivity has been declining accordingly. When the hardwood forest was first converted to softwood plantations, however, yield of wood fiber (first rotation of intensive forestry) (Figure 75) was high, but then it declined steadily. Are our forests following the same pattern because we practice intensive plantation management with the same philosophical foundation as the German foresters from whom we initially learned?

After the native forest is liquidated, we may be deceived by apparently

FIGURE 76   Large fallen trees that decompose and recycle into the forest soil are a reinvestment of nutrient capital and ecological processes in the next forest. For example, the upslope side of a fallen tree oriented along the contour of a slope is filled with humus and inorganic material, which allows invertebrates and small vertebrates to tunnel alongside. The downslope side provides protective cover for larger vertebrates. The wood itself is saturated with water and acts as a reservoir under protection of the forest canopy. (USDA Forest Service photograph by J.F. Franklin.)

successful growth of a first-rotation plantation that lives off the stored, available nutrients and processes embodied in the soil of the liquidated native forest. Without balancing biological withdrawals, investments, and reinvestments, both biological interest and principal are spent, and thus both biological and economic productivity must eventually decline. The dysfunctional "managed forests" (plantations) of Europe, which are biological deserts compared to their native forests, bear testimony to such short-sighted folly.

In our linear product mode, we have learned little or nothing from the history of the European experience of total forest exploitation. The view of the North American timber industry is to utilize everything possible, to squeeze more and more from the forest, with no biological reinvestment into the system and its processes that produce the coveted trees in the first place. Here, the invisible present—impoverishment of the soil by industrial greed—is, in fact, the "grim reaper" of future forests.

Do not mistake planting seedlings (so-called reforestation or afforestation) or fertilization as biological reinvestments; they patently are not reinvestments in the forest! (Reforestation, as I use the term, is far more that just planting seedlings; it is the purposeful, scientific simulation of Nature's forest as it evolved through the cumulative addition of structural diversity that initiates and maintains process diversity, complexity, and stability through time.) Replacing a mature or old-growth native forest with a plantation or applying fertilizers are neither biological reinvestments nor economic reinvestments in either the forest or the soil; they are economic investments in "crop trees."

The outlay of economic capital required to liquidate the inherited forest, plant seedlings on bared land, and fertilize the young stand is the *initial economic investment* in the intended product: the crop tree. However, a forest does not function on economic capital; it functions on *biological capital*—the decomposing organic material (including the wood of large fallen trees) and genetic and biological diversity.

In temperate, coniferous forests, the biological capital is obvious in the soil as organic matter. The story is different in the tropical areas of the world. The resilience of sites, both temperate and tropical, following disturbance is at least partially related to the ability of the soil to retain nutrients and to maintain its structural and biotic integrity during the period that plants are becoming re-established. The capacity of a soil to store nutrients is a function of its age and the climatic regime under which it evolved. In areas without recent (in geological time) tectonic or glacial activity, soils are quite old, and reserves of nutrients are low, particularly if rainfall is high. Such is the case in much of the tropics, particularly interior South America and Africa.

Nutrients are primarily stored in living biomass (living tissue) in these forests, and mechanisms buffering the loss of nutrients are mainly biological rather than geological. Loss of forest cover removes the major sink for nutrients, which is the living tissue of the forest in which they are contained. This loss is compounded because both evapotranspiration (the loss of water to the atmosphere through the leaves of plants) and the covering of the soil are drastically reduced. This, in turn, allows an increased volume of rainwater to flow overland through runoff, which carries the nutrient elements out of the system.

Change in the water balance due to deforestation in the Amazon Basin may be responsible for the increase in annual peak flows of the Amazon River, which averaged 8 percent higher in the 1970s than in the 1960s and which occurred with no increase in rainfall. Centuries may be required to replace the lost nutrient elements. Because the soils have a low capacity to retain elements, conventional fertilization will probably be unsuccessful and in fact may pollute the supplies of local water.

As we redesign the forest, therefore, we will be wise to acknowledge Nature's blueprint, which holds the potential for understanding the whole story.

DEFILE NOT THE LAND

Because economists in our linear capitalistic system refuse to accept intrinsic ecological value as "real" value, the use and management of natural resources are guided only by the cost–benefit analysis of their potential economic value when converted to something else, such as trees cut into boards. This means that the only value economists can see is short-term specialization, a view that is killing the soils of the Earth.

Soil

Many cultures have emphasized the trusteeship of the soil through religion and philosophy. Abraham, in his covenant with God, was instructed to "Defile not therefore the land which ye shall inhabit, wherein I dwell."[76] Confucius saw in the Earth's thin mantle the sustenance of all life and the minerals treasured by human society. And a century later, Aristotle viewed the soil as the central mixing pot of air, fire, and water that formed all things.

In spite of the durability of such beliefs, most people cannot grasp them because they are intangible. The invisibility of the soil is founded in the notion that it is as common as air and therefore is taken for granted as air is. Although to many people the soil seems "invisible" because it is part of the invisible present, humanity is nevertheless somehow tied to the soil for reasons beyond measurable materialistic wealth.

Even though soil protection can be justified economically, our ultimate connection with it escapes many people. One problem is that traditional linear economics deals with short-term tangible commodities (such fast-growing plantations of timber planted to maximize the production of wood fiber) rather than long-term intangible values (such as the prosperity of future generations). However, when we recognize that land, labor, and capital are finite and that every system has a carrying capacity that depends on natural or artificial support, the traditional linear economic system becomes more like a cyclic biological system.

In the late 18th century, Thomas Malthus proposed that the human population would grow faster than the soil's ability to sustain it, but agronomic advances in this century led many short-sighted leaders to dismiss this idea as simplistic and overly pessimistic. Today, however, Malthusian theory seems prophetic when one considers the trends in air pollution that poison the soil, overgrazing by livestock and the growing desertification, global deforestation and the loss of the soil's protective cover of vegetation and its vitality, and the ensuing famines.

Today forest managers often see the protection of the soil as a cost with no benefit. Those who analyze the soil by means of traditional linear economic analyses weigh the net worth of protecting the soil only in terms of the expected short-term revenues from future harvests of timber. They ignore the fact that it is the health of the soil that produces the yields, because the standard method for computing "soil expectation values" and economically optimal crop rotations commonly assumes that the soil's productivity always remains constant or increases but never declines. In reality, however, reducing the productivity of the soil on marginal sites can push the expected present net worth of subsequent harvests of timber below zero.

Given this short-sighted and flawed reasoning, it is not surprising that those who manage the land seldom see protection of its productivity as cost effective. However, if the real effects of management practices on long-term economic yields could be predicted, the invisible costs associated with poor care of the soil might be viewed differently.

One of the first steps along the road to protecting soil fertility is to ask how various management practices affect long-term ecosystem productivity, particularly that of the soil. Understanding the long-term effects of management practices in turn requires knowing something about what keeps the ecosystem stable and productive. With such knowledge, we can turn our often "misplaced genius," as soil scientist Dave Perry rightly calls it, to the task of maintaining the sustainability and resilience of the soil's fertility. Protecting the soil's fertility is buying an ecological insurance policy for future generations.

After all, soil is a bank of nutrient elements and water that provides the matrix for the biological processes involved in the cycling of nutrients. In fact, of the 16 chemical elements required by life, plants obtain all but 2, carbon and oxygen, from the soil. The soil stores essential nutrients in undecomposed litter and in living tissues and recycles them from one reservoir to another at rates determined by a complex of biological processes and climatic factors. In a forest, for example, the losses of nutrients in undisturbed sites are small, but some are lost when timber is harvested. Others may be lost through techniques used to prepare the site for planting trees, reducing the hazard of fire, or controlling unwanted vegetation.

Therefore, the resilience of forested sites following a disturbance like harvesting timber is at least partly related to the ability of the soil to retain nutrients and water and to maintain its structural and biologically functional integrity during the period in which plants are becoming reestablished. Beyond that, the health and fertility of the soil is reflected in the growth of the forest and the quality of the timber harvested now and in the future.

Special Cases and Common Denominators

In his discussion of the predictability of human behavior, Sherlock Holmes touched the core of special cases and common denominators. He saw each person as a special case and therefore unpredictable, but if enough special cases are studied with an eye for their common traits (common denominators), then certain predictions can be made about their behavior.

Psychoanalyst Carl Jung put it differently. Because self-knowledge is a matter of getting to know the individual facts, theories help very little in this respect. The more a theory claims universal validity, the less capable it is of doing justice to individual facts. Any theory, which is based on experience, is necessarily statistical and formulates an ideal average, which abolishes all exceptions at either end of the scale and replaces them with an abstract mean. Although the mean is valid, it does not necessarily occur in reality. Nevertheless, the mean figures as an unassailable fundamental fact in the theory. Exceptions at either extreme, although equally factual, do not appear in the final result because they cancel each other out.

For example, determining the weight of each stone in a bed of pebbles and finding an average weight of 145 grams reveals little about the real nature of the pebbles. If, on the basis of these findings, you try to pick up a pebble weighing 145 grams on the first try, you would be in for a serious disappointment. Indeed, it might well happen that no matter how long you searched, you would not find a single pebble weighing exactly 145 grams.[77]

These same concepts are used in forestry, but we do not derive as much meaning from them as did Sherlock Holmes and Carl Jung. Consider Figure 77, for example. Each numbered shape above the soil represents an individual stand of trees, a sale unit if you will. Note that each sale unit is a discrete numbered entity, and thus, like an individual human being, is a special case.

Each sale unit (special case) is measured to derive the volume of marketable

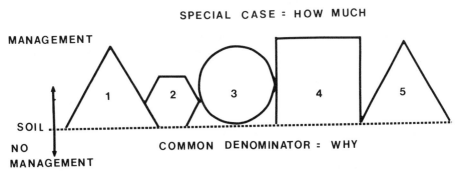

FIGURE 77   The individual numbered shapes above the soil are discrete stands of trees or sale units (special cases). Each is quantified to determine how much wood fiber there is and how much it is worth. Note that all of our management is aboveground. The soil, one of the common denominators in the forest, tells us why a particular stand of trees grows the way it does. Each stand of trees is only a mirror reflection of the ability of the soil to grow it.

wood fiber it has produced that can be harvested for a profit. Because the volume of each tree is averaged, an abstract mean is produced and an absolute dollar value is assigned to each sale, which is valid precisely once.

Once harvested, we expect to plant more trees and derive at least the same volume of wood fiber from those acres (the predictable, artificial, absolute minimum). Whether or not this goal is achieved depends on the "cooperation" and "coordination" of all the variables that constitute the forest.

So long as we think of each stand or sale unit as an isolated special case, we will never understand the forest, because a special case is thought to be out of relationship and out of context to the whole, which is an aggregate of special cases. We must understand how each delineated stand of trees relates to neighboring stands, to the watershed, and to the landscape before we can begin to understand the forest and make any kind of reasonable predictions about future trends in behavior.

We make two serious, fundamental errors in our view of forestry. First, we fail to realize that all we "manage" is what we see aboveground (Figure 77). We do not manage, or even think about or plan for, belowground processes. This brings us to one of the common denominators: the soil. Each tree, each stand of trees, and each forest is only a mirror reflection of the soil's ability to grow that tree, stand, or forest *once*.

Second, we assume that the foundation stones of forestry (the depth and fertility of the soil on which the forest grows, the quality and quantity of the water falling on the forest, the quality of the air infusing the forest, the quality of the sunlight reaching the forest, the biological and genetic diversity driving the forest, and the climate in which the forest lives) are economic constants. Because we assume they are economic constants, they are omitted from our economic and planning models and, even more basically, from our thinking. Each, however, is an ecological variable. Soil is eroded in two ways, chemically and physically, and we are doing both. Water and air are polluted with chemicals. Air pollution has a direct effect on the forest and in turn affects the quality and quantity of the sunlight that energizes the forest. Through our linear economic thinking, we are simplifying forests by converting them to plantations and causing untold extinctions in the process. And today's changes in global climate are affecting everything else. *Nothing in Nature is static.* Nature gives us only variables that are constantly changing.

Let's return to the soil for a moment. Sigmund Freud introduced the notion that our minds contain both conscious and unconscious parts. The soil can be thought of as our management unconscious, not only because we take it for granted but also because that is where we hide our toxic wastes. For example, the meltdown of the nuclear reactor at Chernobyl, in the former Soviet Union, early in 1986 was not potentially as dangerous as the buried nuclear dump that blew up near Chelyabinsk in the southern Ural Mountains in late 1957 or early 1958.

"The land was dead—no villages, no towns, only chimneys of destroyed homes, no cultivated fields or pastures, no herds, no people—nothing. It was like the moon for many hundreds of square kilometers, useless and unproductive for a very long time, many hundreds of years."[78]

We must remember that the soil not only supports all plants growing in it but also supports myriad hidden processes that are necessary for its fertility and for healthy forests. The nuclear accident is only a more drastic, faster version of the global damage we are causing to the soil through our indifference and through the insidious poisons of our management-introduced array of biocides and the toxic wastes of industry and war. We can only sustain healthy forests by learning about and planning for the special cases (stands of trees) in relation to their common denominators: soil, water, air, sunlight, biological and genetic diversity, and climate.

With the preceding in mind, it is wise to consider the observation of soil scientists V.G. Carter and T. Dale, who point out that civilized people despoiled their favorable environment mainly by depleting or destroying the natural resources; cutting down or burning most of the usable timber from the forested hillsides and valleys; overgrazing and denuding the grasslands that fed their livestock, killing most of the wildlife and much of the fish and other water life; permitting erosion to rob their farm land of its productive topsoil; allowing eroded soil to clog the streams and fill their reservoirs, irrigation canals, and harbors with silt; and in many cases, using or wasting most of the easily mined metals or other needed minerals. As a result, their civilization declined in the middle of its despoliation, or they moved to new land.[79]

AIR: THE GLOBAL COMMONS

It is late afternoon on a warm September day. A tiny spider climbs a tall stalk of grass in a subalpine meadow and raises its body into the air, almost standing on its head. From spinnerets on the tip of its abdomen, it ejects a mass of silken threads into the breeze. Suddenly, without visible warning, the spider is jerked off its stalk of grass and is borne skyward, where it joins other spiders riding the warm afternoon air flowing up the mountainside. All cast their fortunes to the wind, like their ancestors in centuries past, floating on air currents from the far corners of the Earth to be the first inhabitants of the newly formed South Seas islands.

Spiders are not the only things carried on currents of air. In 1883, Krakatoa (a small Indonesian island between Java and Sumatra) was virtually obliterated by explosive eruptions that sent volcanic ash high enough above the Earth to travel the world's airways for more than a year, affecting the climate by filtering the sun's light. Air also carries the reproductive spores of fungi and the pollen of various trees and grasses, as well as dust and microscopic organisms. It also carries oxygen and water as well as pollution. Air can therefore be likened to the key in the Chinese proverb: "To every man is given the key to the gates of heaven and the same key opens the gates of hell." In this case, air is the key, which carries both life-giving oxygen and death-dealing pollution. As pollution and deforestation increase, the usable oxygen content decreases.

With the above in mind, we need to consider carefully the counsel of George Santayana: "Those who cannot remember the past are condemned to repeat it." What are Nature's penalties for economically and ecologically disregarding the soil and air? Two obvious penalties are loss of fertile topsoil and polluting the entire world.

In considering topsoil, a lesson can be learned from history, because although the birth of agriculture caused civilizations to rise, it was abusive, linear agricultural practices that destroyed the topsoil and thus caused the collapse and extinction of civilizations. Yet today such things as acid rain exist in many parts of the industrialized world. With all the glaring lessons of history and with all our scientific knowledge and technological skills, we insist on walking the historical path of agricultural ruin and impending social collapse.

The supreme irony is that even as we work to rid the world of nuclear weapons and to establish a lasting peace, we continue to commit genocide by ruining our environment. Through rain and snow, airborne pollutants reach the entire Earth from the tops of the mountains, through the vegetation, down into the soil, and down into the deepest recesses of the sea.

With dirty air, self-destruction is only a matter of time. As we continue to poison our environment, we also destroy the stage—the soil—on which the entire human drama depends for life. Destroy the stage, and the drama is no more, which brings us to water.

13

WATER:
AN INESCAPABLE NECESSITY

Water is a physical necessity of life. The world's supply of quality water is therefore precious beyond compare. Water is the *most important* commodity from the world's forests.

The amount and quality of water available for human use are largely the result of climate and strategies for managing the ecological health of watersheds. In North America, sustaining the ecological health of watersheds is particularly important in high-elevation forests in order to protect the annual snowpack from which the vast majority of all useable water comes. However, protecting the quality and quantity of society's water supply is not a primary consideration in the harvest of timber (Figure 78).

People seldom realize that drinkable water comes from forested watersheds. Even much of the prehistoric ground water that is pumped came from forested watersheds. Water, and therefore much electricity, is a forest product just as surely as is wood fiber.

A curious thing happens, however, when water flows outside the forest boundary: we forget where it came from. We fight over who has the "right" to the last drop and pay little attention to the supply—the health of the forested watersheds. Consider, for example, the following:

> A concrete canal ribbons its way 190 miles over red rock mountains and scorched sand, defying the laws of both gravity and economics, representing both the past and future of watering the West.
>
> The umbilical cord called the central Arizona Project carries water uphill at 4 mph from the Colorado River to bursting, thirsting cities. It is the last and most expensive of all the great federal water works, a $3.6 billion aqueduct conceived as a way to irrigate the desert and hailed as the final answer for Arizona's needs....

FIGURE 78   Clearcut and hot burn show that water quality is not a primary concern when timber is harvested. Note the completely exposed second-order stream at the bottom.

"Most of the West is going to have to pay more," said water director Frank Brooks, the pioneer of Tucson's tactics.

"Historically, what we've done with all our natural products is use the cheap stuff first. It works the same way in a coal mine, the same way with oil. [Note that both examples refer to mining.] I think there's certainly much greater awareness now that *all of our natural resources are exhaustible—even water*" [emphasis added].[80]

Although such development of water projects has obviously contributed greatly to the economic and social growth of the country, in the western United States such developments have been used primarily to promote irrigated agriculture. As a nation with bountiful resources, the United States has rarely faced natural resource limits. However, present trends and experience indicate that every additional drop of water conserved and thus available enables more growth and development, which further raises the demand for more development and more water. Effective management of water will thus necessitate attention to both demand and supply.

The availability of water for agricultural use varies by location and over time. The availability of water depends on such variations in components of the hydrologic cycle as precipitation, evaporation, transpiration, infiltration, and runoff. Because these components are interrelated, a change produced by technology in one component of the cycle will inevitably affect other components.

In the brief history of the United States, there have always been more lands and more resources to develop and a philosophy that technology could supplement natural resources when needed. Today, however, stretching such water resources to accommodate the continuing economic growth of the western United States while protecting existing patterns of water use will require levels of technical development that are increasingly damaging ecologically and no longer feasible economically. Most people do not realize that only a small part of the water used in the United States goes to towns and cities. The overwhelming share is used for irrigation.

For example, withdrawals of water for irrigation range from 80 percent of the total use in Utah to 90 percent in New Mexico. Further, the use of water for irrigation is inefficient at best as shown by the U.S. Geological Survey, which found that the loss of water by seepage from canals was one third of the amount actually delivered to irrigated farms. And this does not include the loss of water to the atmosphere through evaporation in the arid west.

According to Professor Luna B. Leopold, the persistence of the pro-development bias of the U.S. Bureau of Reclamation is increasingly inexcusable. This attitude is still held in spite of the obvious strain on both the quality and the quantity of the supply of water. Leopold goes on to say that, "It is deplorable that the government agency most responsible for managing water in water-short regions continues to be so insensitive to the hydrological continuum and the equity among claimants."[15]

The hydrological continuum, as used by Leopold, is different from the hydrological cycle. The hydrological cycle continues for better or for worse, but the idea of a hydrological continuum implies the maintenance of a quasi-equilibrium operational balance among the processes within the hydrological cycle, which involve the air, water, soil, biosphere, and people. In other words, if withdrawals of water are balanced with Nature's capacity to replenish that which is used, the use of water can be measured in such a way that the available long-term supply is protected from being overtaxed.

There are thus two options in managing the use of water. One is to protect the availability of the long-term supply by disciplining ourselves to use only what is necessary in the most prudent manner. The other is to take water for granted and use all we want with no discipline whatsoever (as we do now) and then wonder what to do when faced with a self-inflicted shortage.

By using all the water we want in a totally undisciplined manner, we are insensitive to both the care we take of the world's watersheds and the speed with which we mine the world's supply of stored, available water. As stated by Professor D.J. Chasan, "One might suppose that people would automatically conserve the only naturally occurring water in a virtual desert, but one would be wrong. Land and farm machinery have capital value. Water in the ground, like salmon in the sea, does not. Just as salmon are worth money only if you catch them, water is worth money only if you pump it."[81] We are therefore damming, diverting, and channeling the world's rivers to "tame" and to "harness" them for short-term use based on poor economics, rather than nurturing the environment to ensure the availability of an adequate long-term supply of water.

For instance, the U.S. Army Corps of Engineers designs and builds structures to control flooding and to improve navigation, but it also issues permits for the alteration of bodies of water, marshlands, and estuaries. The Soil Conservation Service, which was originally a land-management agency, became another engineering organization following World War II. Where earlier policy had been in the hands of agronomists, soil technicians, and managers of rangelands, following the war most of the policymaking positions were filled by engineers.

One aspect of the engineering programs of both the U.S. Army Corps of Engineers and the Soil Conservation Service is the practice of "channel improvement" in streams and rivers. In this program at least 5000 miles of "improvements" have been carried out by the Corps of Engineers and over 3000 miles by the Soil Conservation Service.

Such "improvements" lead to a straightening of a stream's channel and a change of its shape, which in turn destabilizes the channel and causes downstream effects, such as erosion of the banks, alterations of the channel's bed, degradation of the aesthetics, and changes in the composition of the plants and animals that inhabit the stream, which are often considered to be undesirable. From these numerous "improvements," each planned on its own isolated rationale, comes the next, larger-order magnitude of massive flooding, such as occurs in the lower reaches of the Mississippi River.

No studies, according to Leopold, have been conducted by either the Corps of Engineers or the Soil Conservation Service to determine the long-term effects, either on-site or downstream, of such alterations in the channel. The interest of these government agencies is not in the long-term future of the landscape because the computed, but often unrealistic, cost–benefit ratio is on the side of utility.

As with old-growth forests and soil, fresh water is being mined, which is another example of the economics of extinction. A report by the Office of Technology Assessment of the U.S. Congress, entitled "Water-Related Technologies for Sustainable Agriculture in U.S. Arid/Semiarid Lands," states that two major classes of watersheds occur in the western United States. One is the highland watersheds located in the major mountain ranges, which consist of

FIGURE 79   Winter snowpack in the high mountains is the source of most of our water from forested watersheds.

both the unforested "alpine" zone (above the timberline) and the forested "montane" zone. The second is the lowland watersheds, which consist of grass- or brush-covered valleys and plains. Technologies for the management of watersheds are designed to increase surface runoff by removing vegetation, by replacing it with something else, or by other modifications to the surface of the ground.[82]

Nothing in this report is said about managing vegetation in order to prevent erosion, to increase the infiltration of water to recharge the underground aquifers and become purified in the process. In fact, the report simply states that most of the available water in the West comes from winter snowpack in the mountains (Figure 79).

The report makes six other points, most of which reiterate points already made:

1. Almost half of the western United States experiences water-supply problems in relation to demand.

2. Water conservation enables and stimulates more growth and development, which demands more conservation, which enables and stimulates more growth and development, and so on. Such behavior creates

a self-reinforcing feedback loop that ultimately leads to water shortage. Effective water management must therefore consider demand as well as supply.

3. Technological change in one component of the cycle affects all other components.

4. Short-term fluctuations in climate that affect the supply of water can be accommodated in management and planning processes by studying past trends, but there is no reliable method for predicting long-term fluctuations or such potential changes as global warming.

5. The philosophy in the United States has always been that there are more resources to develop and that technology could supplement them where needed when needed.

6. Continued growth of the West, while protecting current patterns of water use, requires levels of technology that are no longer economically feasible.

Is water to become the ultimate economic/environmental club with which we bludgeon each other? This question is appropriate here because as we witness the demise of the old-growth forests, we are also running out of available supplies of potable water. The only solution is an environmental one: sound ecological management of watersheds on a landscape scale that first and foremost nurtures the health of soil and water, lest everything else become unhealthy. Like migratory birds and anadromous fish, environmental crises know no political boundaries. Soil, water, air, and climate form a seamless whole, the thin envelope we call the biosphere, which is all we have in all the Universe.

With the growing realization of the global, ecological interdependency among all living forms and their physical environment, it can hardly be doubted that even "renewable" resources show signs of suffering from the effects of society's unrelenting materialistic demands for more and more. These demands have degraded the renewability of resources in both quality and quantity. Water can be thus characterized, because it is increasingly degraded by soil erosion, increases in temperature, and pollution with chemical wastes, salts from irrigation, and overloads of organic materials. Is it any wonder, therefore, that the hydrological system is under stress?

The rub lies in the available technology. Many farmers, interested only in the short-term production of their own fields, are still plowing uphill and downhill, despite 50 years of soil conservation. The soil eroding from their fields, which is changing the conditions downstream, is not their problem. Similarly, the county agent who advises the farmers is more likely to be concerned with their fields than with the whole river basin. Because most water supply engineers see the hydrological system as a whole as being outside of their domain, they are not immediately concerned with its problems. We therefore rely on our technology to provide safe, chemical-laden water, but we are managing water after the fact.

As with any problem, there are solutions, but we tend to look for solutions only where the symptoms are obvious. For example, I used to live in an old mobile home, the kind that leaked immediately every time a dark cloud appeared on the horizon. Finding where the leak ended—on my desk or over my pillow— was no problem, but finding the source of the leak was often so difficult that I had to repair the entire waterproof coating on the roof—and then hope.

HOW WE THINK ABOUT WATERSHEDS

The problem with watersheds begins with the headwaters, the first-order stream and its watershed (Figure 80). A first-order watershed is always a special case; in fact, it is probably the only part of the land that is manipulated where the

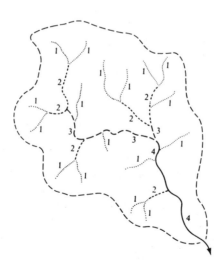

FIGURE 80 Stream order in a typical watershed. The majority of stream mileage is in first- and second-order tributaries in all watersheds in western Oregon and western Washington. (From *Water in Environmental Planning* by Dunne and Leopold. ©1978 by W.H. Freeman and Company. Reprinted with permission.)

hydrology has any semblance of ecological integrity, because it is the headwaters and therefore controls the initial water quality for the whole watershed.

A first-order watershed, by definition, is unique. A second-order watershed is unique among second-order watersheds, but is a common denominator, an integrator, of the first-order watersheds that created it. A third-order watershed is unique among third-order watersheds, but is a common denominator of the first- and second-order watersheds that created it, and so on.

Our thinking, and therefore our view of the world, is generally limited to a kaleidoscope of special cases because we choose to focus on "discrete" parcels of real estate. If we deal only with special cases (a mile of stream, for example), we perpetuate our inability to understand that particular mile of stream, the entire stream, and the watershed as a whole. If, on the other hand, we deal with a particular mile of stream (a special case) in relation to the whole watershed (the common denominator), we enhance our ability to understand both the mile of stream and the watershed because each is defined by its relation to the other.

FIGURE 81 First-order watersheds are the quality control mechanisms of our drinking water, but the water does not hold the same value as wood fiber to the timber industry. (USDA Forest Service photograph.)

Understanding how a reach of stream relates to the whole watershed is like understanding how a single chair relates to a room.

If we stand in the doorway and survey the room, we will see the chair both in the room and in relation to the room, but when we focus only on the chair we can no longer see the room or the chair's relationship to it. Unfortunately we do not see that the first-order watersheds are the initial controllers of water quality for our supplies of domestic water. We therefore cut timber down into the stream bottoms of both first- and second-order streams because the timber is thought to have greater immediate economic value than the water (Figure 81). Moreover, because politically important fish, such as salmon and steelhead, do not live at the high elevations in which most of these small streams occur, they are deemed to be of no visible economic importance.

ROADS AND WATER

In addition to our lack of vision with respect to nurturing watersheds, we are also bleeding water from the land with roads the same way cuts in the bark bleed sap from a sugar maple (Figure 82). The U.S. Forest Service, for example, is constantly expanding its network of roads in the 119 million acres of publicly owned forests. Initially set up to care for the nation's public forests and to protect watersheds, the Forest Service is best understood as the world's largest socialized road-building company. More than 340,000 miles of roads have been constructed under its auspices, which is more than eight times the mileage of the entire U.S. interstate highway system.

A story in *The Oregonian* stated: "The Forest Service's 1985 renewable resource program promises construction of some 77,000 miles of new road[s]…in the 39 national forests in Oregon, Washington, Idaho and Montana." The story goes on to say that, "Some 24,000 miles of the proposed 77,000 miles of the new Northwest forest roads will be constructed in Oregon."[83]

The world at the equator is about 25,000 miles in circumference; thus 340,000 miles is about 13.6 times around the world at the equator. Including the other 77,000 miles (about an additional 3.1 times around the world), this is approximately the equivalent to 16.7 times around the world on forest service roads. This does not even include "temporary" roads. How many acres of land are "permanently" out of timber production to accommodate this road system?

By figuring that 4.5 acres of productive land are lost per mile of road,[83] the existing 340,000 miles of road equals 15,300,000 nonproductive acres. Adding another 77,000 miles of road (3,465,000 acres) brings the total acres out of production to 18,765,000. Adding the 77,000 miles of new road to the 340,000 miles already built would mean that more than one seventh of the Forest Service's

FIGURE 82 Unless water infiltrates deeply into the soil, it runs downhill and reaches a road cut that brings it to the surface, collects it into a ditch, and puts it through a culvert to begin infiltrating again. The water then meets another road cut, and so on. Water is sometimes brought to the surface three, four, or more times. Water is purified by its journey through the deeper soil and not by surface flow. Roads therefore have an impact on the hydrological cycle of a watershed and on the purity of the water that reaches human habitations.

119 million acres are out of production. This does not even include those acres out of production because of private logging roads, logging roads on such other public lands as the Bureau of Land Management, or private, county, state, and national roads and highways through forested lands. How much more can society afford? No wonder it has been said that our national flower is the "concrete cloverleaf."

Why has so much time been devoted here to water, and what does all this have to do with a sustainable forest? The answer is simple. Forest roads are constructed primarily to harvest timber, and these roads affect the quality, quantity, distribution, and effects of water in the soil of the watershed. Enough roads over time can alter the soil–water cycle of entire forested landscapes, and as soil–water cycles are altered, so is the forest's ability to grow. Remember that the quality and quantity of water is an ecological variable that is considered to be an economic constant.

Even if water were a constant, a variable is introduced with construction of the first logging road. The variable is compounded by constructing and maintaining logging and access roads to harvest timber. In addition, logging and plantation management alter the water regime, which affects how the forest grows. Thus, a self-reinforcing feedback loop of watershed degradation is created, altering the soil–water regime, which in turn alters the sustainability of the forest, which in turn affects the soil–water regime, and so on.

One of the main problems is that resource-managing agencies, especially those responsible for water, lack a long-term perspective and suffer from a shortage of public-minded leadership. Such agencies are not guided by an ethos of long-term sustainability. They are instead guided by a plague of special interests and a disdain for equity. As a result, the public is the continual loser.

We can continue to degrade and impoverish our water supply, or we can risk abandoning our conventional thought pattern and—with a strong, concerted commitment—reverse the trend. The choice is ours today in the invisible present, but the consequences belong to future generations when the invisible becomes visible.

In the final analysis, we must remember that only so much water is available, and more cannot be found in the courtroom, no matter how hard we try. It behooves us, therefore, to consider how we manage high-elevation forests.

BALANCING THE STRESSES

The most productive forests in Oregon (those below 4000 feet elevation) were the first to be cut out. To maintain the "sustained yield" from the less productive high-elevation forests (those above 4000 feet), the increase in annual acreage cut has been five times the increase in volume cut during the last 40 years. Are we adding to other ecological blunders of world forestry by mining high-elevation watersheds?

Consider that the industrialists' view of the land is conceived in profits (and they are correct from their point of view); they set about the task of dominating and exploiting the land for all it is worth. This being so, we need to understand money.

Money has no value in and of itself. It is symbolic of the value placed on something else: food, clothing, a home, an automobile, and so on. In fact, money is generally accepted as a medium of exchange—a go-between for things we value, a measure of value for something, or a means of payment for something we want. By the same token, the amount of money—in the form of biological capital—that we are willing to *re*invest in the forests to ensure their sustainability through time is a measure of their perceived value. However, money spent without honoring Nature's design can still kill a forest.

To understand the impact of forest management for profit, let's examine a forest that extends from sea level to the timberline (Figure 83). First, consider the virgin forest (Figure 83A). Note that the amount of natural stress (column 2) is

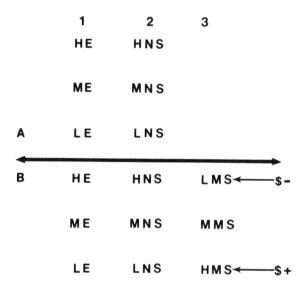

FIGURE 83 (A) A virgin forest going from sea level to timberline. Column 1 represents high elevation (HE), mid elevation (ME), and low elevation (LE). Column 2 represents high natural stress (HNS), moderate natural stress (MNS), and low natural stress (LNS). (B) The same forest under management, where column 3 represents low management stress (LMS), moderate management stress (MMS), and high management stress (HMS). In an unmanaged forest (A), columns 1 and 2 are in balance and the forest is self-sustaining and self-repairing through time. In the managed forest (B), the addition of column 3 (forestry) disrupts the ecological balance. The high-elevation forest is the least inherently resilient. Logging has the greatest impact ecologically here, yet this is where the least money is spent to care for the health of the system that produces the trees because of the perceived small profit margin. Conversely, the low-elevation forest is the most inherently resilient. Logging has the least impact ecologically, yet this is where the most money is spent to care for the health of the system that produces the trees because of the perceived high profit margin.

correlated with elevation (column 1); this means that the higher the elevation, the shorter the growing season, the greater the temperature extremes, the poorer the soil, and the greater the natural stress of survival for the forest. The opposite occurs at low elevations.

If the virgin forest is now managed (Figure 83B), column 3 is introduced and management stress is added to the forest's endeavor to survive through time. The initial clearcutting of the old-growth timber puts a disproportionate stress (management stress) on the forest that correlates with natural stress (column 1) and elevation (column 2). Note where most of the money is spent (low elevation, low natural stress forest) and where the least is spent (high elevation, high natural stress forest).

Because low-elevation forests, with inherently low natural stress, are the most productive in terms of wood fiber, they have the greatest potential to produce profits (Figure 84). Therefore, this represents the vast majority of the perceived value of the forest, and the vast majority of the money (Figure 83B) and the greatest intensity of management activity are centered in these low-

FIGURE 84 Low-elevation forests are high producers of wood fiber and are ecologically resilient. (USDA Forest Service photograph.)

FIGURE 85   High-elevation forests are marginal producers of wood fiber and are ecologically fragile. (USDA Forest Service photograph.)

elevation forests. It is this perceived value that causes an increase in money to be spent to change the forest—to "improve" Nature's design, fast and drastically, based almost entirely on predicted profit margins from short-term economics without hard scientific data.

High-elevation forests, on the other hand, produce little wood fiber and produce it slowly (Figure 85). Therefore, because they have little perceived value once cut, little money is spent on them compared to low-elevation forests. Because they are already near or at the limits of natural stress, any management activity is disproportionately stressful (Figure 83B).

There is, however, little profit to be made from high-elevation forests because they grow slowly and do not reach sufficient size to produce wood fiber at a high enough profit margin per dollar invested. Thus, the perceived value of high-elevation forests has almost exactly the opposite effect as the perceived value of low-elevation forests.

Ironically, however, it is the lack of sensitivity with which high-elevation forests are managed—based on the perceived profit margin from wood fiber produced for dollars spent—that is likely to quickly and drastically lower forest productivity. Such alteration of the forest is based almost entirely on predicted profit margins from short-term economics, but without hard data concerning the consequences of such management.

Elevation is used as an example because that is where snow accumulates and water is stored in the snowpack. How we treat our high-elevation forests is thus how we treat the source of our most important water supply.

If high-elevation forests are clearcut, then the snow melts and runs off early, which may overflow the low-elevation reservoirs and be lost for summer and autumn use. If, on the other hand, the high-elevation forests are healthy, the snow melts more slowly and would-be runoff can infiltrate into the soil, moving downslope in slow-motion storage to be available in streams during late summer and autumn, when it is most needed. Thus, there are forested areas, particularly high-elevation water storage areas, that humility and wisdom dictate should not be cut even once, regardless of the perceived short-term dollar value of the wood fiber. To protect such areas for water storage, however, all available data must be used.

14

PLANNING WITH
HALF-USED DATA

We can either be held hostage by old, rigid belief patterns or we can risk new, innovative ideas about managing forests. Although we may want to think like employers in forest management, we must think like employees of the future because there is always an individual and collective price for what we do, both within the present and among future generations. We also must remember that ultimately there is only one ecosystem—the biosphere—that simultaneously produces and sustains a multitude of products, including ourselves.

In this sense, we talk about our responsibility to the future, but that is not enough. We also need to act in such a way that we ensure, to the greatest extent possible, that the generations of the future have the ability to respond to the legacy of options we leave them. It is therefore imperative that we understand and account for the short-, mid-, and long-term ramifications of our decisions; this can only be done by thinking of them simultaneously, as I learned.

While working in Egypt in 1963–64, I wanted to go to a particular "black hill" I had been told about. The hill was in the desert about 300 miles southwest of Alexandria. The desert in this part of Egypt is flat and sandy with vast areas of desert pavement (Figure 86).

We had traveled by jeep for some time when my Bedouin guide told me to steer about three inches to the right. This sounded ridiculous. What difference could three inches possibly make? He didn't even have a map! Nevertheless, I was finally persuaded to make this "insignificant" correction.

Two days later we were at the black hill (Figure 87), and my guide told me to look at my map. I spread the map on the hood of the jeep and learned about humility. My guide drew a triangle and showed me that a correction of three inches near Alexandria had saved us about 50 miles worth of fuel and water on our way to the black hill (Figure 88). I learned that the further we predict into the

FIGURE 86   Small pebbles in the desert pavement protect the sand from blowing away in the Western Desert of Egypt.

trackless future, the more conscious and clear we must be of our motives, goals, objectives, and our data.

USE *ALL* AVAILABLE DATA

All available data must be used and planning (the invisible present) must be carried out far enough into the future to show the probable consequences our actions have set in motion, as exemplified by my trip to the black hill. In other words, short-term economic planning discloses short-term economic forestry and does not show what the cumulative effects of management actions are likely be through time. Thus, we leave the future blindly to the future.

True, we cannot foresee all the cumulative effects of our actions, and we cannot wait until everything is known before we act. But there definitely are some effects that can be projected, based on available data, which we stubbornly persist in ignoring.

One problem is that because forests, especially those of the Pacific Northwest, are long lived, a forester sees a rotation-age stand on a given acre precisely once in her or his career, even if it is a very short 40-year rotation (Figure 89).

One career forester works with the original old-growth stand (cuts it and

FIGURE 87   "Black hill" in the Western Desert of Egypt.

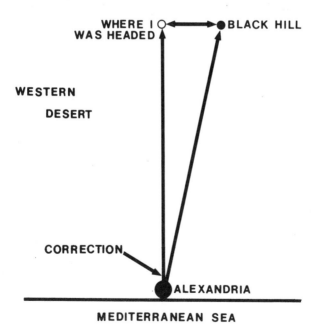

FIGURE 88   A Bedouin guide in Egypt taught me the essential lesson of traveling through the trackless sands of time in the present and future simultaneously. He made a three-inch correction near Alexandria that saved us about 50 miles worth of fuel and water when we reached the black hill 300 miles out into the Western Desert. (Not to scale.)

replants the area) but has no historical knowledge of the stand and cannot foresee what will happen to the future stand, despite all the assumptions, predictions, and data. A second career forester sees the first plantation at harvest time. A third career forester will be involved in harvesting the second short rotation and a fourth will see the third short rotation.

Beyond the first forester (the one involved with the original old-growth forest), every other career forester will see a plantation stand on a given acre only once at harvest time. The forester may or may not have good historical data on the stand and cannot foresee its future.

Select a figure *without any scientific data* that is easy to compute, say 10 percent of the potential recyclable nutrients (Figure 89). Now, let's assume that whatever wood fiber (product) is removed from an acre of forest during the harvest of old growth, subsequent site preparation and planting, commercial thinning, and final harvest will remove 10 percent of the potential recy-

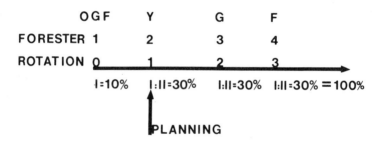

FIGURE 89 Because of the longevity of Pacific Northwest forests, a career forester only sees a stand or rotation once at harvest on any acre, whether old growth (OGF) or young growth (YGF). Let's assume that 10 percent of the potential recyclable nutrients will be removed with the wood fiber taken out at each entry. Each time a human enters the forest to manipulate it in a way that removes wood fiber (I in the figure is equal to an individual entry), 10 percent of the potential recyclable nutrients will be removed. Thus, harvest entry into the old-growth forest represents a 10 percent removal ($I = 10\%$), as does each entry in plantation management that removes wood fiber, such as site preparation and planting ($I = 10\%$), commercial thinning ($I = 10\%$), and final harvest ($I = 10\%$). In precommercial thinning, represented by a colon (:), no nutrients are removed from the site; therefore, the colon is equal to zero. Thus, removal of wood fiber is additive over time: old growth ($I = 10\%$) + first rotation ($I = 10\% + : = 0 + I = 10\% + I = 10\%$, which equals 30 percent per rotation) + second rotation (30%) + third rotation (30%) all equal 100 percent removal of the potential recyclable nutrients. Thus, the soil would be exhausted by the end of the third short rotation. Because we only plan for the next entry (the next 10 years), we neither know, see, or understand nor are accountable for the cumulative effects caused by our actions.

clable nutrients (fertility) that would have remained and been recycled into the next stand had they not been removed with the wood fiber (Figure 89). As illustrated in Figure 89, if each 10 percent removal of nutrients were accounted for, soil nutrients would be depleted by the end of the third short rotation. Thus, the fifth career forester probably would not know what happened or why.

We do not plan far enough ahead; we plan only for the next entry or the next 10-year cycle (Figure 89). We do not plan for even three, short 80-year rotations (three crops of trees) that would take 240 years to grow. It is, of course, not this simple, because some nutrient inputs, disruption of ecological processes, and other cumulative effects have not been accounted for. Nevertheless, what would we find through time if we looked at fertilizers, *or* herbicides, *or* pesticides, *or* soil compaction, *or* alteration of drainage patterns, *or* forest fragmentation...? What would we find if we looked at fertilizers, *and* herbicides, *and* pesticides, *and* soil compaction, *and* alteration of drainage patterns, *and* forest fragmentation...?

Are lack of foresight and emphasis on *economic* rotations rather than *biological* rotations part of the worldwide decline of forests? Is this what the Germans are beginning to learn about their plantations? Perhaps, but we are ignoring the most important aspect of existing data in forest land-use planning: forecasting the possible ecological consequences that could stem from economically oriented actions over three or more short rotations.

THERE IS NO MAGIC HINGE

When the forest industry, forest economists, and forest managers carry the projection of the past into the future, they assume that the young-growth plantation of the future will necessarily produce as much or more wood fiber per unit of area than the original old-growth forest (Figure 90A). That assumption is based on five premises—all of which are false:

1. Each acre can and will grow at least the volume of wood fiber that was represented in the harvested old-growth stand.

2. The first premise is true because the fertility of the soil, the quality and quantity of the water, the quality of the air, the quality of the sunlight, biological and genetic diversity, and the stability of the climate are all ecological constants.

3. Young-growth plantations function the same as the old-growth forests did, only faster and better.

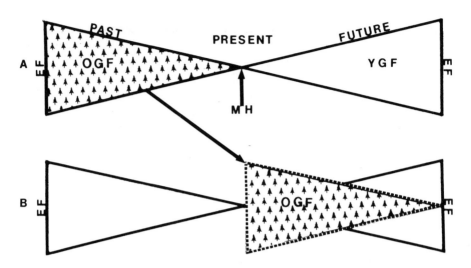

FIGURE 90 The forest industry, forest economists, and forest managers perceive scientific research and technology as a "magic hinge" (MH in part A) in the present that will make the young-growth forest (YGF) "better" than the old-growth forest (OGF). The concept is that future forests can be made sustainable, with the same ecological functions (EF), on a short-rotation basis. This notion is based on experience with the old-growth forest that is economically projected to the young-growth forest (A). Experience in liquidating an old-growth forest is past tense, however, and has little or nothing to do with data on how to redesign and grow a new sustainable, short-rotation forest even when projected into the future (B).

4. A "magic hinge" (scientific research and technology) exists in the present that will make the plantation of the future more predictable and productive of wood fiber than the forest of the past, or at least a mirror reflection of the past yield (Figure 90A).

5. All the variables that affect the growing plantation will, in the future, operate in concert at the same scales of time and space as those of the past, which created the original stand. But what about the impending changes in climate, which can alter everything?

All of these false premises are based on experience with existing old growth, the culmination of the time-space continuum of variables, and not on data with young-growth plantations. Experience with existing old growth reveals nothing about either the ecological functions (processes) or the sustainability of a short-rotation, young-growth plantation that is on a different trajectory of the time-space continuum of variables (Figure 90B).

Speaking of old-growth forests, planning in forestry needs to address the dwindling supply of large wood in the ocean because, among other things, it is an important part of the marine ecosystem's detrital-based food chain. The supply of such wood was once exemplified by vast accumulations of driftwood on beaches around the world. Today, the world's beaches are mostly bare of driftwood because current forestry practices, along with dams in rivers, are quickly eliminating the supply. If this concern seems far-fetched, remember that a forest is an interconnected, interactive ecosystem just as an ocean is, and together they represent a larger interconnected, interactive ecosystem.

FORESTRY PRACTICES AFFECT THE OCEAN*

We (Jim Sedell and I) both grew up in Oregon and over the years spent considerable time at the ocean. As boys we remember the huge piles of driftwood along the beaches, piles that seemed to grow with each winter storm. In fact, one of the challenges of even getting to the sandy shores of the Pacific was having to climb over the jumbled mountains of driftwood. There was so much wood, ranging from small branches to boards to whole trees, that we could build shelters from the wind that easily held 15 or more people. Enormous piles of driftwood were simply taken for granted as part of the beach.

Then we were grown and, suddenly, the driftwood mountains were gone. What happened to them? When did they disappear and why? How could the mountains of driftwood we so clearly remember have vanished without our noticing?

Driftwood is floating trees and parts of trees carried by water from the forest to the sea. It is a critically important source of habitat and food for the marine ecosystem, including the deep-sea floor. Even during its seaward journey, driftwood is both habitat and food source for a multitude of plants and animals, both aquatic and terrestrial. In addition, some driftwood controls stream velocities, stabilizes stream banks, makes waterfalls and pools, and creates and protects fish spawning areas. Other driftwood protects vegetation encroachment on floodplains and allows forests to expand. In short, driftwood makes a vital contribution to the health of streams, rivers, estuaries, and oceans worldwide.

The natural processes by which wood disappears from streams and rivers have positive effects on the ecosystem. Human activities, on the other hand, such as logging to the edge of a stream, salvage logging in riparian zones, cleaning wood out of streams, and firewood cutting, have had negative effects over the last

---

*By Chris Maser and James R. Sedell, excerpted from *From the Forest to the Sea: The Ecology of Wood in Streams, Rivers, Estuaries, and Oceans,* St. Lucie Press, Delray Beach, Fla., 1994.

several decades. The consequences of these actions, however, are both little understood and far reaching.

Streams historically replenished annual supplies of driftwood to the lower portions of river basins and out into the sea, where it washed up on beaches. But the banks of lower rivers and estuaries (the riparian corridor) were probably the common source of large driftwood in the bays.

Substantial amounts of driftwood must have been transported to the sea at the time when most riparian zones were dominated by such large coniferous trees as Douglas-fir, western redcedar, and Sitka spruce and such deciduous trees as black cottonwood, bigleaf maple, Oregon ash, and red alder. Hundreds of millions of board feet of logs and driftwood have entered Puget Sound and Georgia Strait from the rivers draining the Cascade Mountains of Washington and the coastal mountains of British Columbia. They were joined by large numbers of "escapees" from log rafts. Over ten billion board feet of logs are annually stored or travel in the estuaries and the lower segments of rivers in the Pacific Northwest. A 1 percent escape rate would allow over 100 million board feet of driftwood to enter the ocean from this source alone.

A conservative estimate is that in days past, as much as two billion board feet of wood per year was transported to the sea. Two billion board feet per year is a small amount when prorated across the entire North Pacific. Large driftwood, an important ecological component of Pacific Northwest streams and rivers, interfered with human objectives, however, and was summarily removed. In fact, people throughout North America have systematically cleaned driftwood from streams and rivers for over 150 years.

From the 1800s to around 1915, streams and small rivers were cleaned of driftwood so that logs could be floated from the forests to the mills. Several "splash dams" were built on many streams to temporarily augment the flow of water in order to float logs to mills. The net effect of channel clearance and splash damming was to remove large quantities of driftwood from medium to large streams, which is a significant change from the conditions that formerly existed.

Over the last 100 years, millions of drifted trees and other driftwood have been cleared out of streams and rivers to facilitate navigation and reduce flooding. To this end, streams and rivers have been channelized and dammed, and marshes have been drained. In addition, most stream banks have been so altered through logging that they now have dramatically smaller and younger trees of different species than in times past.

Most big western redcedars and Douglas-firs have been logged along Cascade Mountain streams and along coastal streams greater than third order. On private land, more than 70 percent of the coniferous trees greater than 14 inches in diameter at breast height have been logged within 100 feet of fish-bearing streams.

Before the great ecological value of driftwood was known, west coast fishery managers believed that driftwood in streams restricted fish passage, supplied material for driftwood jams, and caused channels to scour during floods. Indeed, during times of flooding such fears might have seemed to be well founded, but we know now that the results of stream cleaning have been ecologically disastrous.

It is now apparent that neither we nor the generations of the future can afford the effects of the loss of driftwood, which connects the forest to the sea and the sea to the forest. The loss of driftwood means the destabilization of streams, rivers, estuaries, complexes of sand dunes, beaches, and sand spits, as well as food chains in the oceans of the world. Sooner or later it means the loss of such jobs and unique cultural ways of life as commercial fishing, because such fish as tuna and salmon benefit from driftwood during various stages of their life cycles.

Nevertheless, driftwood is being prevented from even beginning its journey to the ocean by the removal of as much wood as possible from the forests as a product for human consumption, lest it remain as an "economic waste." In addition, damming of rivers prevents what little driftwood even begins its journey from completing it. Thus the connection between the forest and the sea is severed.

Even today, county sheriffs, port commissions, and recreational boaters still routinely clear driftwood from rivers for safety and personal convenience. As a result, most Pacific Northwest streams and rivers bear little resemblance to their ancestral conditions, when they flowed freely through pristine forests carrying their gift of driftwood to the sea.

Consequently, the supply of driftwood for food on the bottom of the sea off the coast of North America is both dwindling and becoming more erratic. For the first time in the evolutionary history of deep-sea animals, the availability of food has become unpredictable.

If the coastal mangrove forests continue to be destroyed through deforestation, the last direct link of the forest to the sea will be severed. The deep-sea wood-dependent species of the world will then shrink in both numbers and areas they inhabit, and some will become extinct. What does extinction of some species mean in terms of the health of an ocean?

Today, we are substituting for driftwood in the ocean and on beaches such nonwooden human garbage as metal, glass, rubber plastic, oil, bilge, chemical effluents, medical and household wastes, and raw human sewage—none of which can replace Nature's gift of driftwood.[84] We thus face the certainty of grave, uncomfortable uncertainties through our decisions concerning such renewable resources as driftwood by giving economics and technology higher priority than scientific understanding. If healthy streams, rivers, estuaries, and oceans are important for social benefits, then a renewable supply of driftwood (including whole trees) must be incorporated into land-use planning—especially forests.

THE FOREST AS A LIVING TRUST

The situation can be can rectified, however, if we concede that all we have to offer the future is options (which are choices to be made) and that those options, both biological and legal, are held within the forest as a *living trust,* of which we are the legal caretakers or *trustees* for the future. Although the concept of a trustee or a trusteeship seems fairly simple, the concept of a trust is more complex because it embodies more than one connotation. Consider, for example, a legal living trust.

A living trust is a *present* transfer of property (whether real property or personal property, livestock, interests in business, or other property rights), including legal title, into trust. The person who creates the trust can watch it in operation, determine whether it fully satisfies his or her expectations, and, if not, revoke or amend it.

A living trust also allows for delegating administration of the trust to a professional trustee, which is desirable for those who wish to divest themselves of managerial responsibilities. The person or persons who ultimately benefit from the trust are the beneficiaries.

The forest is a "living trust" for the future. A living trust, whether in the sense of a legal document or a living entity entrusted to the present for the future, represents a dynamic process. Human beings inherited the original living trust—the living world of which a forest is but a part—before legal documents were even invented. The Earth as a living organism is the *living trust* of which we are the *trustees* and for which we are all responsible.

Throughout history, administration of our responsibility for the Earth as a living trust has been progressively delegated to professional trustees in the form of elected officials. In so doing, we empower them with our trust (another connotation of the word, which means we have firm reliance, belief, or faith in the integrity, ability, and character of the elected official who is being empowered).

Such empowerment carries with it certain ethical mandates, which in themselves are the seeds of the trust in all of its senses, legal, living, and personal:

1. "We the people" are the beneficiaries and the elected officials are the trustees.
2. We have entrusted our elected officials to follow both the letter *and* the spirit of the law in the highest sense possible.
3. We have entrusted the care of public lands, whether forested or otherwise, to elected officials through professional planners, foresters, and other land managers, all of whom have sworn to accept and uphold their responsibilities and to act as professional trustees in our behalf.

4. We have entrusted to these officials and professionals the living, healthy forest. Through the care of these officials and professionals it is to remain living, healthy, and capable of benefiting both present and future generations.

5. Because we entrusted public lands as *"present transfers"* in the legal sense, we have the right to either revoke or amend the trust (the empowerment) if the trustees do not fulfill their mandates.

6. To revoke or amend the empowerment of our delegated trustees if they do not fulfill their mandates is both our legal right and our moral obligation as individual, hereditary trustees of the Earth, a trusteeship from which we cannot divorce ourselves.

How might this work if we are both beneficiaries of the past and trustees of the future? To answer this question, we must first *assume* that the U.S. Forest Service *is* both a functional and a responsible agency. The ultimate mandate of the Forest Service would then be to pass forward as many of the existing options (the capital of the trust) as possible.

These options would be forwarded to the next ten-year planning and implementation team (in which each individual is a beneficiary who becomes a trustee) to protect and pass forward in turn to yet the next ten-year planning and implementation team (the beneficiaries who become the trustees) and so on. In this way, the maximum array of biologically and culturally sustainable options could be passed forward in perpetuity.

If, however, the professionals in the Forest Service did not fulfill their obligations as trustees to our satisfaction, then their behavior could be critiqued through the judicial system, *assuming* that the judicial system *is* both functional and responsible. The invisible present embodied in our decisions as trustees of today could then create a brighter vision for the generations to come, who are the beneficiaries of the future when they stand in their today. In order for this to happen, however, we must first mandate that the Forest Service and the judicial system be made both functional and responsible, something we have not yet chosen to do.

Finally, in light of recent ecological information, it is imperative that we reexamine the way we treat streams and rivers and the supply of driftwood they no longer carry from the forest to the sea. Our present course impoverishes our waterways and ultimately the ocean. With their impoverishment, we increasingly jeopardize human society. We can change this self-destructive path. Another choice is possible.

Part V

SUSTAINABLE FORESTRY

Two kinds of landscapes are worth looking at—
those that man has never touched,
and those in which man has gained harmony.

Paul B. Sears

Liquidating old-growth forests is not forestry; it is simply spending our inheritance and stealing from our children. Nor is planting a monoculture forestry; it is
simply plantation management, which more often than not is what we are
practicing. Industry is trying very hard to make a gigantic, monotypic plantation
out of most of the forested lands in the United States. In fact, the timber industry
seems to be trying exceedingly hard to make plantations whenever and wherever
they can anywhere in the world. We will practice "forestry" only when we begin
to see the forest and begin to restore its health and integrity—sustainable forestry.
Sustainable forestry is the only true forestry. Sustainable means that the whole
is greater than the sum of its parts. Forestry in this sense is scientific knowledge
guided by a land ethic or ethos in its application to the art and business of
manipulating the forested portion of the ecosystem in a manner that assures the
maintenance and sustainability of biological diversity and ecological productivity throughout the centuries. Inherent in sustainable forestry are intuitive reality
checks and a great deal of humility. The outcome of such forestry will be the
perpetual production of amenities, services, and goods for human use.

In sustainable forestry, we use the forest by removing products, often in the form of biological capital, and then restore its vitality, its sustainability, so that we can remove more products in time without impairing its ability to function. From the time we cut the original old growth, we must continually practice sustainable forestry. Anything else is not forestry. It is simply abuse of the system for short-term economic profit.

15

WHERE ARE WE HEADED?

We are redesigning the world, both wisely and unwisely, through our decisions about everything—from what to purchase as a consumer to which tree to cut as a forester. Before we go much farther, we had best decide what we really want on the landscape. If we do not consider very carefully where we are going, we will end up where we are headed, which may not be sustainable in any useful way. Whatever we do scientifically and/or technologically will have no lasting value if the price of success is the loss of human dignity and the livability of our planet.

The opening paragraphs of a special report in *The Oregonian* sum up where we are headed in the Pacific Northwest, and the situation is worse in parts of Canada and other places in the world:

> The story of the Pacific Northwest's vanishing virgin forests is written on its mountains, in its foothills and along its river valleys.
>
> It's a story best read from the air, where 140 years of logging has torn the deep green carpet that once covered the land into a tattered quilt of large and small clearcuts, threaded together by thousands of miles of logging roads.
>
> In the Northwest, the timber industry is running out of places to cut.
>
> The old growth is gone from private lands and carved up or locked up on federal lands, and most second-growth forests [monocultural plantations] planted in the 1950s won't be ready to harvest for 20 to 30 years.
>
> Over the years, logging practices have contributed to declining fish runs, massive landslides, severe forest fragmentation and ruined streams. Many wildlife species—not just the northern spotted owl—are losing ground.
>
> The timber industry, long the region's economic mainstay and wellspring of political power, is reaping the consequences of a history of overcutting. Many mill owners say their companies are on the verge of

going broke. Since January 1989, 48 mills have closed in Oregon, Washington and Idaho—35 in Oregon alone—and 5,500 workers have lost their jobs.[85]

Forestry professor Julian Dunster indicates that management of forests as distinct from management of timber alone is rare in Canada. He states that a cherished principle is the notion of sustained yield, in which the forest is managed so that an equal volume of merchantable wood fiber is produced annually, which can be projected into perpetuity. However, it now seems that the Canadian timber industry is being sustained solely by the superabundance of existing timber, rather than by carefully created and implemented plans for the management of biologically sustainable forests.

The timber industry has survived by changing its technology and standards of utilization, which has enabled more of the existing timber to be harvested and processed. While this has improved the industry's efficiency of use, it has delayed the apparent need for a critical assessment of the forest's actual condition and has made no provision for the necessities of either the forests or the generations of the future.[86]

If the above scenario does not maintain the kind of landscape in which we want to live, then we have to ask some different questions and make some tough decisions. We simply cannot avoid these questions and decisions any longer. We have run out of time if we want to have anything of desirable quality left.

Whatever we do to move toward the biological sustainability of our forests will take the utmost courage. With the right attitude, any mistakes we make may become the future's strength. But we must act while the Earth still has the strength and the resources to survive in the face of ongoing errors and while there is still the ecological margin to allow a few more mistakes from which to learn.

To assure the potential of future generations to correct our errors and their ability to learn from them, we must remember that all we have to give the future is options. Therefore, each time we make a decision that deals with natural resources we must consider how our decision now, in the invisible present, will either maintain or enhance the options for the future.

We can examine our present choices with respect to natural resources by looking at our relationship with the Earth. For example, before we decide that our technology is better than our endowment from Nature, we must determine that Nature's endowment is not adequate to fulfill our necessities. And before we decide that Nature's endowment is not adequate to fulfill our necessities, we must identify our perceived necessities.

Once this has been determined locally, regionally, nationally, and/or globally, we must translate these necessities into a clearly stated vision, goals, and objectives, without which we cannot establish priorities. Then we must decide if the end justifies the means, particularly if the cost inherent in the means is the degradation of our planet, human dignity, or both.

As human beings, we participate in the creation—change—of the world in which we live, because our very existence and that of every other living thing is involved in this ongoing act of creation. As conscious co-creators with Nature, we are the moral, ecological guides for the future. In this sense, our impressions of our ancestors are the reflections of the care they took of the land that we inherited. And future generations, to whom we are ancestors, will find their impressions of us mirrored in the care we take of the land they must inherit.

If we are to change our image, we must begin now, consciously, to create a new paradigm for our trusteeship of the land, based on a sense of place and permanence, a sense of creation and landscape artistry, a sense of ecological health and sustainability, and a sense of humility and humanity. Although such a harmonious union between people and the Earth is not new in the world, it is new to the western psyche. It is the art of gardening the land with the artistry and the beauty that for so long has lain dormant in our souls.

The images we see on the landscape are but reflections in our social mirror of the way we treat ourselves and one another. As we compete and fight and live in fear, so we destroy the land; as we cooperate and coordinate and live in love, so we heal the land. We see the inner landscape of our beings reflected on the outer landscape of the Earth; we see ourselves reflected in the care we take of the land.

How do we participate in creating our world and to what extent? Do we create a world that is environmentally compatible with human existence, or do we create an unfriendly environment that is hostile to human existence? Do we alienate ourselves from our own planet, or do we accept our responsibility as trustees of Nature's bounty and act accordingly?

One thing is clear. Nothing will change the effect on the collective outer landscape until we first change the cause in our own inner landscapes, until we move toward conscious simplicity in both our inner and our outer lifestyles. And to participate consciously in the process of creating a different forest, indeed a different world, we must reevaluate our understanding of two concepts: natural and native.

WHAT IS NATURAL?

Because people perceive their activities as an "unnatural" contamination of Nature, the connotation of "natural" or "naturalness" becomes pivotal to the very concept of a natural area, say a forest. I originally understood the concept of natural areas when I worked with them, as a pristine piece of the ecosystem that was to be set aside in perpetuity for baseline research. It seemed a noble act to preserve some small pieces of uncontaminated Nature.

After all, *pristine* is of, pertaining to, or typical of the earliest time or condition. Something that is pristine is in its original state and remains pure or uncorrupted. *Preserve* is used in the sense of keeping in a perfect or unaltered condition by maintaining whatever it is in an unchanged form. Yet both definitions imply nonparticipation by human beings, at least post-European-settlement human beings.

Today, however, I have a different sense of the word pristine. It simply means an arbitrarily selected snapshot in time, like arbitrarily selecting a frame within a film loop and saying that this frame is henceforth our measure of purity. Even in an ecological sense, nothing in the world is pristine, if for no other reason than the pervasiveness of air pollution.

Nothing can be preserved, in the sense that the word is used, without first killing it and either sealing it in an air-tight container or preserving it in a fixative medium. The question therefore becomes whether this tiny piece of the world is functioning in a way that seems desirable and within our perception of the healthy state of the ecosystem. If so, is it worthwhile and within our capabilities to protect this tiny piece of the world's ability to function as we perceive it to be functioning? The notion of natural and naturalness must now be taken a step further by introducing the concept of "native."

WHEN IS A NATIVE A NATIVE?

Native means belonging to one by nature. It means originating, growing, or produced in a certain place, indigenous as opposed to exotic or foreign. How do native and natural fit together? To answer this, we must go back in time.

When Europeans first began to settle the islands of the Caribbean in the New World in 1492 and La Florida in 1513, they considered both the forests and the people who inhabited them to be both native and natural. What changed? First, the European invaders saw not the land but only free products, including the "native" or aboriginal peoples from which to grow economically wealthy. To reap such wealth, however, they had to dispose of the aboriginal peoples, which they did. Second, the Europeans brought their Nature-condemning religion with them.

As discussed earlier, the aboriginal Americans viewed the land and all it contained as holy and something to be revered; Europeans, on the other hand, viewed the same land and all it contained (including the indigenous peoples) simply as an object to be subjugated and exchanged for money.

The concept of "native," however, has changed over time. For example, beech, elm, hickory, and sycamore trees were once part of the pre-Pleistocene forests of the Pacific Northwest. If they were reintroduced today from the east coast of North America, would they still be native to the area? If bighorn sheep, extirpated from an area within historical times, were reintroduced from a herd 200 miles away, would they still be native to the area? The aboriginal Americans have been here for about 20,000 years, but George Vancouver, a midshipman with Captain Cook, first saw the coast of Oregon in 1778.

Europeans have therefore been in the Pacific Northwest for less than 275 years, but those whose ancestral roots have grown in the soil for a century also think of themselves as natives, albeit a different kind. Regardless of ancestral heritage, people who were born in a given state consider themselves to be natives of that area, regardless of how old they are.

What we are really talking about is a relative degree of "nativeness," which is thought of in the context of time just as "naturalness" is thought of in the context of human intrusion. In this sense, the further into the past one's ancestry can be traced in a given area, the more "native" one is thought to be.

In order for ecological restoration to become a fully functional tool of sustainable forestry, a system of natural areas needs to be set aside as benchmarks against which the results of ecological restoration can be tested and the connotation of "natural" and "native" in the areas so designated can be reconsidered. As humanity continues to redesign the world, we must grapple with the notions of relative naturalness and relative nativeness, which are constantly changing. If the relativeness of natural and native is not built into the concept now, the definitions of each may become so fixed as to be unworkable simply because particular portions of future landscapes are not natural enough or native enough by today's standard.

In the sense that we are a natural part of the ecosystem and what we do is therefore also natural, although it may be destructive and unwise, "naturalness" is not a definition but rather a descriptive continuum of human interactions, ranging from the most pristine end of the ecological scale to the most culturally altered end of the ecological scale. By the same token, "nativeness" is not a definition but rather a descriptive continuum in time, ranging from the most ancient continuous habitation to the most recent immigration.

Therefore, for the present at least, the standard of naturalness must be based on the best overall data of historical integrity, which also implies the most nativeness, of a plant and animal community that is available for a given location. Such a standard for the characterization of naturalness and nativeness can either be maintained or altered in the future as circumstances dictate without losing the integrity of the concept.

This may prove to be a difficult charge, because the earliest historical records are European. And, depending on one's point of view, it may be erroneous to base the concept of natural or native on pre-European conditions, because even if the earliest Europeans had recorded their perceptions of the environment, it would be suspect because their influence moved more rapidly than they did.

The introduction of the modern horse by the Spanish explorers, for example, caused such a rapid and profound reorganization of aboriginal cultures that the pre-horse ecosystem probably could not be reconstructed. In addition, because European diseases spread so fast and were so virulent, the vulnerable aboriginal population had been severely reduced by the time Europeans documented the landscapes. Finally, enough time had elapsed between the pandemics of disease and the time the Europeans recorded what they saw on the landscape that a case can be made for the recovery of plant and animal communities from the higher levels of utilization by the once abundant aboriginal population.

Nevertheless, the difficult task of characterizing the most natural and native end of the ecological continuum is necessary if we are to have functional concepts of "naturalness" and "nativeness" with which to work. Further, such concepts are imperative if future generations are to have meaningful options from which to choose the types of landscapes they find of desirable quality in which to live.

A REASSESSMENT OF OUR CONTROL OF NATURE

Let's examine which concepts sustainable forestry might cause us to revisit before discussing why it is important and how it might be practiced. Restoration is the act of restoring or returning something to, or in this case toward, its original condition; renewing. We thus have to know what the "original" condition was, which is the value of characterizing successional stages within unmanaged forests as young-growth, mature, and old-growth forest. To do this, some original, unmanaged old-growth forest, mature forest, and young-growth forest must be maintained as a parts catalog, maintenance manual, and service department from which to learn how to practice sustainable forestry.

These unmanaged forests are also living laboratories that simultaneously serve as living dictionaries and living libraries, where such things as productivity of soil for specific types of forest sites can be characterized. Productivity in this case has at least two definitions: "The status of a soil with respect to its ability to supply nutrients essential to plant growth"[87] and "The inherent capacity of a soil for supporting growth of specified plants, plant communities, or sequence of plant communities."[88]

Limitations are necessary, however, because no soil can produce all plant communities or crops with equal success, nor can a single system of management produce the same effect on all soils. The way we use the term productivity emphasizes the capacity of soil to produce crops and is expressed in terms of yields. What does this really mean in terms of sustainable forests with respect to soil type, moisture gradients, elevation, latitude, slope, and aspect within the forests? How will we ever know without a well-distributed system of living laboratories?

Sustainable forestry is the opposite of plantation management practiced today. In plantation management, costs are hidden and deferred to the next rotation or human generation. In sustainable forestry, on the other hand, there are no hidden, deferred costs; it is pay-as-you-go forestry that more closely follows Nature's blueprint for maintaining a self-repairing, self-sustaining forest.

Product extraction is maximized in traditional plantation management and sustainability of the forest is minimized. In sustainable forestry, however, sustainability of the forest is maximized and product extraction is performed at a level and in a way that does not impinge on the sustainability of the forest.

Today we are applying 100 percent of what we know about product extraction and utilization in plantation management and only about 10 to 20 percent of what we know about sustaining the forest. Sustainable forestry demands that we use 100 percent of what we know about sustaining the forest and 100 percent of what we know about product harvest and use, but the former must be maximized and the latter must be carried out with humility (Figure 91).

All of this means that we must reassess the value placed on land, which can be done by examining the amount of money spent for various management practices, because money is symbolic of the value placed on something. For example, many many times more money is spent to extract products from the land (including extractive and utilization research and technology) than is spent to protect and sustain the system that is responsible for the products in the first place. Economic models and management plans on forested lands clearly show that expenditures in extracting products are maximized and expenditures in caring for the land are minimized.

Let's examine some possibilities more closely. Suppose we compare renting a home with buying a stand of old-growth or young-growth timber, which in essence is renting public lands for a profit. In renting a home, an agreement on price is reached and a deposit is paid to cover any possible damage. If damage occurs, an appropriate amount of the deposit is forfeited; if none occurs, the deposit is returned.

Something similar could be done on public lands. A timber company that buys the timber (rents public lands for the profit in the timber) must pay a deposit (say 10 percent of the value in the timber), which is held in escrow until all its logging and other contractual obligations have been fulfilled. If the company lives up to the contract, the deposit is returned; if not, damages are assessed and restitution is made to the public by using whatever portion of the deposit is necessary to repair the damage. The same could be done with contract tree planting or any contractual work on public lands. The point is

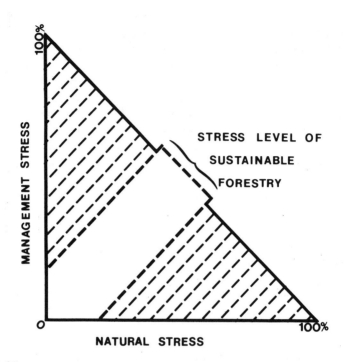

FIGURE 91   The forest can be managed for products on a sustainable basis with sustainable forestry, provided that the stress threshold beyond which the forest may not be repairable is not exceeded. The higher the natural stress, the lower the management stress must be; the lower the natural stress, the higher the management stress can be, but always within the limits of a sustainable forest.

that whoever makes a profit from public lands needs to be totally account-able for any damages they cause and at no cost to the public or future genera-tions. In fact, any cost passed forward represents taxation of the future without representation.

In addition, the public needs to begin sharing part of the financial burden of trusteeship for public lands. This could be done by paying an annual fee for general use of public lands and for such activities as harvesting mushrooms, fishing, hunting, prospecting, and so on. A fee for hunting is already under consideration.

Our understanding of the sustainability of a forest is also skewed by our focus on the theoretical money-product base of plantation management. We simply do not understand or accept that, all too often, parts of forests—such as trees—are not interchangeable or substitutable. Forests are not automobiles that can be tailored artificially by substituting parts. Trees may look similar, but they do not necessarily function the same or even in a similar manner.

For example, one of the problems in Germany, which, according to forestry professor Richard Plochmann, is typical of all pure plantations, is that the ecology of the natural plant associations became unbalanced. The physical condition of the single tree weakens and resistance against enemies decreases when planted in pure stands outside of the natural habitat.

The problem is compounded because in Germany, as everywhere else, all of the ecological factors cannot be controlled when planting trees in a strange environment. Some factors may prove more favorable to enemies of the trees than to the trees. A strange environment in this case does not only mean a foreign country; it also means off-site. Thus, a degree of nativeness comes into play.

To illustrate, when German foresters imported the Douglas-fir from the United States, the trees brought two fungi. These fungi do little damage to Douglas-fir in the United States, but in Germany the destruction was severe. The result was repeated insect and fungi catastrophes, which destroyed large plantations.

The German experience with even three or four generations of pure conifer stands, mainly outside the area of their natural distribution, shows that their cultivation is possible, their management in many ways is easier, and their economic results are even better than those of the natural mixed forests. However, experience also shows that the risks involved are high and that the productive capacity of the soil can be lowered markedly.

The dangers involved can be only partially controlled by artificial means. Artificial fertilization, chemical insecticides and fungicides, and use of the most appropriate seed origin may lower the risk but cannot eliminate it. In addition, the costs incurred will consume large parts of the anticipated profit. This has led German foresters to the conviction that, even from an economic point of view, plantations of pure conifers must be restricted to areas where they occur naturally. This means the best and most stable sites, where the danger of a decreasing production capacity of the soil is minimized to the greatest extent possible.[25]

Yet we continue to ignore data from around the world, for whatever reason, as exemplified in an article in the *Corvallis Gazette-Times*:

> The remote nation of Nepal has a serious deforestation problem. In Oregon there lives a citizen who raises fast-growing hybrid poplar trees. Sen. Mark O. Hatfield, R-Ore., chairs the Appropriations committee. The committee earmarked $2.28 million in its fiscal 1986 foreign operations bill to send 2.5 million Oregon poplar tree cuttings to Nepal.
>
> Nothing new there, you say? Happens all the time? Ah, but in this case, Nepal resisted: These poplars might not grow well there. The Agency for International Development also resisted, listening to one of its foresters who argued that the project would be "completely unrealistic, a waste of money" and "an act of extreme folly" because of transport, land availability and refrigeration problems.

Hatfield was determined. "He really believes that this is an important and valid development project," said aide Rick Rolf....

A feasibility study that AID ordered in January—but which committee aides said was a stalling tactic—reported that Nepal does indeed need "a tree such as a poplar" that can be used as fuel, animal feed, brushwood and a soil holder and windbreak. The study, headed by Argonne National Laboratories, said that other varieties of trees also should be explored.[89]

Nothing was said in the article about using native trees, even though some individuals with the U.S. AID mission were largely responsible for Nepal's problem of deforestation in the first place. As mentioned earlier, I watched the progress of deforestation while working in Nepal during 1966–67.

Here it is necessary to once again discuss the unity of all things before going on to examine some ideas of how a forest functions, which in turn may allow us to understand how to restore a forest and why it is important to do so. We turn again to physicist Fritjof Capra.[8]

Modern physics shows us once again, says Capra, and this time at the macroscopic level, that material objects are not distinct entities, but rather are inseparably linked to their environments. As such, their properties can only be understood in terms of their interaction with the rest of the world.

The bootstrap philosophy constitutes the final rejection of the mechanistic world view, as exemplified in modern physics by Newton's universe. Newton's universe was constructed from a set of basic entities with certain fundamental properties, which because they were created by God were not amenable to further analysis. This notion, in one way or another, was implicit in all theories of natural science until the bootstrap hypothesis stated explicitly that the world cannot be understood as an assemblage of entities that cannot be analyzed further.

In the new world view, the universe is seen as a dynamic web of interrelated events. None of the properties of any part of this web is fundamental; they all follow from the properties of the other parts, and the overall consistency of their mutual interrelations determines the structure of the whole.

The crux of maintaining diversity within a forest is that things must exist before they can be in relation to one another, which is part of the unity of all things. This specifically includes the diversity of processes in the concept of diversity within a forest because diversity—both structural and functional—either maintains or alters the speed and direction of succession and hence the resulting plant community. For example, a forest neither has a single state of equilibrium nor is characterized by a single deterministic pattern of recovery.

Sustainable forestry can therefore be constructed on two premises: (1) within some limits a forest will persist, provided the existing disturbance regime is compatible with its continued existence, and (2) given the chance, a specific condition within an ecosystem, plant community, or successional stage will recur. This means that by accepting the first premise, the second is allowed to fulfill itself, but not on all acres all of the time and perhaps not even on our timetable all of the time.

Bear in mind that we can guide Nature gently, but trying to force Nature will surely result in resistance, unwanted outcomes, and situations over which we have no control. Nature is not our servant, but rather, given a chance, may become our partner.

SUSTAINABLE FORESTRY AS CONSCIOUS EVOLUTION

According to Professor Paul Ehrlich, our species is now at its most important turning point since the Agricultural Revolution. "For the first time, humanity has the knowledge to destroy itself quickly, and for the first time, humanity also has the knowledge to take its own evolution into its hands and to *consciously evolve.*"[90] In this sense, perhaps more than any other, we must come to understand that we, by nature, are participatory partners in and with the landscape, and that our future as a society, even as a species, is inexorably intertwined in this interactive partnership. This being the case, we must also understand that our conscious evolution is based not on the answers we divine, but rather on the questions we ask.

The Questions We Ask

Each question is a key that opens a door to a room filled with mirrors, each of which is a facet of the answer. Only one answer, however, is reflected in all the mirrors in the room. If we want a new answer, we must ask a new question— open a new room with a new key.

We keep asking the same old questions, opening the same old doors and looking at the same old reflections in the same old mirrors. We may polish the old mirrors and hope thereby to find new and different meanings out of the old answers to the same old questions. Or we might think we can pick a lock and steal a mirror from a new and different room in the hope of stumbling onto new and workable answers to the same old questions.

The old questions and the old answers have brought us to the problems we face today and are leading us toward even greater problems tomorrow. We must therefore look long and hard at where we are headed with respect to the quality of the world we leave as a legacy. Only when we are willing to risk asking new questions can we find new answers.

One ostensibly new way of viewing the old problem of continual development is expressed in a report by Norwegian Prime Minister Brundtland prepared for the World Commission on Environment and Development. It calls for sustainable development, which juxtaposes, in one sense, two mutually exclusive concepts: sustainability and development.[91] *Sustainability* is the language of balance and limits over time, whereas *development* is the language of expansion, of always expecting more in some limitless fashion. Brundtland's report thus attempts to see the problem as one of paradox.

If this is the paradoxical way we are going to view the world, says philosopher Ivan Illich, then some of the pressing questions of today include: "After development, what? What concepts? What symbols? What images?" To find an alternative language, to find something that works for today in terms of the future, Illich suggests returning to the past to discover the history around which the mythological "certitudes" that undergird our current thinking were invented. These certitudes encompass need, growth, development, and the like and together form the organizational core of our modern experience.[92]

There is another way of looking at sustainable development, however, one in which development is truly sustainable: cultural/technological evolution. The concept behind cultural/technological evolution is that society can evolve through conscious choice to control both what it introduces into the environment and how it uses natural resources—destructively with arrogance or gently with humility.

Technology can be designed in an environmentally friendly manner. In this sense, sustainable development means working consciously and continuously toward a higher level of awareness and responsibility toward one another and the environment—all of it. It means thinking within the context of an interconnected, interactive system in which a desired outcome will depend on the questions we ask.

The answer to a problem is only as good as the question and the means used to derive that answer. Scientist do not always ask good questions. Professor Walter McDougall, of the University of California at Berkeley, has in fact declared that "Big Science" has become destructive to the scientific endeavor as a whole. The lure of large grants aimed at predestined results is deadly to the most important ingredient in science: the ability to freely ask a new question. The overall goal of scientific endeavor must remain the pursuit of pure knowledge, which by definition demands that the pursuit be totally unencumbered and forever open-ended.

To keep the search for truth on its own credible track, we must first recognize that we tend not only to form a single hypothesis, but also tend to become so attached to it that any criticism of or challenge to our methods raises our defenses. This means that the moment a person has derived what seems to be a satisfactory explanation for a phenomenon, the attachment to his or her intellectual child springs into existence. The more the explanation grows into a definite theory, the more near and dear it becomes. Then the theory is massaged (as I have often heard it called in government agencies) to fit the data and the data to fit the theory.

In addition, we tend to become "method-oriented" rather than "problem-oriented" in our thinking and therefore in many of the questions we ask. It is important to recognize that we become method-oriented in our questions, because we tend to think that through our experiments—our methods—we are learning the "truth" about Nature, when in fact we are learning only about our experimental designs—again, our methods—and our assumptions and expectations.

It is impossible to accurately represent Nature through science, because scientific knowledge is not only a socially negotiated, rigid construct, but is also a product of the personal lens through which a scientist peers. Scientists may attempt to detach themselves from Nature and become "objective," but they are never completely successful. They are part of and must participate with Nature in order to study Nature.

In addition, every scientist sees through his or her lens but dimly, first because we cannot detach ourselves from Nature and second because all we can judge as fact are our own perceptions, which are always colored by our personal lenses. We may polish and wipe them, but appearance—and not reality—is all we can ever hope to see. Therefore, it is appearance to which we often unknowingly direct our questions.

The truth about scientific research is that nothing can be proven—only disproven; nothing can be known—only unknown. This being the case, we can never "know" anything in terms of knowledge. We can "know" it only in terms of intuition, which is the knowing beyond knowledge that is inadmissible as evidence in modern science. Whatever truth is, it can only be intuited and approached, never caught and pinned down.

The irony is that knowledge, which is external to a person, is not "knowable," and intuition, which is internal to a person, is not knowledge and therefore is not subject to disproof. Intuition is inner sight—individualized, inner knowing—for which proof is unnecessary and explanation impossible. Knowledge, on the other hand, is the collective outer experience of humanity's and society's subjective judgments, the truth of which cannot be known and therefore is explainable only in the illusions of its appearance.

Thus the actual objects of our inquiries, the formulations of our questions and definitions, and the mythic structures of our scientific theories and facts are social constructs. All aspects of our scientific theories, facts, and practices—including "scientific method" itself—are but expressions of contemporary socio/political/economic interests, cultural themes and metaphors, personal biases, and personal/professional negotiations for the power to control, albeit minutely, the scientific knowledge of the world.

Over the 20 some years that I worked in science, I learned that truth is an inner phenomenon of faith, which is absolute but unshareable and unprovable. Illusion, on the other hand, is an outer phenomenon of knowledge, which is shareable but only relative and disprovable—the social construct of science.

Facts that scientists construe to be statistically true statements about Nature are demonstrated to be concrete, deified, magical outcomes of the social process of fabricating statements about the world so as to distinguish order from chaos.[93] Thus, instead of scientific consensus being achieved when the "facts" reach the state of speaking for themselves, scientists come to a consensus when the political, professional, and economic costs of refuting them make further negotiation untenable.

There is, however, no single reality, but rather a multiplicity of realities, the representation of which depends on one's position in the process of negotiating an acceptable social view of reality. Thus, regardless of the question, the reproducibility of the experimental design and methods does not mean that the results represent anything about Nature. The reproducibility of the experimental design shows merely that a particular negotiation of reality is reproducible under a certain set of conditions.

The results of every experiment may thus be valid, if unprovable, only because the experimental design tells us nothing about the results. It tells us only that the reproducibility of the experimental design is socially acceptable according to a consensus of scientific opinions.

If, therefore, we are going to ask intelligent questions about the future of the Earth and our place in the scheme of things, we must be free of scientific opinions based on "acceptable" interpretations of scientific knowledge. In addition, we would be wise to approach life with a beginner's mind, which is simply open to the wonders and mysteries of the Universe.

A beginner sees only what the answers might be and knows not what they should be. An expert, on the other hand, presumes to know what the answers should be and can no longer see what they might be. The beginner is free to explore and to discover a multiplicity of realities, while the expert grows rigid in a self-created prison of a single pet reality, which often turns into an obsession to be protected at any cost. Thus the beginner understands the question better than does the expert.

We must, if we are going to ask intelligent questions, be open to multiple hypotheses and explanations, and we must be willing to accept a challenge to our ideas in the spirit of learning, rather than as an invitation to combat. One of the most important scientific questions we can ask is: "How small and elegant an experiment can I perform?"[94] The greatest triumphs in science are not, after all, triumphs of facts, but rather triumphs of new ways of perceiving, thinking, and asking questions.

Such triumphs come through knowing which questions to ask and through the willingness to risk what most people think of as failure. According to university president Harold Shapiro, the avoidance of risk, is, in the end, "an acceptance of mediocrity and an abdication of leadership."[95]

True success or failure is a personal view, however, and lies not in the event itself but in its interpretation. When Thomas Edison's 10,000 experiments with a storage battery failed to produce results, for example (and society would surely have deemed that a failure), he said, "I have not failed. I've just found 10,000 ways that won't work."

Before we can arrive at fundamentally new answers, we must be willing to risk asking fundamentally new questions. This means that we must look long and hard at where we are headed and the legacy we are leaving with respect to the quality of our environment. Keep in mind that the old questions and the old answers have gotten us to where we are today and, unless we change our thinking, will guide us to where we will be tomorrow.

Heretofore we have been more concerned with getting politically right answers than we have been with asking morally right questions. Politically right answers validate our preconceived economic/political desires. Morally right questions would lead us toward a future in which environmental options are left open so that generations to come may define their own ideas of a "quality environment" from an array of possibilities.

A good question, which may be valid for a century or more, is a bridge of continuity among generations. We may develop a different answer every decade, but the answer does the only thing an answer can do: it brings a greater understanding of the question. Because an answer cannot exist without a question, the answer depends not on the information we derive from the illusion of having answered the question, but on the question we ask.

In the final analysis, the questions we ask guide the evolution of humanity and its society. It is the questions we ask, and not the answers we derive, that determine the options we bequeath to the future. Answers are fleeting, but questions may be valid over the long term. Questions are flexible and open-ended, whereas answers are rigid, illusionary cul-de-sacs. The future, therefore, is a question to be defined by questions.

The Hope We Plant

How does a society treat the forests it has ravaged in a way that allows them to heal and is most likely to sustain their health and vitality through time? How does a society harvest products from healthy forests in a way that sustains their health and vitality through time? A partial answer lies in sustainable forestry.

Sustainable forestry offers the hope of sustainable forests and healthy watersheds to the world. Before Nature will cooperate as our partner in this venture, however, we must meet Nature's conditions:

PLANTATION

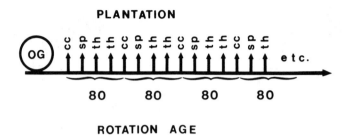

80   80   80   80

ROTATION AGE

FIGURE 92   Linear plantation management in which the old growth (OG) is clearcut (cc). The clearcut site is then prepared and planted (sp); the young-growth forest is precommercially thinned once (first th), commercially thinned once (second th), and finally clearcut (cc) again at the end of 80 years. The plantation cycle then starts over, with the focus always on maximizing wood fiber production.

1. We must shift our focus from products of the forest (Figures 92 and 93) to the processes of the forest.

2. We must balance the energy removed from the forest in products by allowing the forest time to repair its processes and reinvest such things as large woody materials ("biological capital") into itself in a way that is available to the next forest (Figure 94A). We must simultaneously minimize and account for cumulative effects we cause or we will alter plant communities in unacceptable ways (Figure 94B).

3. We must learn to accept management of forests as a set of open-ended long-term trends in which industry must be flexible and accountable to the forest that produces its raw materials.

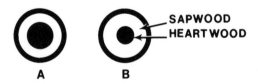

A   B

FIGURE 93   Schematic representation of the heartwood:sapwood ratio in a cross-section of a slowly grown, old-growth Douglas-fir tree (A) and an equal-diameter fast-grown, plantation Douglas-fir tree (B). This shift in heartwood (sapwood ratio between the old-growth tree and the young-growth plantation tree) alters the dynamics of decomposition, nutrient cycling, habitat values, and many other processes that make plantations less stable ecologically over time. The sapwood, which is rich in carbohydrates, is rapidly consumed and decomposed by organisms; the heartwood, composed mostly of lignin, is long-lasting and important in the formation of soil humus and mycorrhizae.

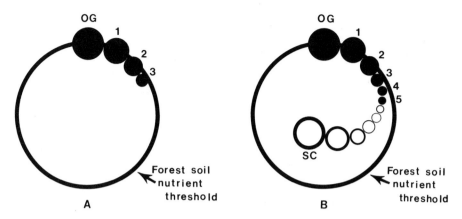

FIGURE 94   (A) The circle represents the forest soil nutrient threshold (maintenance budget) required to maintain the site in a forest community through time. When the old-growth forest (OG) is cut, the first plantation (1) does very well because it draws on the surplus budget of stored, available nutrients (refer to Figures 54 and 70) and relatively intact belowground ecological processes. The second plantation (2) does not do as well because there is less nutrient capital from which to draw, and the ecological processes suffer greater disruption. The third plantation (3) does poorly because the nutrient budget is becoming depleted and the belowground ecological processes have become severely disrupted by intensive management. If at this point the forest is replanted and left to grow *without further human intervention* until it reaches the old-growth stage, the soil nutrient capital and the belowground ecological processes can repair themselves, and the forest again becomes self-sustaining. The forest can then be harvested and a new plantation cycle initiated. The maintenance budget of forest soil nutrients must retain its integrity in order to make this management scenario possible. (B) The same principles function here as in Figure 93A, but instead of letting the forest (solid circles) repair itself after the third (3) rotation, a fourth (4) and fifth (5) rotation are attempted. The soil nutrient capital is well into a deficit budget by the fourth rotation and is definitely irretrievable by the fifth rotation; the belowground ecological processes are irretrievably altered by the fifth rotation, and the forest community becomes a shrub community (open circles, SC), which may last decades to centuries, depending on the severity of management damage to the site. Note that intensive management brings the forest community to an earlier, simpler, successional stage shrub community, which is a predictable consequence of intensive product-oriented plantation management that reduces the forest soil nutrient capital into the deficit budget. The forest soil deficit budget, however, is a surplus budget for the shrub community, which over time can reestablish the maintenance budget for the gradual return of the forest community.

4. Given a chance, the most desirable successional stage will probably recur. The probability is higher in some areas than in others, but this depends on what we do, how we do it, and why. It is therefore wise to carefully select areas for different intensities of management based on the sustainability of the forest at the desired level of harvest.

5. We must develop practices appropriate for the management of our own forest and cease trying to force incompatible practices on the forests of North America, practices that have proven to be detrimental even to the forests in which they were developed.

6. We must practice landscape management with all the tools available to us.

7. In humility, we must accept Nature as our teacher.

Assuming we meet these conditions, how does sustainable forestry work? Sustainable forestry combines two thought processes (cyclic and linear) and operates on the principle that as we alter the forest from the soil surface upward, we simultaneously alter the forest from the soil surface downward. We must also understand and accept that the variability within and among forest stages is infinite, because the variability is composed of ever-changing small, short cycles within larger, longer cycles within larger, longer cycles, and so on. Every cycle ultimately completes itself because time is a human construct and not relevant to a forest (Figure 95).

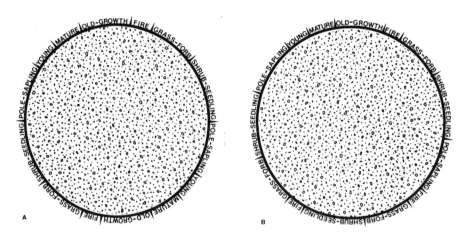

FIGURE 95 The following scenarios are only two possibilities among infinite variations at any given site. (A) Classic forest succession takes place, but the length of time in each stage can vary from years to decades to centuries, depending on a myriad of conditions. (B) Fire starts succession over before the old-growth stage is reached; if fires are frequent enough, several centuries may pass before old growth again occupies the site.

Sustainable forestry is thus born in the concept of variable rotation ages that simulate Nature's blueprint over a landscape. Such forestry strives to achieve the goals and objectives of both biological sustainability and sustainable yields of products through time. It can also be thought of as satellite forestry in that satellites are used to see the whole forest simultaneously. A changing mix of plantations and native forest—including old growth—can thus be managed over a landscape in a way that simulates Nature's irregularity and allows every acre to become a forest often enough and long enough to be self-sustaining and self-repairing while maintaining a viable pattern of connected habitats across the landscape (Figures 96 and 97).

Intensive plantation management is a child of short-term economics in which the entire focus is on fast-grown wood fiber, and anything diverted to another

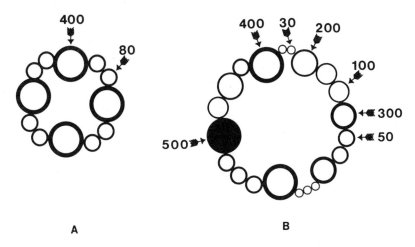

A   **B**

FIGURE 96   A forest/plantation management scheme for the same acres through time. (A) This management scheme represents four forest stages (400 years old) interspersed with eight plantations (80 years old). Each plantation is used to maximize production of low-quality, fast-grown wood fiber, which rapidly draws on stored soil nutrient reserves and impacts ecological processes, both aboveground and belowground, through management activities. Each forest stage allows soil nutrient capital to be replaced and ecological processes to be repaired; this is the time of reinvestment in the forest system in preparation for the next plantations. While the forest is repairing itself, high-quality, slowly grown wood fiber is being produced. The formula for this management scheme is f–p–p–f–p–p–f–p–p–f–p–f (f = forest, which is more than 100 years old; p = plantation, which is less than 100 years old) and takes 2240 years to complete. (B) This example, which takes 3050 years, shows the wide variation of possible options in wood fiber production while simultaneously producing other products and amenities, such as clean water and spotted owls.

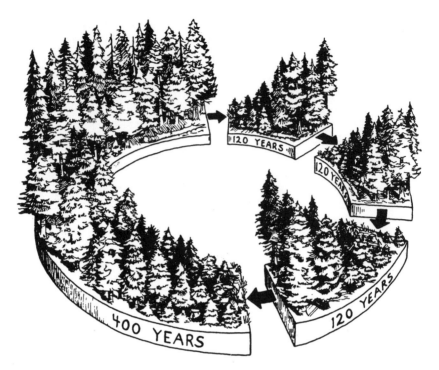

FIGURE 97   After two or three cycles of cutting plantations of Douglas-fir ranging from 80 to 120 years of age, an extended rotation of 300 to 500 years is put in place and allowed once again to become old-growth forest. Such an extended rotation allows the biological processes to heal and also allows the reinvestment of biological capital in the soil to recapitalize the organic material and the soil's savings account of available nutrients in the soil. With the completion of the extended cycle as a forest, the land can be cropped once again in two or three short cycles as plantations, before the next extended rotation as a forest is necessary. (Reprinted with permission from Hunter, M.L. Jr. 1990. *Wildlife, Forests and Forestry,* Prentice-Hall, Englewood Cliffs, N.J.)

product is considered an economic failure. Such management is mutually exclusive of virtually all other human values because rigid constraints of time and a single-species monoculture define the plantation. Spotted owls, for example, cannot survive in a continuous 80-year-old plantation. Water quality and soil fertility are severely impaired with the cumulative effects of erosion (both chemical and physical), compaction, and the loss of soil organic matter, such as large woody debris. In addition, the human values of recreation and spiritual renewal are also missing.

Sustainable forestry, on the other hand, relaxes the constraints of time and a

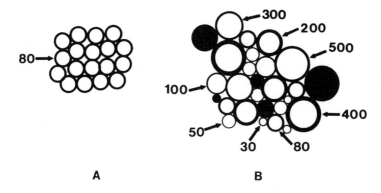

FIGURE 98   The circles represent forest stands. Open circles are managed stands. Solid circles are living laboratories of unmanaged stands of various ages that have been set aside in order to learn sustainable holistic forestry. Numbers represent age at harvest of managed stands. (A) Plantation management where age of harvest is 80 years. Such plantation management is mutually exclusive of almost all human values except fast-grown wood fiber and is not sustainable on a continual basis. (B) Sustainable forestry. Note that plantation management for fast-grown wood fiber (stands 30 to 100 years old) is a part of sustainable forestry. When constraints of time and single-species monocultures are relaxed (stands 200 to 500 years old), the forest can replenish its soil nutrient and process capital; restore the quality and quantity of ground water, streams, and riparian zones; recreate spotted owl, pileated woodpecker, and marten habitats; and simulate old-growth forest conditions for human recreation and spiritual renewal. Throughout the process of restoring an ecologically functional forest, the quality and quantity of wood fiber are maintained on a sustained basis for a diverse market. Public user fees and fees for hunting and fishing can generate revenue annually prior to timber harvest.

single-species monoculture on forest acres. The result is that all values exist on some acres all of the time (Figure 98). Whereas intensive plantation management is cost-deferred, short-term, single-product exploitation, sustainable forestry is a pay-as-you-go sustainable forest that simultaneously produces a multitude of products and amenities over centuries.

One of the reasons we appear to have so much difficulty grasping long-term ideas, such as sustainable forestry, is the brevity of our lives. If our lifetimes are so short, how can we learn to manage such long-term forests? There are several options.

One option is to form a small group of a highly qualified mix of forest scientists, managers, economists, and sociologists and send them around the world for a year or two to study special cases and common denominators of world forestry practices and problems. Their mission would be to figure out what works, what does not, and why. The U.S. Soil Conservation Service learned much about soil erosion by using this approach in the 1930s.

Another option is to determine the current most pressing ecological management problems and then hire a highly skilled group of people per problem to gather all the literature and write a synthesis. This would be a thorough, integrated, up-to-date report of what we think we understand and what we think we need to learn. We could literally catapult ourselves forward in the knowledge of forests and forest management (or other management problems) if we did only these two things.

Simultaneously, of course, we can continue research, both the usual short-term research and innovative long-term research, such as the 220-year study of decomposing logs in the H.J. Andrews Experimental Forest in the Cascade Range of western Oregon.[96] Long-term research projects can also be designed and executed on public lands (and private lands for that matter) as a carefully monitored part of ongoing management practices. What we do today could be interpreted two, three, or more generations from now; this is the only way some critical answers will be derived.

Still another approach to research is to consider Christmas tree farms as the simplest form of forest plantation management. We will learn much about the forest if we can make Christmas tree farms sustainable over time. By studying both ends of the forest—old growth (the most complex) and Christmas tree farms (the simplest)—we should learn much about the hidden thresholds of forest processes that drive the system. Much research also needs to be done on the ecological sustainability of young-growth plantations managed for wood fiber over one, two, or three short rotations.

The richest and as yet the healthiest conifer forest in the world is in western Oregon, western Washington, and southwestern British Columbia. The climate is benign, the air and water are relatively clean, the soils are fertile, and the society has had a severe impact on the forest only within the last century. In addition, the ecology of this forest is probably better understood than that of any other forest in the world. All this gives us the time, if we are committed and we begin now, to learn how to sustain a forest through sustainable forestry (Figure 99).

In sad truth, we may have the *only* chance left in the world! Only when we have accomplished this can we call ourselves "foresters." Only then can we help the people of the world to restore the health of their watersheds, produce a sustainable variety of forest products, and retain their other amenities.

Sustainable forestry is a way to correct our past errors. It will work because the very process of restoring the land to health is the process through which we become attuned with Nature and, through Nature, with ourselves. Ecological restoration is therefore both the means and the end, for as we learn how to restore the land, we heal the ecosystem, and as we heal the ecosystem, we heal ourselves.

FIGURE 99   Sustainable forestry is possible while maintaining carefully and well-distributed stands of Nature's old-growth, mid-age, and young-growth trees across the landscape, which can serve as a model in learning how to sustain the forest.

We also simultaneously restore our options for both products and amenities from the land and we perpetuate these options for the generations that follow. It is not that we do not know enough to grow sustainable forests; we in the United States have simply chosen not to.

16

SUSTAINABLE FORESTRY THROUGH ADAPTIVE ECOSYSTEM MANAGEMENT IS AN OPEN-ENDED EXPERIMENT*

RATIONALE BEHIND THE EXPERIMENT

Scientists can describe, albeit imperfectly, what ecological conditions seem necessary to sustain the production of given commodities, but they cannot describe a management system to sustain the notion of ecosystem integrity. If ecosystem integrity means that all the components necessary to maintain a desired condition are intact and functioning "normally," scientists would need to know what those components are and what normal functioning means with respect to the particular condition.

Scientists feel sure about such critical components of ecosystem integrity as soil, water, air, sunlight, climate, and biodiversity, and they know much about others, such as nutrient cycling and the importance of fungi and insects in releasing nutrients. In addition, scientists are increasingly aware of such things as the effects of fire suppression and toxic substances on the health and sustainability of forests. And they also know much about a variety of organisms that are useful or even essential to the functioning of a forest, but little or nothing about many others.

---

*By Chris Maser, Bernard T. Bormann, Martha H. Brookes, A. Ross Kiester, and James F. Weigand.[97]

Some years ago, for example, five people (including Maser) were studying the survival of nitrogen-fixing bacteria in the guts of small mammals and the effect of nitrogen fixation in soil once the bacteria passed through the mammals' intestinal tracts.[98] The first problem they encountered was an initial inability to isolate pure bacterial cultures and, once isolated, to identify them because much of the taxonomy had not been done. The same problem arose with yeasts interacting with the nitrogen-fixing bacteria within the mammalian intestines. After scouring the United States for someone who really understood yeast classification to help them, they finally found a hospital administrator whose hobby was yeast taxonomy.

When the studies had been completed and written up, they had a terrible time getting the results published because they had bridged five distinct scientific fields in their interdisciplinary approach to the investigation. In one paper, for instance, they stated that, to the best of their knowledge, this was the first time nitrogen-fixing bacteria were found in association with the tissues and spores of mycorrhizal fungi and yeasts, all interacting and surviving passage through mammalian intestinal tracts. When they submitted the paper for publication, one referee (a microbiologist) rejected the paper, saying: prove it—cite another study with the same results!

All but one journal rejected their studies out of hand because, being interdisciplinary and thoroughly integrated, the studies could not be "pigeonholed." They were told by more than one journal that their studies were too complicated; the individual threads could not be traced by reviewers, and they were accordingly instructed to separate the pieces. When they complied and resubmitted the papers, they were told that their work was nonsense. With this kind of thinking, is it any wonder that we know so little about so much and so much about so little?

There are, nevertheless, clear indications that ecosystems are dynamic and constantly changing or becoming "degraded" by some human standard of commodity production. Some indicators are recognized, such as loss of biological and genetic diversity, disappearance of humus, altered hydrologic cycles, and increased pollution and erosion. However, we are still unaware of many subtle processes or tiny, ubiquitous organisms, the loss of which can alter the sustainability of an ecosystem, even under well-intentioned management.

It is therefore impossible to assign either biological or societal standards to ecosystem integrity because it has not been determined what constitutes the "health," or "wholeness," or "sustainability," or even "stability" of an ecosystem. In addition, because ecosystems change constantly, with or without human activities, future conditions cannot be predicted. We can only treat an ecosystem in some manner and hope it will respond in a way that is favorable to us. Ecosystem sustainability can thus only be approximated as a component of a dynamic landscape, the evolutionary adaptability of which is both changing and being affected by our environmental tinkerings.

British philosopher James Allen said it all when he wrote:

> Mind is the Master power that moulds and makes,
> And Man is Mind, and evermore he takes
> The tool of Thought, and, shaping what he wills,
> Brings forth a thousand joys, a thousand ills:—
> He thinks in secret, and it comes to pass:
> Environment is but his looking-glass.[99]

"Management": A Combination of Ecological Principles and Societal Values

It is imperative to understand that "management" to many people connotes "control," but we *cannot control* any ecosystem, not even an urban one. The connotation of management does not, however, need to be controlled. Therefore, the use of management as a concept throughout the rest of this chapter really implies that we are *manipulating* some portion of the ecosystem in an attempt to cause it to behave in a particular way that we find satisfying to our societal values. Management practices and societal values, demands, and concerns are thus closely interrelated.

Consider, for example, that the high sustained rates of timber harvest, livestock grazing, and water diversion have been demanded for high profits as well as cheap wood, red meat, wool, water, and food. Concern also comes from citizens convinced that management practices, such as prescribed fire and its smoke or applications of herbicides to control unwanted vegetation, are harmful to them and/or the environment.

Biological and societal components of forest issues cannot be separated. Society and forests are inexorably intertwined, and a wide variety of human influences and unpredictable events on all scales, from the local area to the global commons, compound each forest issue. In addition, "health" and "sustainability" are human concepts, and ideas differ among people. Nevertheless, in order for ecosystem management to be a viable concept, it must rest on the fundamental belief that people, as an inseparable part of the system, can contribute to or detract from ecosystem sustainability through their behavior—for which they are accountable.

Placing total responsibility for forest health and sustainability on the forest itself, its processes, and the inadequacy of forest science ignores the underlying issues, the triggering mechanisms of forest decline embedded in society's failure to understand and/or accept the limits under which ecosystems can produce the goods, services, and conditions society demands. But the crux of the entire problem—both historically and now—is society's failure to take responsibility for understanding and controlling itself. Analyzing the options available to society for directing its own behavior in promoting ecosystem health and sustainability is at least as important as, if not more important than, research in natural sciences and land management.

Fundamental changes in economic motives, thinking, and behavior; changes in societal institutions; the understanding, acceptance, and application of scientific information; and changes in society's psychological and philosophical underpinnings are all necessary to realize sustainable ecosystem management. How people, especially those entrusted with land management, have contributed to the declining health of our forests must be analyzed, understood, accepted, and acted on. This includes imperfect understanding of and/or willfully ignoring social and ecosystem processes, the resulting unintended consequences of decisions, and resisting the acceptance of changing social values.

Given a clearly expressed vision and compatible goals, natural science can provide assessments of which, if any, management alternatives are likely to be ecologically sustainable and the possible consequences of those that are not. There is a continuing proviso, however: all management concepts and practices, no matter how attractive, must be viewed as they really are—experimental. It is paramount, therefore, that close attention be paid to their effects and, if necessary, modifying the offending concept and/or practice as results are collected and interpreted.

Many people are concerned about and distrust the motives guiding the decision-making process. Many feel they have no real voice in creating the alternatives and in choosing the one selected. This frustration includes local people as well as those from across the country.

Successful ecosystem management requires a continual process through which conflicts can either be avoided or resolved. In either case, the process of creating a vision toward which to work collectively transcends conflict. Keep in mind, however, that conflicts are moving targets, conditions in the stream of life. One no sooner comes into view and is swept away when another comes 'round the bend, and each in its turn stimulates debate.

This continual debate is essential in a democracy. It not only gives people a voice and defines the framework in which ecosystem management will take place, but also nurtures the continual growth in social consciousness and the evolution of core values.

Sustainable Forestry Is More than a Local Issue

The organizational structure of land-management agencies has led to their dysfunctional aspects and is not conducive to change. In dysfunctional agencies, homeostasis prevents the adaptation of any management activities, local or otherwise, if such activities are perceived as a threat to an agency's survival. Public policy, formed in the political arena, is a primary tool through which homeostasis is maintained in dysfunctional agencies and in the minds of many in the Congress of the United States, which sets its own policy and that of land-management agencies.

Because of increasing congressional and agency dysfunction, people are becoming more insistent that their voices be heard in the decision-making process affecting land management. As the human population increases, diversifies, and moves from one area to another, conflicts accelerate over what should be done, where, when, how, and why. Neighbors disagree, communities are divided, voices become shrill as tempers flare, and lobbyists and lawyers prosper. Scientists increasingly find themselves arguing questions of science in court and before congressional subcommittees, in a political arena so charged with emotion that rational logic is all but precluded.

Effective public communication of both scientific and management concepts is essential, however, in building rationally informed communities. As these communities become educated and grow, it is imperative that their ideas are truly and justly heard, understood, and incorporated in both good management alternatives and good choices among them.

In addition, scientists must transcend their specialties and learn to communicate effectively among disciplines, especially between social and natural sciences. Finally, natural scientists and social scientists must both contribute to formulating public policy within a framework of a future-oriented ethos while fully recognizing the influence of their own personal and professional biases. (Future-oriented ethos, as used here, is the uniquely human morality that is consciously acted on in the present for the benefit of both the present and the future; it is the necessary protective weave of our cultural fabric if society is to evolve in a sustainable manner.)

Despite society's best efforts at planning, however, uncertainty and unpredictability of both ecological events and changing societal values affect ecosystem sustainability when it comes to producing desired commodities. The risk of catastrophic change, both ecological and social, underlines the necessity to maintain options for future generations as a primary goal in managing ecosystems—including forests. Whether or not to maintain options is "our" choice, meaning this generation at this time; however we choose, we must do so consciously and know what our decision is, for we will be held accountable as we have held those before us accountable.

Maintaining options is paramount because people's values and behavior, and thus their necessities and desires, change both within and among generations. Considering human behavior at multiple geographic and temporal scales expands the kaleidoscopic variation of human interactions with ecosystems beyond the bounds of easy analysis.

Although people believe they have choices in how they influence ecosystems, society has seldom elected to treat ecosystems in a way that even ensures the production of what was originally identified as desirable. More often than not, society has failed to see the consequences of its own expectations from forests until the biological capacity of a forest to fulfill societal demands (not to mention necessities) is severely compromised and the cumulative effects of human actions produce a crisis.

Society's approach to these crises is to overreact. A widely held assumption is that crises, especially big ones, will disappear if both science and money are "thrown" at them in an effort to find the magic technological quick fix that will preclude society from having to change how it behaves. What is really needed, however, is natural scientists, social scientists, citizens, and policymakers all working together in an effort to resolve the crisis within the ethos of the democratic social system and the biological capabilities of the ecosystem.

Despite the human capacity to perceive major problems and conduct research, analyze data, and apply information to natural resource policy in the United States, the responses of policymakers have seldom been timely, appropriate, or adequate. This is primarily because constraints have been imposed to protect the dysfunction of both our flawed economic theory and/or practice and the survival of our dysfunctional agencies and Congress.

Delay in responding to crises is not the result of inadequate expenditure of either money or effort. Rather, it is the dazzling array of dysfunctional coping mechanisms that are marshalled to maintain informed denial, while keeping the connections between science, society, and policy both nebulous and tenuous, which is possible because of our confused use of language.

A Clarification of Terms

For sustainable forestry to work, we must both understand and accept that a forest is but one component of the landscape. The landscape is ultimately a social landscape if for no other reason than we humans have affected and therefore altered the entire planet. However, before we can step into the future with a reasonable assurance that a sustainable society and therefore sustainable forestry is at some level possible, we must renegotiate our frame of reference, as symbolized by the phonetic language with which we communicate.

To move forward demands a clarification of terminology, because our current definitions of basic concepts have brought us as far as they can. The following terminology reflects the best expression of that which we think we understand.

Ecosystem is a system of organisms and their physical environment interacting so that a flow of energy leads to a clearly defined trophic structure (foodweb) and material cycling among the nonliving and living parts of the system. Ecosystem boundaries are designated to address specific concerns; an ecosystem can thus be as small as the surface of a leaf or as large as the Earth and beyond.

Through movement of energy and material across boundaries, ecosystems affect and are affected by other ecosystems. Because people directly and indirectly influence all of the Earth's ecosystems, and because people obtain sustenance from and make demands on ecosystems, people are recognized as both important and inseparable parts of all ecosystems. Societal processes are thus important ecosystem mechanisms. (See *societal processes.*)

Ecosystem approach means that the "system" in ecosystem embodies three fundamental concepts: designating the physical boundary of the system and its components, understanding the interaction of its parts as a functioning whole, and understanding the relation between the system and its context. Context in this sense means both the external factors that influence the system as well as the internal information that must by synthesized at the scale of the defined system if we are to have any understanding of it. For a continental ecosystem, global air pollution and population growth are examples of external context and local politics and endangered species are examples of internal context.

Ecosystem management is a system of making, implementing, and evaluating decisions based on the ecosystem approach, which recognizes that ecosystems and society are inexorably linked and always changing. Ecosystem management is based on the understanding that ecosystems are constantly changing independently of society and human influence.

Disturbance, succession, and natural selection are important mechanisms of change within ecosystems; parallel mechanisms create change in society. Ecosystem management must therefore be able to incorporate change in societal processes, in our understanding of natural and social phenomena, and in the capacity of a system to produce that which society desires.

Thus far, management is still largely focused on individual parts of an ecosystem (such as timber, water, forage, game fish, big game, and wilderness) in relative isolation from one another, with little attempt to see how the parts fit together as a functional whole. Ecosystem management treats these parts as interrelated components of an interactive system, rather than just as independent and isolated variables.

Ecosystem sustainability, on the one hand, is the degree of overlap between what is ecologically possible and what is socially desired by the current generation and, on the other hand, is the ability of both the ecosystem and the management system to adapt to changing environmental conditions over centuries (Figure 100). Within this concept, the current generation must protect both the social and ecological options for the future, because both will change in some related and probably unpredictable fashion.

Managing for ecosystem sustainability encompasses the following ideals: (1) providing a continuing balance through which society obtains a desirable yield of goods, services, and conditions from an ecosystem in the present, while maintaining the ecosystem's capability to produce desirable yields of goods, services, and conditions for society in the future; (2) seeking to increase overlap

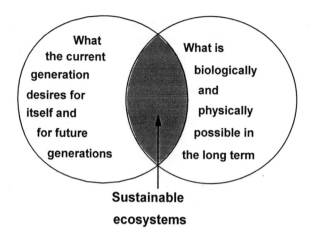

FIGURE 100   Ecosystem sustainability is the degree of overlap between what is ecologically possible and what is currently desired by the present generation. The degree of overlap is dynamic because both ecological capacity and societal values are continually changing. (From Bormann et al.[109])

between societal desires and ecological possibilities wherever and whenever possible; (3) developing a method for making a quantitative, objective assessment of sustainability for producing very different ecosystem products from very different ecosystems over varying periods; and (4) restoring ecosystems, as necessary, to some former productive capacity so they can sustainably produce socially desirable good, services, and conditions.

To manage an ecosystem sustainably, people must come to terms with continually conflicting changes in societal values and must alter the way society makes decisions about the goods, services, and conditions it demands. Under ecosystem management, society must focus on the conditions necessary to produce a set of goods, services, and conditions sustainably rather than focusing on the goods, services, and conditions themselves strictly for the present.

Achieving a desired condition for an ecosystem will place constraints on the types, combinations, and quantities of obtainable ecosystem products as long as there is growing resource scarcity caused by unbridled consumption and an exploding human population. Sustainable ecosystem management addresses and deals with the problem of ensuring that an ecosystem can deliver as yet unspecified and even unknown products in the future for the future.[11] (See *Global Imperative*[11] for a thorough discussion of sustainability.)

The following steps are required to achieve sustainability in ecosystem management: (1) select candidate goods, services, and conditions desired by society; (2) determine ecosystem patterns and processes thought to be necessary for producing the products in question; and (3) simultaneously evaluate and set priorities among societal demands and ecosystem patterns and processes.

Ecosystem health is qualitatively synonymous with ecosystem sustainability.

Ecosystem degradation is the reduction in a system's sustainability with respect to a socially desired condition.

Ecosystem restoration is the return of a degraded system to an approximation of a former socially desirable condition.

Products are all possible goods, services, and conditions that society desires from an ecosystem. Goods include such commodities as lumber, and services include such things as recreational opportunities and clean air. Examples of conditions are attractive landscapes and abstract entities like biodiversity. Production is the creation and flow of products. Conditions result from ecosystem processes and societal values and actions.

Societal processes in the sense of sustainable ecosystems are how people collectively regard, affect, and are affected by ecosystems.

These are but a few of the terms that need to be updated as we move into the next century. The importance of clarifying the language we use is too often underrated and dismissed as "semantic drivel." In reality, however, it is the lifeline of communication and thus the survival of society as we know it. In addition to clarifying terminology, there are a number of fundamental premises that must be taken into account in outlining the experiment.

FRAMING THE EXPERIMENT

Emphasis in land management has thus far focused on defining management practices (such as fire suppression, livestock grazing systems, and silvicultural treatments) to produce immediate goods with the greatest possible profit margin, as well as desired conditions for increased production of goods in the future. The effects of commodity management on society have been little studied, even though society is the customer for ecosystem products and the inheritor of future environmental conditions.

The current management system also has multiple externally and internally conflicting influences. The resolution of these conflicts, through decisions of trade-offs, often starts as top-down directives. Determining which goals are achievable, with which trade-offs, is difficult in managing large, complex systems. It is not even possible to anticipate and plan for all outcomes when designing a complex system, especially in ecosystem management, with its many unpredictable elements.

Attempts to plan and control products may work for simple production systems through a top-down decision pyramid, where expected procedures and outputs are determined at the top and expanded downward through a bureaucratic chain of command. Such management leaves little room, however, for analysis by the "hands-on" people at lower levels in the chain of command, which causes them to use their creativity and produce "downstream" compensations for system errors. Top-down planning is therefore neither efficient nor effective and does not allow flexibility in correcting errors or accommodating changes in societal values.

People can judge the feasibility, effort, and trade-offs of achieving various goals and objectives only by physically working with the resource itself. Bottom-up knowledge must flow to the decision-makers at broad geographical scales in a repetitious process so that the correct balance can be maintained with respect to what is desired, what is achievable, and what the costs, benefits, and trade-offs are.

Four strategies will help integrate the top-down and bottom-up aspects of management:

1. **Manage simultaneously at multiple geographical scales.** National goals can be achieved when regional and local feedback on feasibilities and trade-offs is taken into account. In turn, local goals can be achieved when regional and national feedback on feasibilities and trade-offs is taken into account.

2. **Manage with trust.** Especially in ecosystem management, specific techniques to achieve goals and objectives will vary in time and space with biological, social, and technological conditions and knowledge. Management decisions must therefore be made at the appropriate scale.

   Although the specific goals of managing for certain ecosystem products need to be broadly stated, means of achieving them will differ dramatically. Management can remain efficient and effective only if it is flexible enough to change methods of achieving goals as conditions change.

3. **Control the management system itself.** Ecosystem management can only be maintained if management itself remains both efficient and effective. If the system contains many ad hoc downstream compensations, the compensations themselves will inhibit both efficient and effective management as conditions change.

   A primary focus of management must be self-control. If self-control is designed into a management system and is working efficiently and effectively, then, by definition, the system will produce the desired and expected outcomes.

4. **Keep policy-makers informed.** To ensure realistic decisions about the consequences, trade-offs, and efforts spent to achieve various stated goals

and objectives, policymakers, along with society, managers, and scientists, must communicate and keep one another informed. In addition, trade-offs can be incorporated into legislation by setting priorities for possibly conflicting goals or limited funds.

Developing a theory and appropriate practice of land management for ecosystem sustainability demands that people understand how they regard, affect, and are affected by ecosystems. Because people are an inseparable part of the forest, defining goals for ecosystem management is a societal process, which means that sustainable forestry requires social science expertise combined with knowledge of ecological interactions. Societal interactions, along with science and technology, form the policy process that translates acceptable goals and practices into ecosystem management.

Technology, often considered a manifestation of science, is best thought of as an outworking of the interaction between society and science. The multiple (and often unexpected) effects of technology on society are similar to those of science, but usually more striking. Changes in technology, especially information technology, can, for example, alter societal demands by allowing people to recognize an increasingly wide range of available choices through such strategies as reducing costs and providing substitutes.

Technology, however, has unexpected effects that may be either beneficial or detrimental to a wider set of goals than the one it was originally designed to address. Although the effects of a given technology on an particular ecosystem may be direct if unanticipated, its effects on society may be difficult to trace. The net outcome may be to change society's attitude toward technology as a whole, a change that must be incorporated into ecosystem management.

Many people believe that science and technology can overcome whatever has gone awry in ecosystems, but both are oversold as cures for environmental problems. Ecosystem managers must understand their limits and place both—but especially science—in the broader context of all activities leading to ecosystem management.

Although no one fully understands ecosystem management, the following premises (summarized in Table 2) will help clarify our current view of the relation between people and the ecosystems of which they are a part. (To understand Table 2, integrate the columns both vertically and horizontally.) The premises constitute a framework for understanding the origins of societal goals for ecosystem management and form a basis for evaluating the breadth and durability of popular support for those particular goals.

They also form a basis for evaluating the social effects of practices used in ecosystem management. Such ongoing evaluation is necessary because, in the final analysis, the land and people are one. As the land prospers, so the people prosper; as the land suffers, so the people suffer.

TABLE 2   Translation of Fundamental, Societal, and Scientific Premises into Principles for Management for Sustainable Ecosystems

Fundamental premises	Societal premises	Scientific premises	Management principles
Society and science are both influenced by individual and societal values, sometimes recognized but often forgotten.	People choose goals for future ecosystem goods, services, and states based on their perceptions of need.	Natural sciences recognize people as part of the ecosystem; social sciences recognize biological and physical constraints.	Manage. No management is poor management, although minimal management activity may be desirable.
	People group themselves into communities of common interests to advance their personal goals.	Science has roles both as forecaster and as conscience.	Use an ecosystems approach; integration of the parts of a system and its relation to its context are keys to improved efficiency and to understanding the system as a whole.
	Diversity, complexity, and the changing nature of human communities create uncertainty about demands and priorities, both now and in the future.	Limitations of science must be understood; predictions should focus on ranges, rather than on specific outcomes.	Consider information a primary resource; success of a management system depends on a decision process that includes adequate societal and scientific knowledge.
Good can come from management.	In a society of diverse and sometimes discordant communities, too many goals or conflicts over goals may develop; some people may not get what they want.	Ecosystems are artificial constructs and therefore must be selected and classified with caution.	Develop communities of interest. A process is needed to draw all interested communities together into a larger planning community to develop and implement management policy.

Ecosystems are fundamentally complex and inherently difficult to predict.	Ecosystem management decisions benefit from societal goals that are clear, informed, and integrated and that recognize that ecosystems are complex.	Diversity is essential to adaptability—of organisms, rural and national economies, and public employees.	Apply general information carefully; considerable specific information and involvement of local human populations are needed to effectively implement management.
The entire system must be understood in its context; study of systems must include external factors and embedded processes.	Scientific and societal information are integrated and codified in laws, treaties, bureaucracies, planning processes, and budgets.	Ecosystem patterns and processes appear, and must be studied, at different geographic and time scales.	Manage across boundaries; to the extent that adjacent landowners have a common vision for ecosystem management, achieving multiple objectives will be easier.
	Policy will always be decided under conditions of uncertainty and ignorance.	Local conditions may override or obscure general patterns and processes; the general may not contain the particular.	Manage for change; diversification is the primary method for reducing risks of unexpected changes in future ecological conditions and societal demands (maintain options for future products).
	Unexpected ecosystem events and shifts in societal demands may require changes in societal institutions, and these changes will bring changes to society.	Ecosystem science at large scales relies on ecosystem management for empirical evidence; small, short-term studies do not extrapolate well.	Manage as an experiment; knowledge is needed from each management action by using a scientific approach that describes anticipated outcomes of an array of treatments and compares them to actual outcomes.

Scientific Premises in Ecosystem Management

Science can describe to some degree what happens in ecosystems, why it happens, when it happens, where, and how much. Based on such information, science can describe constraints on ecosystem management, but it cannot specify management actions in precise detail.

Relatively speaking, we know much about ecosystems and their components, but ecosystem management must be founded on the integration of a wide array of natural and social science disciplines. To this end, the following premises form the integrated basis for ecosystem management.

Premise 1. Both science and society are influenced by individual and societal values that are sometimes recognized but often forgotten

Ecosystem management requires greater participation and more scientific information than does managing for intellectually isolated commodities and benefits. A critical component of any approach to ecosystem management is recognizing how the beliefs and biases of citizens and scientists influence their willingness to participate in land-use planning and their ensuing effectiveness.

An understanding of these beliefs and biases lies partly in basic societal and scientific assumptions that need to be acknowledged, examined, and reconciled if the required societal consensus for land-use management, especially on public lands, is to be forthcoming.

Societal assumptions: What individuals and groups want and demand from ecosystems is understandably diverse, but the closer one comes to the absolute necessities of life, such as available drinkable water, the greater is the societal consensus. In these times of growing resource scarcity, it is essential that the basic societal necessities are both recognized and planned for prior to attempting accommodation of societal desires and demands, which really mean "wants."

The first step in understanding and transcending the conflict of what is actually necessary and what is desirable in a social sense is an open discussion leading to a common set of premises under which the individuals and groups agree to operate. There are a variety of conflict-resolution and consensus-building strategies that can support a visioning and goal-setting process for ecosystem management. Using such a process is critical because the ability of land managers to function in a professional manner requires a clearly defined vision and set of goals.

Scientific assumptions: Scientists also operate with sets of underlying assumptions, although people (most often scientists themselves) are less able or willing to recognize and accept this subjective "human" quality.

Science has many disciplines, each with a set of underlying assumptions that often are passed unquestioningly to new generations of students. Although basic scientific assumptions can be corrected, it is difficult to correct the subsets of assumptions unique to each discipline. Ecosystem management requires information from a multitude of disciplines, which to be useful must be integrated. Differences in the subsets of assumptions are often serious barriers to such integration.

Science is divided into natural or ecological science and social science. Natural science is split into botany, zoology, geology, physics, and chemistry, which are further divided into such subdisciplines as plant pathology, plant physiology, entomology, herpetology, ornithology, mammalogy, geomorphology, hydrology, and so on. Social science is split into economics, sociology, history, and psychology, which are further divided into such subdisciplines as microeconomics, macroeconomics, and so on. Each discipline, and even its subdiscipline, is then fragmented into schools of thought and various contending camps with differing perspectives, each valid under its own paradigm.[100]

The size, complexity, and especially the internal fragmentation of science usually preclude the kind of research necessary for ecosystem management, which focuses on wholeness, integration, and simultaneous consideration of data and concepts from all these various disciplines.[100] Although integrated research has often been tried, it commonly fails because disciplines use different vocabularies, work at differing scales of space and time, often have different funding sources with vastly different economic interests, and fail to understand the appropriate subsets of underlying assumptions. In addition, scientists—stimulated by the current reward system—are more inclined toward interpersonal and interdisciplinary competition for funding and recognition than they are toward cooperative, coordinated team efforts, where egos must be put aside for a common goal.

For ecosystem management to work, scientists must step across disciplinary boundaries and undertake the truly interdisciplinary research that ecosystem management requires. They need to challenge their own biases, expose and question the assumptions of their own disciplines, and demand that their colleagues in other disciplines do the same.

At the same time, inter- and intra-university administrations must mature into integrated programs as the foundations from which socially responsible ecosystem management can grow. Such maturation once again demands putting aside egos and revamping the award system to stimulate interdisciplinary collaboration across broad sweeps of disciplines and universities.

Beyond this, there is a widely perceived notion of science as value-free and completely objective, but it is only an illusion. Scientists are as subjective and biased as anyone else, which means that science is also biased. Ultimately, all science, whether "data-based" or a matter of "expert opinion," is an article of faith.[11]

If society has the courage to take science off its pedestal and allow—even insist—that scientists acknowledge their subjectivity while they strive to be as objective as possible, science will be far more honest and relevant than it is today. In order for ecosystem management to work, science must be as relevant as possible, which means interdisciplinary work must be based on revisiting underlying assumptions. Only then will society and its environment come into a more harmonious and sustainable relationship.

Premise 2. Natural sciences recognize people as part of the ecosystem; social sciences recognize biological and physical constraints

People choose their use of scarce resources, and these choices are both measurable and interpretable. The study of these choices, through surveys and other methods of inquiry, provides an initial estimate of what society wants and can help planners formulate a preliminary set of options for societal goals for ecosystem management. The actual choices, however, must be made through political processes.

Regardless of people's choices, physical and biological laws determine what is possible. Therefore, general rules derived from fundamental theories and scientific laws must be used to forecast future ecosystem conditions. The idea is to have robust guidelines, which specify when an ecosystem is probably being used beyond its sustainable capacity.

We are only now beginning to identify the rules that may be useful in ecosystem management, such as the limits imposed by physical laws, which are impossible to avoid. No management action is going to repeal the law of gravity, for example, and any plan that implies an outcome in violation of this law is easily rejected. By the same token, the law of gravity gives no positive counsel on how to manage an ecosystem. We therefore need principles for which violations are more or less obvious and which give positive hints on how to manage sustainably:

1. **Conservation of matter and energy:** The laws of conservation of matter and energy set up an accounting framework for the physics and chemistry of ecosystem processes. However, it is important to realize both that the *dynamics* of matter and energy are embodied in all ecosystem components and that the matter and energy contained with any ecosystem are finite.

2. **Laws of thermodynamics:** The laws of thermodynamics play an important role in ecology because they set limits on the efficiency of energy

transfer. At each step of energy transfer within food chains, the efficiency is only about 2 to 20 percent.

3. **Fundamental theorem of adaptation:**   The ability of plants and animals to adapt to changing conditions depends on genetic diversity and natural selection.

4. **Limitations imposed by forever imperfect knowledge:**   Any assessment and management of social desires and ecosystems relies on sampled data. The laws of statistics, when used with a great deal of humility, can provide guidelines as to whether and when we can actually collect enough data to make reasonable statements about ecosystems. Here we must remember, however, that averaging hides the truth about the role of variability in designing ecosystem dynamics.

Implicit in the preceding list is the need to continually question the underlying assumptions of all science. In addition, if we conceive of ecosystem management as setting and testing hypotheses, we must use our intuition as the final reality check before making decisions and acting on them.

Premise 3. Science has roles as both forecaster and conscience

It is increasingly stated that forest management must be founded on scientific principles, which puts science more and more in the role of both condition forecaster and social conscience. In its role as forecaster, science provides the machinery necessary for predicting the future condition of an ecosystem, given knowledge of its present state and any management practices applied to it. If, however, ecosystem forecasting is to be possible, we must have a theory for defining ecosystem components and the dynamics among those components, as well as the data necessary to estimate parameters.

One reason why ecology does not forecast well is that the models used to predict ecological conditions are linear, while the ecosystem itself is cyclic. Another reason why ecological conditions are difficult to forecast is that most causes have multiple effects and most effects have multiple causes. Thus, even when a correct prediction is made, we have no way of knowing whether to interpret it as insightful forecasting or apparent coincidence.

In addition to being forecaster, science must play an increasingly important role as social conscience, guiding society to make ecologically sustainable decisions for present and future generations. Science must continually make people aware that ecosystems operate on a series of thresholds and will collapse (even if the exact mechanism or timing is not understood) if pushed too hard in an insatiable, myopic drive for short-term products.

Premise 4. The structure and use of science limits its application for ecosystem management

Ecosystem management is more like managing a stock portfolio than an oil refinery. Stock values are uncertain, but an oil refinery can be made to produce predictable products in predictable amounts. In ecosystems as in stock markets, the very process of setting goals must consider the element of uncertainty.

The practice of ecosystem management must therefore develop tools for managing uncertainty, including forms of hedging and arbitrage. Ecosystem processes are somewhat predictable, however, and management can be tailored to each circumstance when it is understood that the world is a network of relationships rather than a collection of objects.

An advantage of recognizing the cautions of limited predictability in goal setting is that expectations of what managers can actually accomplish are more realistic, and managers therefore are more likely to succeed. The important part of recognizing and accepting uncertainty is that managers must, through good information and frequent interactions, "manage" the expectations of those concerned with the results of ecosystem management. They must also ensure that the public understands that a goal for ecosystem management is more like a constantly changing cloud than a fixed point.

Premise 5. Ecosystems are artificial constructs and therefore must be selected and classified with caution

In the absence of a strong theory of ecosystem evolution, the definition of an ecosystem is subject to wide variation, depending on the perspective of the definer. Although we have no set criteria for defining ecosystems per se, we can develop criteria for defining them operationally to help set and achieve management goals.

We do, on the other hand, have a strong, complex theory about the constitution of species and how they evolved, which makes the species an important component in ecosystem management because it is understood as a critical unit in the evolutionary process. Nevertheless, the concept of species, like the taxonomic entities above species (genera, families, and so on) and below species (subspecies), must always be recognized and accepted as working hypotheses.

In fact, we think it safe to say that, because our understanding of ecosystem components is only in the form of working hypotheses, ecosystem management itself is an open-ended experiment in which both the identification of components and their dynamics beg continual investigation. Experimental results therefore imply something about changes in ecosystem components and something about the usefulness of considering their actual existence.

Forecasting requires identifying the ecosystem variables we are interested in and then devising a model with which to forecast them. The problem lies in knowing which variables to choose. Some, such as chemical elements and species, have well-developed theoretical bases that tell us how to identify and separate them from all other variables. Others, however, are the result of artificial classifications without theoretical bases, and their separation is nebulous and/or temporary at best.

An example of the limitations of artificially created ecosystem variables is a clearcut that is planted in a single-species monoculture in the midst of an all-aged, mixed-species forest. Although the forest itself cannot be subdivided into obvious natural stands, the process of clearcutting and establishing one or more monocultural plantations transforms the forest into a set of obviously artificial stands. Clearly, this form of classification, which lacks any grounding theory, produces temporarily identifiable entities but simultaneously reduces the possibility of alternative definitions.

Premise 6. Ecosystems are fundamentally complex beyond our understanding and at best difficult to predict

The science of evolution focuses on the variability of Nature and how that variability is generated and maintained. Much of that variation has a large, often dominating, and seemingly random component. Most, if not all, ecosystem processes vary incredibly over scales of space and time. Thus, a world view, an understanding and acceptance of ongoing change and the uncertainty it fosters, is necessary for ecosystem management.

Even when variability in Nature seems restricted, unpredictability is high and surprises are limitless. With this in mind, statistics must be applied with a great deal of humility to any given circumstance. Humility is important because most often it is the variability of the variability that designs the process and thus the system in question, as discussed earlier in relation to fire and ponderosa pine regeneration. As previously stated, there is no such thing in Nature as an "independent variable."

Premise 7. Diversity is essential to adaptability

The concept of diversity is central to the framework of ecosystem management and extends across all system components, both social and biological.[11] Benefits of diversity, within some limits, are assumed to increase the adaptability to changing conditions of such things as organisms, landscapes, human communities, rural and national economies, and management agencies in our ever-evolving world.

The fundamental theorem of adaptation provides the basic biological constraint on ecosystem management. In general terms, the rate of adaptability within a system is proportional to the amount of variation within the system, which establishes the absolute importance of genetic and biological diversity per se in ecosystem management.

Premise 8. Ecosystem patterns and processes appear, and must be studied, at different geographic and time scales; reconciling these differences is difficult

Many processes operate at scales of individual plants and animals. The oxygen concentration around the root of a pine tree, for example, determines the extent of nitrogen fixation in the rhizosphere, and the cumulative nitrogen fixation influences the ability of the ecosystem to capture solar energy.[101] Evolution also controls ecosystem processes and patterns at greatly expanded scales of space and time and needs to be accounted for in ecosystem management.

The study of evolution is the study of both life's history on Earth and the processes by which evolution occurs. Accounting for evolution—an ecological and social necessity—leads to broader scales of space and longer scales of time than is customary in traditional forest management.

Premise 9. Local conditions may override or obscure general patterns and processes; the general may not contain the specific

Theories that generally prove true may be difficult to apply in specific instances where the particular condition and history of a site obscures the function of a principle. This being the case, theories, principles, or models may not be easily and/or directly applicable to a particular case in ecosystem management. Therefore, as individual decisions are made in a given management activity, such decisions must be measured by how well they approximate the general theory. Although the approximation may be relatively poor, it shows us where we can improve our management performance.

Because general principles may be difficult to apply, or not even be applicable, to a particular case, reasoned judgment and intuition must be recognized and accepted as crucial components of ecosystem management. And everyone involved in ecosystem management must be open and honest about the role of judgment and intuition.[11]

Attempting to be scientific, forest managers often try replacing judgment and intuition with a theory or model. Although theories and models are necessary components of decision-making, they can never replace the wisdom of sound judgment and intuition.

Premise 10. Ecosystem science at large scales relies on ecosystem management for empirical evidence

Research has long been hindered by the difficulty and expense of large-scale, long-range experiments to understand how and why ecosystems respond to change. Watershed studies, the best large-scale, long-range experiments, have proven to be expensive and have accordingly declined in number and scope. Without experimental data from the scales of landscapes and years, ecosystem management will be weakly founded on untested hypotheses drawn from incidental observations and data from small-scale, short-range experiments that are most often isolated from one another in space and time.

It *is* possible to provide large-scale, long-range data with which to test ecosystem theory: (1) if management is laid out in an experimental design, (2) if the concept of "best practices" is changed to allow several treatments to be simultaneously compared, (3) if we both recognize and accept that controls are impossible at broad scales and long time frames because of inherent complexity, and (4) if standard, small-scale, short-range, more controlled experiments are nested within large-scale, long-range experiments.

Premise 11. The entire system must be managed within its context

Once an ecosystem has been delineated for purposes of management, which includes study of management-related questions, the world outside the ecosystem must also be accounted for. All ecosystems are open and thus affected by the outside world. Consider air pollution, for example, in which a pollutant from far away is carried aloft and affects an ecosystem that does not itself produce that pollutant.

Having identified an ecosystem, management must proceed in such a way that the system, at a minimum, survives as identified through time. Beyond this, scientists and managers need to trace the patterns of relationship among the system's components to understand the way in which management actions affect the system as a whole, including social impacts.

Premise 12. Good can come from management

Management, like any other concept, is a tool of thought, which in and of itself is neither good nor bad. A positive or negative outcome depends on several things: (1) partly on how the tool is used, which is a matter of clarity and purity of motives; (2) partly on the validity and interpretation of information forming the basis of decisions; (3) partly on clearly stated vision and goals; and (4) partly on the novelties and vagaries of the ecosystem being manipulated. Thus, with honest motives, sound information, and clearly stated vision and goals, all of which a manager must *feel* good about, he or she can be optimistic that his or her management decisions and actions will lead to beneficial effects for both society and the ecosystem.

Underlying the notion that good can come from management is the assumption that managers enjoy two forms of trust: trust in themselves as individuals and practitioners of their craft and public trust, that of the people. Trust on all levels must be continually earned, because once lost it is exceeding difficult, and often impossible, to regain, which is the thrust behind the earlier discussion of forests (or any ecosystem for that matter) as living trusts.

Societal Premises in Ecosystem Management

Because ecosystem management must be a marriage between societal necessities, desires, and demands[11] and the biological capacity of the ecosystem to produce them over time, societal processes must be included in forestry if it is to be sustainable. People themselves must integrate research in social and natural sciences and then consciously use the results to understand the effect of social patterns and processes on the patterns and processes of forest ecosystems.

Premise 1. People choose goals for ecosystem goods, services, and conditions based on their strong desires, which they perceive as needs; goals are set and evolve by cultural and political processes

People have basic necessities for survival and well being, such as food, water, shelter, safety, love, esteem, self-realization, comprehension, and aesthetics.[102] Ecosystems fulfill immediate human necessities by generating goods (timber, water, food, medicines), services (air), and conditions (biological diversity). Under certain conditions, ecosystems can also fulfill immediate human desires by generating other goods (mushrooms, truffles), services (recreation, including hunting and fishing), and conditions (old-growth forests, wilderness). Everyone has his or her own list of perceived necessities and desires.

Something in the ecosystem is valued as a resource only when it is perceived as a resource, usually something to be converted into money. An example is the Pacific yew tree. It was considered a weed until taxol was discovered in its bark, which caused the yew to become an instant resource because taxol is a drug used as an experimental treatment for ovarian, and more recently lung and breast, cancer in women. Resource recognition, which often takes place on the heels of scarcity, may also occur as cultural values change and the ecosystem is seen anew.

As some wants are satisfied, others emerge, and wholly new and unexpected demands arise. As people strive to tailor an ecosystem's processes to their advantage, they change the system so that its appearance and functioning are as much altered by their efforts to extract what they want as by processes inherent to the system itself.

Goals for ecosystem management are set *neither* through science *nor* technology, although attempts are often made to attain them through one or the other or both. The great physicist Albert Einstein observed: "Perfection of means and confusion of goals seems, in my opinion, to characterize our age." His statement emphasizes the impatience with which Western society deals with life. We find it extremely difficult to slow down enough to do the crucial work of crafting our vision for the future and establishing goals through which to achieve the vision. We instead cling tenaciously to our impatience by letting action overwhelm our wisdom.

Premise 2. People group themselves into communities of common or complementary values to further their personal goals

People with similar cultural and economic values band together in common cause to strengthen their sense of safety and self-esteem and to acquire those things that set them apart as a community within society. Aggregations of individuals into one or more communities and aggregations of smaller communities into one or more larger communities give rise to complex kaleidoscopic scales of social organization. These scales of community organization exist simultaneously in time and space alongside and within ecosystems and social hierarchies of political and geographical organization.

People in communities advocate general philosophies to explain and justify their use of ecosystems to satisfy their demands. Examples are subsistence hunting and trapping, mining, single-use timber extraction, wilderness preservation for spiritual reasons, and such ecosystem conditions as old-growth forests.

In addition, notions of land ethics (total utilization, stewardship, trusteeship, husbandry, or conservation of the productive capacity of an ecosystem) establish individual and communal concepts of reciprocity and responsibility of people to ecosystems, of the present to the future—a social/ecological contract, if you will. Regardless of what they call it, however, people believe they are obliged in varying degrees and in various ways to participate with ecosystems, usually through the notion of some degree of control embodied in the term "management."

Thus far, most communities of people tend to create de facto visions of the future by focusing on the individual products, services, and conditions they want now. In so doing, they strike different balances between ecosystem use and trusteeship, between regard for ecological constraints and self-restraint in their efforts to meet their demands though management. Out of this conglomerate comes a de facto desired future condition.

Premise 3. Diversity, complexity, and the changing nature of human communities create uncertainty about societal priorities and demands for ecosystem products, both now and in the future

Because human necessities and perceived needs, goals, and behavior in ecosystems are diverse, the range of ecosystem goods, services, and conditions demanded by people is highly variable. Managing forest ecosystems has focused historically on a narrow range of products for a narrowly defined constituency. It is crucial to successful ecosystem management, however, that we recognize the disparate cultural and class values placed on natural resources by different ethnic groups, urban and rural dwellers, and the rich and poor.

How an expanded awareness of diverse priorities and demands translates into managing specific ecosystems, now and in the future, is uncertain. Who participates in the decision process and who actually makes the decisions is deservedly a subject for debate. Diverse and conflicting resource demands require balancing individual and collective preferences within political and ecosystem organizational processes across the scales of space and time.

Individuals or communities may have different priorities for different ecosystems or may set conflicting priorities about values from a single ecosystem. Most European Americans, for example, consider lampreys, which are parasitic fishes, "undesirable" because they attack and suck blood from other "desirable" fishes, such as salmon. The aboriginal peoples of the mid-Columbia River Basin, however, prize lampreys as a traditional food.[103] As expected, therefore, individuals or communities are inconsistent about their perceived goals toward managing forest ecosystems.

Many Americans feel profoundly attached to the wilderness experience. At the same time, many other Americans have great expectations for consumptive uses of ecosystems that are inconsistent with, and even counter to, maintaining wilderness and its values. The array of goods, services, and conditions to be derived from a particular ecosystem is thus often unclear in the minds of individuals, communities, and/or society as a whole.

Public land management currently flounders because rapid societal changes come up against resistance to those same changes from conservative and dysfunctional policymakers and ecosystem managers in dysfunctional agencies. The greater the agency dysfunction, the greater the resistance to change. That notwithstanding, ecosystem management must address societal changes, no matter how stressful.

Culture, values, and communities change at somewhat predictable, poorly predictable, and unpredictable rates. Because many changes are poorly predictable, ecosystem management needs a process to continually monitor and respond to changing perceptions of societal goals.

The expression of societal goals for ecosystem management will become more diverse, complex, and changeable in the future because of increases in: (1) the number of consumers competing for ecosystem resources; (2) the demand for access to the political policymaking process for resource management; (3) the breadth of societal demands for ecosystem goods, services, and conditions, including intergenerational and even interspecific equality; (4) public demand for access to burgeoning ecosystem information; and (5) legal constraints to private property rights.

This diversity of human perspectives, like biological diversity, promotes adaptability, prevents a single philosophy from dominating within the constraints of a given society, and hedges against cultural loss. Diversity is also thought to maintain adaptability, or perhaps even increase it up to some point, and thereby also maintain cultural sustainability in our changing world.

Premise 4. In a society of diverse and often discordant communities, conflicts are likely to develop when too many goals for ecosystem management evolve in competition with one another, resulting in some people not getting what they want

Goals expressed for ecosystem management by local, regional, national, and global communities may not be compatible when applied to the same ecosystem. Individual or collective goals not only may exceed the capacity of the ecosystem but may also be mutually exclusive. Societal conflicts must therefore by sufficiently resolved so that ecosystem management can proceed with clarity of purpose.

An essential human necessity is for conflict resolution so that order might impose sense in the world. Society determines the ground rules concerning interpersonal conduct, such as justice and respect for minorities In this way, civic discourse becomes established at multiple forums and people resolve conflicts, which momentarily restores societal order and its contract with a given ecosystem. For society to maintain internal order and its contract with the ecosystem, personal goals must defer to a democratic goal-setting process by the public at large.

When the process of representative government is perceived as inadequate to satisfy the necessities of some people, many strategies are available to alleviate people's suffering and increase their sense of trust. Society as a whole makes concessions to people in minority communities or to those incapable of representing themselves, such as future generations. An informed society is most able to accept and accommodate change through compromise, compensation, and alteration of individual, community, and societal goals.

Several approaches are available to resolve conflicts among communities. Approaches differ in their degree of openness in decision-making, the locality for decision-making, the assumptions about who is right or wrong, who is a winner or loser, and whether or not the silent parties (the ecosystem in question and future generations) are represented. Three approaches are as follows.

Management by professionals. Civil service professionals have authority and may be considered experts. They decide how societal demands should be translated into public land-management goals and practices. Various degrees of information might be integrated among special resource interests, depending on legal mandates. Such management tends to be specific over small areas and general over large ones.

Management by prescription. Courts, legislative bodies, and executive decrees establish the program of management based on the rule of law. Management tends to be prescriptive over large areas. This process relies on the expertise of lobbyists, lawyers, and politicians to solicit ecosystem experts to convincingly support their cases.

Management by collaboration. Communities with common interest in an ecosystem but dissimilar sets of values and demands from that system collaborate to establish binding, site-specific goals through trade-offs, compromise, and consensus. No one side is right by rule of law or expert opinion. The process of goal-setting often includes a mediator or facilitator, lay people, experts, and litigators.

There is at present no single approach to decision-making for policy and management of the national forest system in the United States. The complexity of decision-making is, in part, the result of multiple societal scales at which decisions about ecosystem management are made and the dysfunctional premises under which they are made.

Premise 5. Ecosystem management decisions benefit from societal vision and goals that are clear, concise, informed, and integrated and that recognize and accept that ecosystems are complex beyond our comprehension

Public debate and conflict reveal the spectrum of demands from ecosystems. Being heard depends on access to policy processes, and access depends on the ability to both acquire money and organize for effective advancement of ideas through lobbying and testifying. Legislative authority establishes a body of "policymakers," whose composition is more or less open to interested community members, professionally involved civil servants, and social and natural scientists. Policymakers may call on the general public, social scientists, and natural scientists to provide substantive information for integration into policy decisions about societal goals.

Policymakers need evidence of the public's preferences for ecosystem management. One method of gathering information is public hearings at various geographic and governmental scales. Hearings depend on self-selection of interested individuals and communities; bias is thus inherent. Individuals and communities with no awareness of or access to hearings are disenfranchised. Such people hold unstated and unheard opinions, and problems often result from their exclusion.

An important question for society is to what extent policymakers and society need to consciously gather perceptions from all communities, especially those seldom heard. Social scientists use the most objective methods possible to provide such data for policymakers. These methods include analyzing the present expression and anticipating the expression of future societal demands, particularly among communities with no previous access to decision-making processes, and estimating the present and future value of ecosystem products, services, and conditions that satisfy societal demands for ecosystem management.

Natural scientists provide society with information concerning ecosystem composition, structure, and function as well as the compatibility of various sets of ecosystem goods, services, and conditions. Although they encourage both the public and policymakers to reexamine priorities and goals in land management, their primary task is to describe, analyze, and explain trends, emerging problems, and potential solutions.

Effective scientists are experts in their disciplines and influence policy for managing ecosystems, but rarely are the only participants in policymaking. Policymakers and the public base their decisions on scenario planning for the most predictable outcomes of future management, and the primary scenario is usually an overly optimistic version of the status quo projected into the future. Other scenarios reflect management for an array of goods, services, and conditions demanded by various communities.

Ecosystem managers require clear mandates from policymakers to implement ecologically realistic goals. Mandates, in turn, require the best available ecosystem and societal information, clear and unencumbered communication between policymakers and ecosystem managers, trust in managers to exercise well their professional expertise, and recognition and acceptance of the need for a flexible response to unpredictable ecosystem events.

Ecosystem managers often lose public trust when the dysfunctional agencies they represent refuse to change management direction and practices despite changing societal values and goals. But even decisive managers doing their best to comply with changing societal values and goals can lose public trust if society itself does not invest adequately to acquire needed ecosystem information and to express clear goals.

Premise 6. Policymakers translate scientific information about ecosystems and societal goals into treaties, laws, government bureaucracies, planning, and budgets

Codification organizes land management into spatial and temporal patterns consistent with ecosystem properties and societal goals. Society allocates property rights, use or extraction of goods and services, and the acquisition of monetary profit through formal contracts among users and between present and future generations. Formal planning processes help ensure that management properly recognizes ecosystem limitations and societal goals and achieves the desired effects for both society and ecosystems.

Regulation of ecosystem use is a political effort to enhance human well-being by providing reliable and efficient use of property, goods, services, and ecosystem conditions. Regulation is aimed at preventing users from causing net losses through private actions for personal gain.

Treaties, property rights, and use and extraction rights are inviolate under the terms and duration of societal contracts, and any illegally withdrawn rights must be restored. Government cannot withdraw rights from a holder without just and acceptable compensation to the holder of the rights. American government treaties with aboriginal American nations, for example, have the full force of international treaties.

Ideally, bureaucracies are a means of streamlining communication: collecting information and disseminating it to the proper levels of decision-making, so that tasks can be identified and assigned to those most qualified to carry them out.[104] Land-management bureaucracies are currently dysfunctional, however, and require continual monitoring to ensure that they: (1) follow acceptable guidelines for generating, standardizing, conserving, and distributing information as needed to interested communities; (2) ensure that societal goals are in fact achievable within present and future ecosystem constraints; (3) remove institutional barriers to implementing societal goals that are compatible with ecosystem management; (4) reduce jurisdictional conflict in decision-making; (5) improve both organization and technology of information generation, storage, and dissemination in decision-making processes; (6) delegate some responsibility for decision-making away from centralized bureaucracies to ecosystem managers in local, peripheral settings; and (7) arrange for sufficient appropriations, appropriate them, and then manage them effectively for implementing ecosystem management.

Premise 7. No matter how people attempt to anticipate and control outcomes and effects of decisions, policy for ecosystem management will always be determined under conditions of uncertainty and ignorance

Science emphasizes probability, not certainty. This emphasis runs counter to existing societal dogmas, which insist on predeterminism of the future. Some communities refuse to accept uncertainty in scientific and societal frameworks for ecosystem management, thereby creating social discord and, at times, ecosystem impairment.

Unfortunately, policymakers usually interpret the lack of certainty in science as justification for inaction ecologically and/or overly optimistic action economically, based on the pseudo-certainty of economic theory and predictions. Such justifications are inappropriate for delaying planning and action to curb such human activities as those affecting global warming, ozone depletion, control of human population growth, deforestation, desertification, and premature depletion of fossil fuels.

People do not perceive the diversity, complexity, and changing nature of society and ecosystems around them, because people operate, as discussed earlier, in the "invisible present"—that scale of change to which we humans are oblivious. Information is thus inevitably inadequate for complete identification of human necessities and opportunities to fulfill them.

Although technology may not exist to manage and use certain information, scientists and bureaucrats may also simply choose to ignore available information about societal and ecosystem responses to management practices. In turn, certain information about societal and ecosystem responses to management practices may not be interpreted for policymakers and the public, or they may simply choose to ignore it.

Incomplete and/or mismanaged information leads to increased dysfunction in an already dysfunctional system of land management. As dysfunction increases, it becomes easier for people to further corrupt the policymaking process and its use.

Monitoring decision outcomes by scientists and other interested communities provides a means of drawing attention to ignorance, uncertainty, and agency dysfunction. Lack of investment in scientific monitoring (a chronic problem) subverts the credibility of both science and ecosystem management. In addition, reactive monitoring by disaffected communities may produce better information and simultaneously discredit policymakers and government agencies entrusted with ecosystem management.

Policymakers and ecosystem managers must understand and manage various types of responses to uncertainty and ignorance, when possible to their advantage, because people unaware of their ignorance perceive no need for additional information or research. Unpleasant surprises often provide sudden awareness of previously unimagined ignorance.

The recognition of ignorance that comes after experiencing an unpleasant surprise leads to a point where ignorance recognizes itself and arrives at the awareness needed for knowledge. Although attempts to overcome ignorance with knowledge may at times reduce immediate personal or collective ignorance, the greatest danger for society is the resulting illusion of the certainty of its knowledge, not the scope of its ignorance. Then again, ignorance may not be overcome, which is always grounds for humility.

Wisdom therefore dictates that, recognizing the limits of knowledge and the prevalence of uncertainty, policymakers and ecosystem managers must plan for the unforeseeable with enough options and flexibility to switch direction quickly for the benefit of society—present and future—and the ecosystems on which it depends.

Premise 8. Unexpected ecosystem events and shifts in societal demands for ecosystem products require changes in societal institutions, which will in turn bring changes to society

Policymakers deliberately choose whether or not to respond to surprises in the ecosystem or in society. They also choose to respond effectively or ineffectively. Intentions without actions are an ineffective response, which clearly shows that policy rhetoric and subsequent action must be consistent.

Without timely and decisive responses to change, policymakers, at whatever level, lose credibility with society. Then again, society itself can overreact when ecosystem conditions and societal values are changing rapidly or unexpectedly. Public policy, on the other hand, is timid and its mandates often vague. It is therefore imperative, following policy rhetoric advocating sustainable ecosystem management, that adequate investment actually be made in ecosystem research and monitoring, ecosystem management personnel, and market research concerning social necessities.

Society increasingly demands to know the cost of ecosystem management and its future risks and benefits. And people certainly have differing attitudes toward investing and incurring risk to achieve admittedly uncertain future net benefits through ecosystem management. Additionally, in an era of high national deficits and public demands to cut federal spending, potentially high start-up costs for achieving ecosystem sustainability may meet with public disapproval and political resistance. Yet despite high national deficits, some citizens will recognize the direct benefits to be accomplished through ecosystem management and/or the immediate and long-term consequences of failure to account for ecosystem sustainability.

Society must recognize and accept that achieving desired ecosystem conditions will require public investment and that such investment must compete with other demands for scarce financial resources. Nevertheless, as with any investment decision, inaction or insufficient funding for ecosystem management will incur even greater risks and costs in the future and will result in fewer options and lower net benefits in the future. We thus risk saddling future generations with a form of taxation without representation.

Policymakers are obligated to estimate and make known to the general public as explicitly and honestly as possible the societal costs, risks, and net benefits of ecosystem management. They must also choose and justify to various communities the allocations of costs and risks necessary to achieve a new mix of ecosystem benefits under sustainable ecosystem management.

Changes in composition and allocation of ecosystem benefits (who pays, who gains) shift power both within and among communities. Local, specialized communities that depend directly on ecosystem goods for livelihoods but do not control the resources are especially vulnerable to power shifts.

Because most Americans live in cities, demographic and economic power resides in urban areas. Policy decisions must therefore determine issues of justice and equality in ecosystem management, paying particular attention to rural communities, which, because they perceive no gain from ecosystem management, may feel alienated and create new sources of societal conflict. To avoid unnecessary conflict, policymakers must clearly demonstrate the net gains of ecosystem management for both rural and urban communities, although the type and degree of gain will differ among communities.

Efforts to avert ecosystem catastrophe by implementing ecosystem management must be done carefully so as not to unwittingly initiate any more stressful or undesirable societal changes than absolutely necessary. Even so, the rate of societal change induced by alterations in the availability and allocation of ecosystem resources may exceed the thresholds of some communities to cope with such changes, particularly those specialized, resource-dependent communities that steadfastly resist moving into the future.

In the end, when all is said and done, we must recognize and accept that ecosystem management is an open-ended experiment through which society is continually learning how to participate sustainably with its environment.

CONDUCTING THE EXPERIMENT

Having framed the experiment, it is time to consider how it will be conducted, which brings us to the "adaptive" in adaptive ecosystem management.

What Is Adaptive Ecosystem Management?

Adaptive ecosystem management is, at a minimum, a continuing cycle of action based on studying, planning, monitoring, evaluating, and adjusting (Figure 101). The "adaptive" part of ecosystem management focuses on using current scientific and societal information to inventory, plan, act, and accumulate new information as an integral part of management with which to improve future decisions. The "adaptive" concept underscores that knowledge and societal values are changing ever more rapidly and that management must keep abreast of these changes.

Adaptive ecosystem management is therefore designed as a twofold process: one part for rapid and effective learning by scientists, managers, society, and policymakers and the other for rapid change. This twofold process is central to managing for ecosystem sustainability.

Sustainable ecosystems for present and future generations is the ultimate goal of adaptive ecosystem management. What people want over the long run and what they can actually have depends on what is ecologically possible within the sustainable capacity of the ecosystem.

Achieving the overlap between societal desires and ecological possibility (see Figure 100), that condition which is sustainable, requires continual integration of social values with ecological capacity. This integration demands a better understanding of changing societal values and the ecological and social processes that shape ecological capacity. Recognizing and accepting that societal processes have profound effects on ecosystems is the core of ecosystem sustainability. Adaptive ecosystem management is the means of

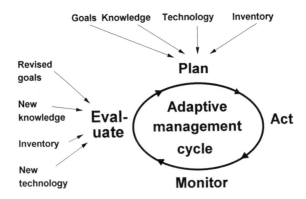

FIGURE 101   The "adaptive" in adaptive ecosystem management denotes a series of steps that promote management's *rapid* awareness of and *quick* response to changing ecological knowledge and societal values and goals. (Adapted from Birch et al.[110])

achieving such recognition, acceptance, and integration and, ultimately, sustainability.

Management goals and objectives must be set by society because successful implementation of management—adaptive, sustainable, or otherwise—depends on societal approval. Broad management goals and concise objectives can be derived from the directives given by President Clinton at the 1992 Forest Conference in Portland, Oregon: (1) protect the long-term health of forests, wildlife, and waterways; (2) be scientifically sound, ecologically credible, and legally defensible; (3) plans must seek sustainability and predictability; (4) always remember the human and economic dimensions; and (5) make the federal government work together for you.

Taken together, these directives describe a vision of the overlap between what people want for themselves and future generations and what is ecologically possible in the long run. A systematic, scientifically valid, and socially acceptable approach to adaptive ecosystem management is necessary in attempting to implement all of these directives simultaneously.

Why Is the "Adaptive" in Ecosystem Management Necessary?

The need for the "adaptive" in ecosystem management is based on the following premises: (1) the synergistic social/ecological changes in the world are becoming more rapid and ever more tightly linked in self-reinforcing feedback loops, which makes anticipating them extremely difficult; (2) practices previously thought to be ecologically and socially desirable often turn out to be undesirable or even detrimental when viewed historically; (3) the concepts of "best management practices" and "cookbook management" are invalid, especially when applied over broad or diverse areas; (4) policymakers will never have fully adequate knowledge on which to base policies, and managers will never have fully adequate knowledge on which to base decisions; and (5) the rate of learning, along with the acceptance and application of new knowledge, controls the frequency of desirable outcomes.

Although people are beginning to recognize the speed with which politics, governments, social values, and environmental conditions are changing, even more rapid ecological and social changes are likely as the world population continues to increase and environmental effects continue to intensify. As people come to better understand the incredible complexities of large ecosystems and how they respond to human actions, they will begin to realize their inability to accurately predict ecosystem changes and their outcomes.

The basic need for adaptive ecosystem (forest) management comes from recognizing and accepting that the current concept of "best management practice" is invalid. A "best practice" assumes that we know the ramifications of current practices and their scientific underpinnings—which we do not!

Best practices have been the political basis for developing technical, cookbook prescriptions that do not work over the long run or over ecologically and socially diverse landscapes. Best practices and cookbook management have given people an excuse for not thinking about what they are doing or not evaluating the consequences of their management choices.

People are continually and unpleasantly surprised by previously unrecognized flaws in past "best practices" because it is impossible to know if a "best guess" will be correct. Managers must therefore concentrate more and more on detecting errors and making quick course corrections, which is the essence of adaptive ecosystem management.

If sustainability requires that errors be not only quickly detected but also quickly corrected, the rate with which sustainability is achieved should be proportional to the rate of learning, accepting, and using new knowledge. If adaptive ecosystem management is learning from management experience, and if sustainability is in fact proportional to how fast we learn and apply what we have learned, then adaptive ecosystem management must be structured for rapid learning, which requires two major steps: (1) designing decision-making and planning processes and institutions to ensure that they use the best available societal and scientific information to achieve management goals and (2) designing and analyzing the management experiment to improve the information needed for future decisions.

Applying the concept of management as an experiment and the knowledge derived as a working hypothesis is one way to increase the rate of learning. In theory, arrays of management treatments can be designed to allow for statistically valid comparisons that, along with judgment and intuition, can be relied on for correcting management practices. At a minimum, a range of management practices can be applied to an ecosystem, but the differences between them must be distinct enough to detect any relative success.

In addition, the following must be kept in mind:[105]

1. Manage where you are, because although ecosystem management is site specific, it transcends all human-conceived boundaries.

2. All global ecosystems are subject to human influences, known and unknown, purposeful and accidental. Ecosystem management must therefore begin with a careful evaluation of social desires, influences, and responsibilities because every system definable in biological and physical terms interrelates and interacts with an ever-expanding network of human values, uses, and other social constructs.

3. Ecosystem management transcends all defined boundaries at one time or another. Such transcendence must be accommodated through a complex

process of joint goal-setting, compromise, regulation, and incentives among joint landowners.

4. The key to continuously improving ecosystem management is an ever-growing and applied knowledge of specific processes and interactions responsible for system activities and outcomes. Terrestrial systems are so complex that they can seldom be managed by enumerating and tracking all components. The only path to predicting system behavior and outcomes is thus a thorough knowledge of why things occur as they do, which requires reliance not only on sampling but also on knowledge gleaned elsewhere and tested in and adapted to the system at hand.

5. Ecosystem management requires that all known contents of the system in question be accounted for when the decisions and manipulations are made.

Then, armed with new knowledge and a good forum for exchanging information, trust and understanding can continually develop among scientists, managers, communities, and policymakers.

Decision and Planning Processes

We must begin adaptive ecosystem management by understanding more clearly than ever what people want and by including different points of view and ideas in a variety of experimental treatments as a way of involving more than just the majority opinion. The public, representing social diversity, must be continually involved with scientists and land managers in the process of defining their vision, goals, and objectives and in helping to interpret the results of management experiments. Continual, open learning from one another is a prerequisite to developing innovative, consensus-based resolutions to as many management conflicts as possible.

In addition, decision-makers need quantitative assessments of societal necessities and wants. Businesses in a market economy live and die on their ability to know what customers want and are willing to pay for. Many of these same techniques can be adapted for use in natural resource management.

The best and latest ecological knowledge must also be available in a form useful to decision-makers. To understand what is ecologically possible in the long run, a significant investment in obtaining and synthesizing high-quality ecological information at all scales is required. Further, ecological knowledge, like societal knowledge, needs to be local in the sense that it is known to apply to the area being managed, regardless of its size.

The decision-making process must be designed to maintain and/or build sustainability or the overlap of what people want and what is ecologically possible in both the short and long term. This approach is the essence of the

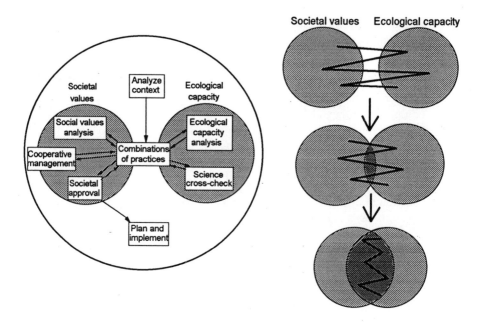

FIGURE 102 The lacing model is a conceptual approach to building overlap between what people want for themselves and future generations and what is ecologically possible in both the short and long run. Collective information on what society wants from the forest is passed to scientists and managers, who assess what is ecologically and operationally possible. The scientific and managerial assessment, now passed back to the public, may influence society's assessment of its necessities and wants and may stimulate creative solutions to problems and conflicts. In this way, information about societal necessities and wants and ecological possibilities is passed back and forth in a continual, repetitive process until it is thoroughly integrated and ready for the planning process. (From Bormann et al.[109])

"lacing model," in which societal and ecological knowledge is repeatedly exchanged and integrated before plans are constructed and implemented (Figure 102). This model assumes two conditions: (1) that the knowledge and expertise exist with scientists and managers to implement sustainable ecosystem management to some degree and (2) that all people participate fully and equally, sharing responsibility, continually learning from one another, and collectively developing creative means of resolving conflicts.

A decision-making process that fully and openly permits, even encourages or requires, public participation and shared responsibility will necessitate a new legal framework. In addition, federal and state agencies will require new mandates and restructuring to accommodate and facilitate a new decision-making process and continual learning by all employees. Finally, sustainability will be difficult to achieve until society at large both accepts and applies the concept of rapid, continual learning.

Design and Analysis of the Management Experiment

When will we know whether a management strategy framed as an experiment has been a success or a failure? When will we have sufficient data to change our management or policy? What criteria will we use to make this decision, and how will we account for unexpected occurrences that may fool us into thinking that an experiment has succeeded or failed? These questions are pivotal to designing an adaptive ecosystem management process.

Determining what evaluation criteria to use must be part of the decision-making process. Scientists and statisticians can help in the selection, evaluation, and interpretation of these criteria. They can also help by designing management experiments to produce needed information.

Management experiments, if properly designed and analyzed, can efficiently and effectively produce information of known quality. Appropriate statistics will vary with the scale and complexity of management treatment.

Traditional experiments in forestry research are relatively small, assumed to be highly controlled, and use agricultural-based parametric statistics. At the larger scale of whole watersheds, complexity increases and no two watersheds can be easily compared, which means that replication is not usually possible and that other statistics, such as time-series analysis, must be used. If the scale is further increased to a river basin or regional experiment, statistical considerations will probably have to be changed again.

Management experiments will have to occur at a wide range of geographic and time scales. On the geographic scale, for example, planners might conceive of an experimental design that forms a regional framework for management experiments in different forest types (say, coastal Sitka spruce–lodgepole pine, low-elevation Douglas-fir–western hemlock, and high-elevation noble fir–silver fir–mountain hemlock), the goal of which is to evaluate strategies for achieving ecosystem sustainability. Monitoring must be thought of as an integral part of evaluating the experiment. The extent of local adaptation, public involvement, and the approaches to research and planning would be compared among the forest types.

Nested within this regional-scale experiment are a multitude of small-scale experiments designed to test, among other things: (1) such assumptions as the necessity for riparian buffer zones along intermittent streams, (2) approaches to recreating such late successional conditions as old-growth forest, and (3) techniques for restoring damaged ecosystems.

Most experiments within forest ecosystems take considerable time to yield reliable information. Thus, smaller-scale, shorter-duration "research" experiments must also be nested within the longer-duration, regional-scale experiment. Past practices must also be evaluated because such studies will quickly produce valuable information.

Carefully conceived, high-quality management and research experiments and studies nested within a regional management experiment will work synergistically to provide the best possible information. The more linkages that can be established among experiments of differing scales and purposes, the more likely it is that meaningful information will be forthcoming on ecosystem sustainability.

Where should we begin adaptive ecosystem management as an open-ended experiment? We can begin by: (1) building community involvement in and developing new approaches to public participation in the process of deciding what people want from public lands; (2) developing strategies for building ecosystem sustainability while taking into account local societal, managerial, and scientific knowledge, as well as such legal obligations as treaties; (3) exploring how an array of social values can be accounted for in alternative, sustainable practices that will allow comparison in management experiments; (4) discussing how current dysfunctional institutions need to change so that they can rationally participate in adaptive ecosystem management; and (5) installing long-term experiments to begin testing our long-held scientific, managerial, and societal assumptions as they apply to ecosystem sustainability.

Regardless of what society ends up doing with respect to the adaptive ecosystem management experiment, every decision—whatever it is—will have either positive or negative effects on the future. A decision simply cannot be neutral!

17

TODAY'S DECISIONS, TOMORROW'S CONSEQUENCES

As stated in the preface, if we dare to dream boldly enough, we can have a sustainable forest, not only in the Pacific Northwest but also in the world. These forests can include old-growth trees, wood fiber, wild areas, wildlife (including fish), clean water, and so on. But as they say in old western movies, we will have to check our guns at the door. We will have to change our attitudes, our thinking, and our perception of the forest, and we will have to transcend our own special interests and encompass all interests in the forest as a whole.

Changing our attitudes and thinking means that we will have to accept that we, as product consumers, are the problem, and this being so, we are also the solution. As Sitting Bull, the great Sioux chief said, "Let us put our minds together and see what life we will make for our children."

In 1987, as the speaker at the annual banquet of a conservation organization, I spoke about a dream large enough to encompass the various interests of people in a sustainable forest in a way that maintains and nurtures human dignity. When I was finished, two young men questioned both my honesty and my faith in humanity. They steadfastly insisted that there had to be enemies out there to hate and that I was dishonest if I did not agree.

I could see why they thought this way after reading an article in which John Stockwell, an ex-Central Intelligence Agency (CIA) employee, was quoted as saying that "American foreign policy [is] a result of [a] need for enemies."[106] On the other hand, I can only see two possible reasons for "needing" an enemy. The first is to have someone onto whom I can project my fear and thereby feel more in control internally. The second is the illusion that if I project my fear onto an "enemy," I am somehow absolved of all responsibility for my thoughts and actions.

My alternative is not to see people as enemies in the first place, but rather to see them as candles. I have chosen the metaphor of a lighted candle because it represents the inner process of growth and change required of each of us in our own, unique way and timing.

As we each discover our inner source of strength, radiance, creativity, and love, we can begin to make the gradual, outer shifts, which affect not only our personal relationships but also the environment beyond our immediate selves. As each person's candle burns brighter, it is able to light the way for others whose flame as yet is dim and struggling. In helping one another to see, the whole world can be alight with the dancing flames of brotherhood, love, and peace, because one candle loses none of its brightness by lighting another.

Another consideration is the power of an idea. An idea that is acted on can change the world. Look, for example, to the Middle East where Judaism, Christianity, and Islam, each attributed to a separate individual, all arose in an area about the size of Oregon. The growth of these ideas into religions, each in its turn, changed the world.

There is a lesson in this: if you want to own "your idea," to get credit for it, then keep it to yourself and it will change nothing. If, on the other hand, you want an idea to effect change, you must give it away freely, so that everyone who wishes to may own it. Even if you do not care who gets the credit for "your idea," keep in mind that large organizations and societies change with seemingly glacial speed, but the speed with which change takes place is not as important as the kind of change that occurs.

The kind of change that is important is what professor and capitalist revolutionary Ed Deming calls "Total Quality Management." Deming maintains that in order for total quality management to work, you must continually improve whatever it is you are doing, not in great leaps forward, but a little here and a little there—sustainable development. It never gets any easier, according to Deming, because you have to change the way you think, you have to unlearn everything you have been taught about management, and you have to develop an understanding of statistics and psychology, of how people learn and what makes them change, of people's need to feel pride and joy in their work.

How many steps are there? No one knows. How long will it take? The rest of your life. The good news, however, is that you should start seeing results within three to five years. Why do it? Because people who have tried it say it works.[107]

Still another consideration is change itself. There is much insistence today that either we need not change because science and technology will give us the answer or that we dare not change *until* science and technology have given us the answer. Keep in mind that science, with the aid of technology, can tell us what happens, to some extent why it happens, how it happens, where it happens, how much it happens, maybe when it happens, and perhaps even what we need to do to alter a predicted outcome. But neither science nor technology can change it for us; we must figure out how to do that. In other words, scientific data without a context—the dream, the vision—in which to frame it is like a jigsaw puzzle without a pattern or border.

To change anything in a meaningful sense, we must first know what is there; then we must decide what we want to be there and figure out the options available to achieve what we desire. This means that we must struggle with defining our vision, goals, and objectives. I can, for instance, define "sustainable forest" for you, but then it is my definition—not yours. In order for the Pacific Northwest (or any other geographical area) to be managed on the basis of a sustainable forest, we must struggle with the definition together, so we will *all* "own" the outcome, including people from public land-management agencies, industrialists, conservationists, and anyone else interested in the forest. We the people must define our vision and choose to strive for it.

Our forests are certain to decline through time with present management attitudes and practices. Once our vision is defined, however, and the choice is made to pursue it, we have the chance to achieve it and the flexibility to see and take advantage of heretofore unknown options.

To have a sustainable forest, we must maintain some stands of unmanaged old-growth, mid-age-growth, and young-growth forest from which to learn. If we are willing to maintain these unmanaged stands, we suddenly have flexibility we did not have before.

To illustrate, if we cut all the old growth simply to maintain the present volume of wood fiber for industry, society will be jolted suddenly when the old growth is gone and jobs with it. We will lose all the potential options that old growth afforded. On the other hand, if we are committed to saving enough of the unmanaged stands from which to learn how to sustain a forest, then industrial changes will be more immediate, but future generations will have a greater ecological chance to live in an environment that fosters human dignity. In addition, old-growth stands have recently revealed, through their fire history, that they are more like selective logged forests than clearcuts, which provides another option for future management of long rotations in a sustainable forest.

Our dream—a sustainable forest—must be bold enough to allow change not only in the forest but also in our thinking, because the land is not to be conquered but rather is to be nurtured. We must also understand, accept, and remember that the world is always in a state of becoming, in a state of change, and therefore nothing is ever "finished." Trying to hold things constant, like yesterday's timber values projected into tomorrow's plantations, is like driving through life looking in the rear-view mirror. Today's decisions will design and sustain or destroy the forests of tomorrow.

We too often pursue our science and our technology in intellectual isolation of their long-term consequences to the environment. If we are to have a sustainable environment for ourselves and our children, however, we must remember the Kenyan proverb: "Treat the Earth well. It was not given to us by our parents. It is loaned to us by our children."

Philosopher Ivan Illich puts it differently. "I distinguish," he says, "between the attitudes of hope and expectation. 'Expectation' is based on a belief in instruments and the naive acceptance of socially constructed certitudes. 'Hope' is based on historically-rooted experience." To face the future freely, Illich feels we must be neither optimistic nor pessimistic, but we must place *no trust* in the tools of human kind. Instead, he says, we must place *all our hope* in one another as human beings.[92] I believe what Illich says, because I know that we can do anything we set our minds to if we are willing to work as hard as we dream.

Thus, as we the people elevate our personal, environmental, and social consciousness (the constant human struggle), we begin to take our rightful place in the Universe, not as conquerors, for we have conquered nothing, but as custodians of the world and the Universe for those who must follow. Democratic governance works, however, only when it is being used, and the will of the people is effective and responsive only in conjunction with justice, fairness, equality, and an ethos in the administration: a set of guiding beliefs in decision-making even though not written into law. This ethos is the intuitive "gut feeling" that the resources of the planet and the nations are worthy of respect and care, husbandry if you will, and that they are equally essential to the long-term well-being of all people in all cultures.

According to Luna B. Leopold, however, there is a striking disregard for the public interest and a net loss of the local, national, and global wealth, both intellectual and monetary. Yet the allocation of advantage to the few at the expense of the many persists despite the growing alarm over pending shortages, increasing pollution, and ongoing, irreversible degradation of the quality and quantity of our resources.

Here it must be kept in mind that, in most instances, Congress has and continues to set the stage for the environmental degradation the world is experiencing, not only by writing legislation that favors special interests but also by giving no counteractive instructions for the protection of the public.

Thus, the public, lacking information and insight, fails to react. Even when reaction does come, it is only after considerable damage has already befallen the environment. For example, the public only now has begun to perceive that maintaining clean air, clean water, and the shield of ozone; decreasing acid rain and deforestation; and controlling the greenhouse gases are commonalties, which sooner or later will affect all people on Earth.

It thus falls to the citizenry to express the ethos of both the governed and the government. If the citizenry becomes divorced from concern for the common good, then the government follows in the same path. Citizens can become so divorced if they are not informed, if they do not see the consequences of neglecting the general welfare, and if they are given no insight into the operational details of how their own interests and those of their children are being taken care of and protected. Note that while they may be late in expressing their ethos, it is the public that is forcing Congress to take heed.[15]

If *we the people* really want a sustainable forest in the Pacific Northwest or anywhere else, then we must understand that the government of the United States is not in Washington, the White House, or the Capitol. The government of the United States resides in us, *the people.* Only the *administration* of our government resides in Washington.

We, individually and collectively, must be willing to risk change if we are to grow enough to envision ourselves as representatives of our own government, dealing in foreign policy beyond the physical borders of the United States. It is inevitable that we must deal in foreign policy because we are part of a larger global community that desperately needs to have the forests of the world in a biologically sustainable condition—sustainable forestry.

Consider, for a moment, that the most important terrestrial ecosystems are those of our global forests. They produce much or most of the oxygen we breathe. They store and absorb much carbon dioxide, which helps to reduce the buildup of greenhouse gases. And they supply and store most of the available potable water on which society depends. Because oxygen, carbon dioxide as a greenhouse gas, and water are global commons in that they not only recognize no political boundaries but also affect the entire world, it is time to recognize and accept that the forests of the world belong to the global community as a matter of societal necessity for its biological survival and social well-being.

Therefore, if we in the United States are to benefit from the oxygen supplied by the forests of South America, Europe, and Africa, to name a few, we have both a moral obligation and a biological mandate to see that the South Americans, Europeans, and Africans can benefit from our forests in return. This means that we must become one another's keepers if society, indeed humanity, is to survive well beyond the 21st century. That is our moral responsibility as human beings because each irreversible option we foreclose represents the limitation—the impoverishment—of the future. The generations to come have no choice but to respond to our decisions, which will have become their inherited circumstances. Because the decisions we make today will inexorably create the circumstances of tomorrow, in sober reality the future is today.

How we individually behave determines the collective condition of the world for every living creature, because we—the largely self-appointed human trustees—are ultimately responsible for our own thoughts and actions and how they affect the global condition. The world is but the outer mirror of our inner selves.

This being the case, we the global community desperately need to review our understanding of the world's forests and come to grips with the fact that each acre of forest, regardless of "ownership," has in concert with every other acre a cumulative effect on our home planet and therefore our global community. Human society and human life simply cannot survive without healthy, biologically sustainable forests. I propose, therefore, that the 21st century officially be made the century of healing the world's forests. If our children are to inherit a livable world environment, we adults must attack the problem of forest health and sustainability with the same strong commitment that we devote to the luxury of war!

Starting now, we the people of the United States can become the first to consciously commit ourselves to healing the forests of the world, beginning with our own. Rather than pointing fingers and fixing blame on other nations, we need to mend our own fences. We not only need to stop deforesting the United States but also need to stop American companies from deforesting other countries.

To heal the forests of the world, we must care about one another. We must be willing to openly and freely share all of our knowledge, ecologically appropriate technology, and personal expertise with the single aim of helping our fellow human beings to heal their forests, from which we all will benefit. We can do this. It is only a choice, but what a choice it is!

If we have the courage to set aside, even for a while, our competitive view of the world and in greater self-interest cooperate and coordinate, working together to heal the forests of the world could be the first unequivocal step toward world peace. With healed biologically sustainable forests as the vision of what is possible, we the global community can come together and invest in one another as never before. We could in fact become one another's keepers!

Time is of the essence. We must start **now**, because we are simultaneously the hope and the limitation of the future. Neil Postman said, "Children are the living messages we send to a time we will not see." What shall we send with them into that distant time—fear and depravation or hope and dignity? The choice is ours. To our children, their children, and their children's children we bequeath the consequences.

I say this because I remember that as a child, whenever I expressed my opinion about what I thought the world needed to be like or how I wanted it to be, I was inevitably told to be quiet, that I was just a child who did not know what was good for me, and I was therefore summarily dismissed. But consider for a moment that the children must inherit the world and its environment as adults leave it for them (Figure 103). Our choices, our generosity or greed, our morality or licentiousness will determine the circumstances that must become their reality.

Why do adults assume that they know what is best for the children, including their own, when it is adults (many of whom have children) who are destroying the world with greed and competitiveness? Why are children never asked what they want adults to leave for them in terms of environmental quality and forest

FIGURE 103 While children plant a forest with their hearts and a beginner's mind, adults plant trees with their expert's intellect and call them a forest.

health? Why are they never asked what kinds of choices they would like to be able to make when they grow up?

In the society of the future, it is going to be increasingly important to listen to what the children say because they represent that which is to come. Children have beginners' minds and, not knowing what the answers "should be," can see what the answers might be. To children all things are possible until adults with narrow minds, who have forgotten how to dream, put fences around their imaginations.

Adults, on the other hand, too often think they know what the answers should be and can no longer see what they might be. To adults, who have forgotten how to dream, all things have rigid limits of impossibility. We would do well, therefore, to consider carefully not only what children see as possible in the future but also what they want. The future, after all, is theirs. What legacy shall we leave them?

Appendix

COMMON AND SCIENTIFIC NAMES

Common name	Scientific name
FUNGI	
Blackstain root rot	*Verticicladiella wageneri*
Laminate root rot	*Phellinus weiri*
Red ring rot	*Phellinus pini*
PLANTS	
Alaska-yellow-cedar	*Chamaecyparis nootkatensis*
Alder	*Alnus* spp.
Beech	*Fagus* spp.
Bigleaf maple	*Acer macrophyllum*
Birch	*Betula* spp.
Black cottonwood	*Populus trichocarpa*
Black spruce	*Picea mariana*
Bristlecone pine	*Pinus aristata*
Cacao	*Theobroma cacao*
Coast redwood	*Sequoia sempervirens*
Common hornbeam	*Carpinus betulus*
Cryptomeria	*Cryptomeria japonica*
Douglas-fir	*Pseudotsuga menziesii*
Elm	*Ulmus* spp.
Engelmann spruce	*Picea engelmannii*

Common name	Scientific name
European beech	*Fagus sylvatica*
Fir	*Abies* spp.
Hemlock	*Tsuga* spp.
Hickory	*Carya* spp.
Incense-cedar	*Libocedrus decurrens*
Japanese cypress	*Chamaecyparis obtusa*
Juniper	*Juniperus* spp.
Larch	*Larix* spp.
Lodgepole pine	*Pinus contorta*
Mahogany	Meliaceae
Maize	*Zea mays*
Mountain hemlock	*Tsuga mertensiana*
Noble fir	*Abies procera*
Norway spruce	*Picea abies*
Oak	*Quercus* spp.
Oregon ash	*Fraxinus latifolia*
Pacific silver fir	*Abies amabilis*
Pine	*Pinus* spp.
Pinyon pine	*Pinus edulis*
Ponderosa pine	*Pinus ponderosa*
Port-Orford-cedar	*Chamaecyparis lawsoniana*
Quaking aspen	*Populus tremuloides*
Red alder	*Alnus rubra*
Red spruce	*Picea rubens*
Scotch pine	*Pinus sylvestris*
Silver fir	*Abies amabilis*
Sitka spruce	*Picea sitchensis*
Spruce	*Picea* spp.
Sugar maple	*Acer sacharrum*
Sugar pine	*Pinus lambertiana*
Sycamore	*Plantanus* spp.
Western hemlock	*Tsuga heterophylla*
Western larch	*Larix occidentalis*

Common name	Scientific name
Western redcedar	*Thuja plicata*
Western white pine	*Pinus monticola*

INVERTEBRATES

Carpenter ant	*Camponotus* spp.
Douglas-fir tussock moth	*Orgyia pseudotsugata*
Gribble	*Limnoria* spp.
Western spruce budworm	*Choristoneura fumiferana*
(two species)	*Choristoneura occidentalis*

BIRDS

Horned lark	*Eremophila alpestris*
Marbled murrelet	*Brachyramphus marmoratus*
Pileated woodpecker	*Dryocopus pileatus*
Red-cockaded woodpecker	*Picoides borealis*
Sage grouse	*Centrocercus urophasianus*
Sooty tern	*Sterna fuscata*
Spotted owl	*Strix occidentalis*

MAMMALS

African elephant	*Elephas maximus*
Black bear	*Ursus americanus*
Bushy-tailed woodrat	*Neotoma cinerea*
Chipmunk	*Eutamias* spp.
Deer	*Odocoileus hemionus*
Deer mouse	*Peromyscus maniculatus*
Douglas squirrel	*Tamiasicurus douglasi*
Ground squirrels	*Spermophilus* spp.
Long-legged bat	*Myotis volans*
Mantled ground squirrel	*Spermophilus lateralis*
Northern flying squirrel	*Glaucomys sabrinus*
Red-backed vole	*Clethrionomys* spp.
Red tree vole	*Arborimus longicaudus*
Western gray squirrel	*Sciurus griseus*
Western red-backed vole	*Clethrionomys californicus*
White-footed mouse	*Peromyscus leucopus*

ENDNOTES

1. *Forestry Terminology.* 1950. Society of American Foresters, Washington, D.C., 93 pp.
2. *Webster's Third New International Dictionary of the English Language Unabridged.* 1961. G. and C. Merriam, Springfield, Mass., 2662 pp.
3. International Union of Forestry Research Organizations. 1977. *Terminology of Forest Science, Technology, Practice, and Products, Addendum Number One,* Society of American Foresters, Washington, D.C., 348 pp.
4. Bak, Per and Kan Chen. 1991. Self-organizing criticality. *Scientific American* pp. 46–53.
5. Stevens, William K. 1990. New eye on nature: The real constant is eternal turmoil. *The New York Times* July 31.
6. Magnuson, John J. 1990. Long-term ecological research and the invisible present. *BioScience* 40:495–501.
7. Freud, Sigmund. 1961. *The Future of an Illusion,* W.W. Norton, New York, 63 pp.
8. Capra, Fritjof. 1975. *The Tao of Physics,* Shambhala, Berkeley, Calif., 330 pp.
9. Pinchot, Gifford. 1947. *Breaking New Ground,* Harcourt, Brace, New York, 522 pp.
10. Covington, Wally W. and M.M. Moore. 1991. Changes in Forest Conditions and Multiresource Yields from Ponderosa Pine Forests Since European Settlement, unpublished report submitted to J. Keane, Water Resources Operations, Salt River Project, Phoenix, 50 pp.
11. Maser, Chris. 1992. *Global Imperative: Harmonizing Culture and Nature,* Stillpoint Publishing, Walpole, N.H., 267 pp.
12. Lowdermilk, W.C. 1975. Conquest of the Land through Seven Thousand Years, Agricultural Information Bulletin No. 99, Soil Conservation Service, U.S. Department of Agriculture, U.S. Government Printing Office, Washington, D.C., 30 pp.
13. Maddock, T. III, H. Banks, R. DeHan, R. Harris, J.H. Kneese, J.H. Lehr, P. McCarty, J. Mercer, D.W. Miller, M.L. Munts, M.A. Pierle, A.Z. Roisman, L. Swanson, and J.T.B. Tripp. 1984. Protecting the Nation's Groundwater from Contamination, OTA-0-233, Office of Technology Assessment, U.S. Congress, Washington, D.C., 244 pp.

14. Hand, D. 1990. Breadbasket ecology. *Yoga Journal* May/June:23–24.
15. Leopold, Luna B. 1990. Ethos, equity, and the water resource. *Environment* 2: 16–42; Maddock, T. III, H. Banks, R. DeHan, R. Harris, J.H. Kneese, J.H. Lehr, P. McCarty, J. Mercer, D.W. Miller, M.L. Munts, M.A. Pierle, A.Z. Roisman, L. Swanson, and J.T.B. Tripp. 1984. Protecting the Nation's Groundwater from Contamination, OTA-0-233, Office of Technology Assessment, U.S. Congress, Washington, D.C., 244 pp.; Hand, D. 1990. Breadbasket ecology. *Yoga Journal* May/June:23–24.
16. Harris, Larry D. 1984. *The Fragmented Forest,* University of Chicago Press, Chicago, 211 pp.; Harris, Larry D., Chris Maser, and Arthur McKee. 1982. Patterns of old-growth harvest and implications for Cascade wildlife. *Transactions of the North American Wildlife and Natural Resources Conference* 47:374–392.
17. Harris, Larry D. and Chris Maser. 1984. Animal community characteristics. in *The Fragmented Forest,* Larry D. Harris (Ed.), University of Chicago Press, Chicago, pp. 44–68.
18. Emmons, Louise H. 1989. Tropical rain forests: Why they have so many species, and how we may lose this biodiversity without cutting a single tree. *Orion* 8:8–14.
19. Bukkyo Dendo Kyokai. 1985. *The Teaching of Buddha,* 110th rev. ed., Kosaido, Tokyo, 307 pp.
20. Castellano, Michael A., James M. Trappe, Zane Maser, and Chris Maser. 1989. *Key to Spores of the Genera of Hypogeous Fungi of North Temperate Forests, with Special Reference to Animal Mycophagy,* Mad River Press, Eureka, Calif., 186 pp.
21. Perry, David A. 1988. Landscape pattern and forest pests. *Northwest Environmental Journal* 4:213–228.
22. Hesse, Hermann. 1971. *Siddhartha,* Bantom Books, New York, 152 pp.
23. Hardin, Garrett. 1986. Cultural carrying capacity: A biological approach to human problems. *BioScience* 36:599–606.
24. Martin, Clyde S. 1940. Forest resources, cutting practices, and utilization problems in the pine region of the Pacific Northwest. *Journal of Forestry* 38:681–685.
25. Plochmann, Richard. 1968. *Forestry in the Federal Republic of Germany,* Hill Family Foundation Series, School of Forestry, Oregon State University, Corvallis, 52 pp.
26. Plochmann, Richard. 1989. *The Forests of Central Europe: A Changing View,* 1989 Starker Lectures, Forestry Research Laboratory, College of Forestry, Oregon State University, Corvallis.
27. Pinchot, G. 1947. *Breaking New Ground,* Island Press, Washington, D.C., 522 pp.
28. Wilkinson, C.F. 1985. *The Greatest Good for the Greatest Number in the Long Run: The National Forests in the Next Generation,* College of Forestry and Natural Resources, Colorado State University, Fort Collins, 12 pp.
29. Clark, T.W. 1986. Professional excellence in wildlife and natural resource organizations. *Renewable Resources Journal* 4:8–13; Jones, C.O. 1970. *An Introduction to the Study of Public Policy,* Duxbury Press, Belmont, Calif., 70 pp.; Wilson, J.Q. (Ed.). 1980. *The Politics of Regulation,* Basic Books, New York, 364 pp.
30. The Associated Press. 1987. Whistle blower's actions branded disloyal by Army. *The Oregonian* (Portland, Ore.), November 10.

31. United Press International. 1988. Hatfield slams environmentalists. *The Daily Barometer* (Oregon State University, Corvallis), December 5.
32. The Associated Press. 1989. Government seeks to ban apple growth chemical. *Corvallis Gazette-Times* (Corvallis, Ore.), February 2.
33. Wilkinson, C.F. 1987. The end of multiple use. *High Country News* March 30, pp. 15–16.
34. Stoessinger, John G. 1974. *Why Nations Go to War,* St. Martin's Press, New York.
35. Kennedy, John F. 1961. *Profiles in Courage,* Harper & Row, New York, 266 pp.
36. Taylor, Duncan M. 1986. Nature as a mirror of changing human values. *The American Theosophist* 74:333–337.
37. Allen, J. 1981. *As a Man Thinketh,* Grosset & Dunlap, New York.
38. Carroll, Lewis. 1933. *Alice's Adventures in Wonderland,* Doubleday, Doran, New York, 162 pp.
39. The Daily Barometer. 1986. The Millennium Grove massacre. *The Daily Barometer* (Oregon State University, Corvallis), April 7.
40. Arden, H. 1987. The fire that never dies. *National Geographic* 172:375–403; Mohawk, J.C. 1986. in Wallace, P.A.W. 1986. *The White Roots of Peace,* The Chauncy Press, New York.
41. *The Holy Bible,* Authorized King James Version, World Bible Publishers, Iowa Falls, Iowa, Proverbs 15:1.
42. Kübler-Ross, Elisabeth. 1969. *On Death and Dying,* Macmillan, New York, 289 pp.
43. Nierenberg, G.I. 1981. *The Art of Negotiating,* Pocket Books, New York, 254 pp.
44. Maser, Chris. 1989. *Forest Primeval, The Natural History of an Ancient Forest,* Sierra Club Books, San Francisco, 282 pp.
45. Kiefer, Michael. Fall of the Garden of Eden, *International Wildlife* July–August: 38–43.
46. Chen, Allan. 1987. Unraveling another Mayan mystery. *Discover* June:40, 44, 46, 48–49.
47. Savonen, Carol. 1990. Ashes in the Amazon. *Journal of Forestry* 88:20–25.
48. Swift, Jamie. 1983. *Cut and Run, The Assault on Canada's Forests,* Between the Lines, Toronto, 283 pp.
49. Perlin, John. 1989. *A Forest Journey: The Role of Wood in the Development of Civilization,* W.W. Norton, New York, 445 pp.
50. Roosevelt, Theodore. 1990. *The Sunday Oregonian* (Portland), July 22.
51. Pinchot, Gifford. 1914. The Training of a Forester, J.B. Lippincott, Philadelphia, 149 pp.
52. Rubner, H. 1985. Greek thought and forest science. *Environmental Review* 9:277–295.
53. Thirgood, Jack V. 1981. *Man and the Mediterranean Forest, A History of Resource Depletion,* Academic Press, New York, 194 pp.; Butzer, Karl W. 1961. Climatic change in arid regions since the Pliocene. in *A History of Land Use in Arid Regions,* UNESCO, Paris.
54. Society of American Foresters. 1984. *Acidic Deposition and Forests,* SAF Policy Series, Bethesda, Md., 48 pp.

55. U.S. laws, statutes, etc., Public Law 94-588 [S. 3091]. Oct. 22, 1976. National Forest Management Act of 1976. in U.S. Code Congressional and Administrative News, 94th Congress 2nd Session, 16 U.S.C. Section 1600, Vol. 2, 1976, pp. 2949–2963, West Publishing, St. Paul, Minn.; U.S. laws, statutes, etc., Public Law 86-517 [H.R. 10572]. June 12, 1960. An act to authorize and direct that the national forests be managed under principles of multiple use and to produce a sustained yield of products and services, and for other purposes. in U.S. statutes at large, Vol. 74, p. 215, U.S. Government Printing Office, Washington, D.C., 1961 [16 U.S.C. sec. 528–531 (1976)].

56. Schütt, Peter and E.B. Cowling. 1985. Waldsterben, a general decline of forests in central Europe: Symptoms, development and possible causes. *Plant Disease* 69:548–558.

57. Forrester, S. 1986. The invisible shadow in the forests. *Register Guard* (Eugene, Ore.), June 8.

58. Kauppi, Pekka E., Kari Mielikäinen, and Kullervo Kuusela. 1992. Biomass and carbon budget of European forests, 1971 to 1990. *Science* 256:70–74.

59. Worster, Donald. 1979. *Dust Bowl, The Southern Plains in the 1930's,* Oxford University Press, New York, 277 pp.

60. Dixit, Kunda. 1986. *The Oregonian* (Portland), February 16.

61. Gold, Gerald and Richard Attenborough. 1983. The unfinished revolution. *Heart* Autumn:17–19, 108–112.

62. Goeller, D. 1986. Southern pine forest decline worrisome. *The Oregonian* (Portland), January 14.

63. Shigo, Alex L. 1985. Wounded forests, starving trees. *Journal of Forestry* 83:668–673

64. Crowell, John B. Jr. 1986. More, not less, timber should be cut. *The Oregonian* (Portland), June 11.

65. Kulp, J. Laurence 1985. *Corvallis Gazette-Times* (Corvallis, Ore.), June 19.

66. Molozhnikov, Valadimir. 1990. Advice to Americans. *Corvallis Gazette-Times* (Corvallis, Ore.), December.

67. Smuts, Jan Christian. 1926. *Holism and Evolution,* Macmillan, London. 361 pp.

68. Ledig, Thomas F. (in press). Human impacts on genetic diversity in forest ecosystems. *Oikos.*

69. Susuki, David and P Knudtson. 1988. *Genethics, The Ethics of Engineering Life,* Stoddart Publishing, Toronto, 384 pp.

70. Thomas, Jack Ward, Eric D. Forsman, Joseph B. Lint, E. Charles Meslow, B.R. Noon, and Jared Verner. 1990. A Conservation Strategy for the Northern Spotted Owl, Report of the Interagency Scientific Committee to Address the Conservation of the Northern Spotted Owl, U.S. Government Printing Office, Washington, D.C., 427 pp.

71. Berry, Wendell. 1990. The road and the wheel. *Earth Ethics* 1:8–9.

72. Clawson, Marion. 1979. Forests in the long sweep of American history. *Science* 204:1168–1174.

73. Neihardt, John G. 1961. *Black Elk Speaks,* Bison Book, University of Nebraska Press, Lincoln, 280 pp.

74. Durbin, Kathie. 1990. Private timberlands show promise. *The Oregonian* (Portland), December 20.
75. Dwyer, William. 1989. *The Forest Voice* (Eugene, Ore.), September:1, 9.
76. *The Holy Bible,* Authorized King James Version, World Bible Publishers, Iowa Falls, Iowa, Numbers 35:34.
77. Jung, Carl G. 1958. *The Undiscovered Self,* A Mentor Book, New York, 125 pp.
78. Tumerman, Leo. 1986. quoted in *Corvallis Gazette-Times* (Corvallis, Ore.), April 29.
79. Carter, Verne Gross and Timothy Dale. 1974. *Topsoil and Civilization,* rev. ed., University of Oklahoma Press, Norman.
80. McCartney, S. 1986. Watering the west. Part 3: Growing demand, decreasing supply send costs soaring. *The Oregonian* (Portland), September 30.
81. Chasan, D.J. 1977. *Up for Grabs, Inquiries into Who Wants What,* Madrona Publishers, Seattle, 133 pp.
82. Kendrick, J.B. Jr. (Chairman), A.A. Adams, Jr., T.G. Bahr, W.H. Blackburn, et al. 1983. Water-Related Technologies for Sustainable Agriculture in U.S. Arid/Semi-arid Lands, OTA-F-212, Office of Technology Assessment, U.S. Congress, U.S. Government Printing Office, Washington, D.C., 412 pp.
83. Blumm, M.C. 1987. Fewer roads fast way to improve elk habitat. *The Oregonian* (Portland), June 10.
84. Maser, Chris and James R. Sedell. (in press). *From the Forest to the Sea, The Ecology of Wood in Streams, Rivers, Estuaries, and Oceans,* St. Lucie Press, Delray Beach, Fla.; Maser, Chris, Robert F. Tarrant, James M. Trappe, and Jerry F. Franklin. 1988. From the Forest to the Sea: A Story of Fallen Trees, USDA Forest Service General Technical Report PNW-229, Pacific Northwest Research Station, Portland, Ore., 153 pp.
85. Durbin, Kathie and P. Koberstein. 1990. Forests in distress. *The Oregonian* (Portland), October 15.
86. Dunster, Julian A. 1990. Forest conservation strategies in Canada, a challenge for the nineties. *Alternatives* 16:44–51.
87. Soil Science Society of America. 1984. *Glossary of Soil Science Terms,* Soil Science Society of America, Madison, Wisc., 38 pp.
88. U.S. Department of Agriculture, Forest Service. 1986. USDA Forest Service Manual—2500-Watershed.
89. *Corvallis Gazette-Times* (Corvallis, Ore.). 1986. Nepal tree-planting fight provokes new scrutiny. June 6.
90. Ehrlich, Paul. 1990. Changing our minds. *Earth Ethics* 1:6–7.
91. World Commission on Environment and Development. 1987. *Our Common Future,* Oxford University Press, New York.
92. Illich, Ivan. 1990. The shadow our future throws. *Earth Ethics* 1:3–5.
93. Bird, Elizabeth Ann R. 1987. The social construction of nature: Theoretical approaches to the history of environmental problems. *Environmental Review* 11:255–264.
94. Platt, John R. 1964. Strong inference. *Science* 146:347–353.
95. Shapiro, Harold T. 1990. The willingness to risk failure. *Science* 250:609.

96. Wyant, D. 1986. 220-year study tracks log rot. *The Register-Guard* (Eugene, Ore.), March 3.

97. The following discussion is based heavily on Bormann, Bernard T., Martha H. Brookes, E. David Ford, A. Ross Kiester, Chadwick D. Oliver, and James F. Weigand. 1994. A Framework for Sustainable-Ecosystem Management, USDA Forest Service General Technical Report PNW 319, Pacific Northwest Research Station, Portland, Ore.; the part on adaptive management is based on Birch, Kevin, Roger Blair, Bill Bradley, Bernard Bormann, Arnie Browning, Michael Collopy, Ken Franklin, Van Manning, and Jerry Monesmith. 1993. Interim Concepts for Developing an Adaptive Management Process, unpublished report, 8 pp.

98. Li, C.Y., Chris Maser, Zane Maser, and Bruce A. Caldwell. 1986. Role of three rodents in forest nitrogen fixation in western Oregon: Another aspect of mammal–mycorrhizal fungus–tree mutualism. *Great Basin Naturalist* 46:411–414; Li, C.Y., Chris Maser, and Harlen Fay. 1986. Initial survey of acetylene reduction of selected microorganisms in the feces of 19 species of mammals. *Great Basin Naturalist* 46(4):646–650.

99. Allen, James. 1981. *As a Man Thinketh,* Grosset & Dunlap, New York, 72 pp.

100. National Research Council. 1990. in *Forestry Research. A Mandate for Change,* G. Caitlin (Ed.), National Academy Press, Washington D.C., 84 pp.

101. Bormann, B T., F. H. Bormann, W.B. Bowden, R.S. Pierce, S.P. Hamburg, D. Wang, M.C. Snyder, C.Y. Li., and R.C. Ingersoll. 1993. Rapid N_2 fixation in pines, alder, and locust: Evidence from the sandbox ecosystem study. *Ecology* 65:394–402.

102. Maslow, Abraham H. 1970. *A Theory of Human Motivation,* Harper and Row, New York, pp. 35–58.

103. Hunn, E.S. 1990. *Nch'i-wana, "The Big River": Mid-Columbia Indians and Their Land,* University of Washington Press, Seattle, 378 pp.

104. Burch, William R. Jr. 1971. *Daydreams and Nightmares: A Sociological Essay on the American Environments,* Harper & Row, New York, 175 pp.

105. Gordon, John C. 1993. The New Face of Forestry: Exploring a Discontinuity and the Need for a Vision, unpublished manuscript, Yale University, New Haven, Conn.

106. Bennett, T. 1987. American foreign policy a result of need for enemies, Stockwell says. *The Daily Barometer* (Oregon State University, Corvallis), May 28.

107. Dobyns, Lloyd 1990. Because better costs less. *Smithsonian* 21:74–82.

108. Cramer, H.H. 1984. On the predisposition to disorders of middle European forests. *Pflanzenschutz-Nachrichten* 37:98–207.

109. Bormann, B.T., Martha H. Brookes, E. David Ford, A. Ross Kiester, Chadwick D. Oliver, and James F. Weigand. 1994. A Framework for Sustainable-Ecosystem Management, USDA Forest Service General Technical Report PNW-GTR-319, Pacific Northwest Research Station, Portland, Ore.

110. Birch, Kevin, Roger Blair, Bill Bradley, Bernard Bormann, Arnie Browning, Michael Collopy, Ken Franklin, Van Manning, and Jerry Monesmith. 1993. Interim Concepts for Developing an Adaptive Management Process, unpublished report, 8 pp.

REFERENCES

Allen, G.M. and E.M. Gould, Jr. 1986. Complexity, wickedness, and public forests. *Journal of Forestry* 84:20–23.

Amaranthus, M.P. and D.A. Perry. 1987. The effect of soil transfers on ectomycorrhizal formation and the survival and growth of conifer seedlings on old, nonforested clearcuts. *Canadian Journal of Forest Research* 17:944–950.

Bak, Per and Kan Chen. 1991. Self-organizing criticality. *Scientific American* pp. 46–53.

Barrett, T.S. 1990. The sky's the limit. *Earth Ethics* 1:1.

Bella, David A. 1987. Organizations and systematic distortion of information. *Journal of Professional Issues in Engineering* 113:360–370.

Bella, David A. 1987. Engineering and the erosion of trust. *Journal of Professional Issues in Engineering* 113:117–129.

Bella, David A. and W. Scott Overton. 1972. Environmental planning and ecological possibilities. *Journal of Sanitary Engineering Division ASCE* 98:579–592.

Bella, David A., C.D. Mosher, and S.N. Calvo. 1988. Technocracy and trust: Nuclear waste controversy. *Journal of Professional Issues in Engineering* 114:27–39.

Bella, David A., C.D. Mosher, and S.N. Calvo. 1988. Establishing trust: Nuclear waste disposal. *Journal of Professional Issues in Engineering* 114:40–50.

Birch, Kevin, Roger Blair, Bill Bradley, Bernard Bormann, Arnie Browning, Michael Collopy, Ken Franklin, Van Manning, and Jerry Monesmith. 1993. Interim Concepts for Developing an Adaptive Management Process, unpublished report, 8 pp.

Bird, Elizabeth Ann R. 1987. The social construction of nature: Theoretical approaches to the history of environmental problems. *Environmental Review* 11:255–264.

Blaschke, Helmut and W. Bäumler. 1986. Über die Rolle der Biogeozönose im Wurzelbereich von Waldbäumen. *Forstwissenschaft. Centralb.* 2:122–130.

BML (Bundesministerium fur Ernahrung, Landwirtschaft und Forsten.). 1976. *Wald, forst-und Holzwirtshaft, Jagd in der Bundesrepublik Deutschland,* AID, Munster-Hiltrup, 184 pp.

Bormann, Bernard T. and Sidle, R.C. 1990. Changes in productivity and distribution of nutrients in a chronosequence at Glacier Bay National Park, Alaska. *Journal of Ecology* 78:561–578.

Bormann, Bernard T., Martha H. Brookes, E. David Ford, A. Ross Kiester, Chadwick D. Oliver, and James F. Weigand. 1994. A Framework for Sustainable-Ecosystem Management, USDA Forest Service General Technical Report PNW-GTR-319, Pacific Northwest Research Station, Portland, Ore.

Botkin, Daniel B. 1979. A grandfather clock down the staircase: Stability and disturbance in natural ecosystems. in *Forests: Fresh Perspectives from Ecosystem Analysis,* Proceedings of the 40th Annual Biological Colloquium, Richard H. Waring (Ed.), Oregon State University Press, Corvallis, pp. 1–10.

Brauns, A. 1955. Applied soil biology and plant protection. in *Soil Zoology,* D.K. McE. Kevan (Ed.), Butterworths, London, pp. 231–240.

Brown, Lester R. 1981. World population growth, soil erosion, and food security. *Science* 214:995–1002.

Bruck, R.I. and W.P. Robarge. 1984. Observations of boreal montane forest decline in the southern Appalachian Mountains: Soil and vegetation studies. in Aquatic Effects Task Group (E) and Terrestrial Effects Task Group (F) Research Summaries, North Carolina State Acid Deposition Program, Raleigh, 425 pp.

Buber, Martin. 1970. *I and Thou,* Charles Scribner's Sons, New York, 185 pp.

Bukkyo Dendo Kyokai. 1985. *The Teaching of Buddha,* 110th rev. ed., Kosaido, Tokyo, 307 pp.

Burgess, R.L. and Sharpe, D.M. 1981. *Forest Island Dynamics in Man-Dominated Landscapes,* Springer-Verlag, New York.

Butzke, H. 1984. Untersuchungsergebnisse aus Waldböden Nordrhein-Westfalens zur Frage der Bodenversauerung durch Immissionen. *Wissenschaft und Umwelt.* 2:80–88.

Calvo, Javier G.P., Zane Maser, and Chris Maser. 1989. A note on fungi in small mammals from the *Nothofagus* forest in Argentina. *Great Basin Naturalist* 49:618–620.

Campbell, Joseph. 1988. *The Power of Myth,* Doubleday, New York, 233 pp.

Capra, Fritjof. 1975. *The Tao of Physics,* Shambhala, Berkeley, Calif., 330 pp.

Carter, Verne Gross and Timothy Dale. 1974. *Topsoil and Civilization,* rev. ed., University of Oklahoma Press, Norman, 292 pp.

Chasan, D.J. 1977. *Up for Grabs, Inquiries into Who Wants What,* Madrona Publishers, Seattle, 133 pp.

Clark, Tim W. 1986. Professional excellence in wildlife and natural resource organizations. *Renewable Resources Journal* 4:8–13.

Cox, T.R., R.S. Maxwell, P.D. Thomas, and J.J. Malone. 1985. *This Well-Wood Land, Americans and Their Forests from Colonial Times to the Present,* University of Nebraska Press, Lincoln, 325 pp.

Cramer, H.H. 1984. On the predisposition to disorders of middle European forests. *Pflanzenschutz-Nachrichten.* 37:98–207.

Cramer, H.H. and M. Cramer-Middendorf. 1984. Studies on the relationships between periods of damage and factors of climate in the middle European forests since 1851. *Pflanzenschutz-Nachrichten.* 37:208–334.

Crosson, P. 1979. Agricultural land use: A technological and energy perspective. in *Farmland Food and the Future,* M. Schnepf (Ed.), Soil Conservation Service of America, Ankeny, Iowa, pp. 99–111.

Crow, T.R. 1990. Old growth and biological diversity: A basis for sustainable forestry. in *Old Growth Forests,* Canadian Scholars' Press, Toronto, pp 49–62.

Crutzen, P.J. and M.O. Andreae. 1990. Biomass burning in the tropics: Impact on atmospheric chemistry and biogeochemical cycles. *Science* 250:1669–1678.

Davis, M.B. and C. Zabinski. (in press). Changes in geographical range resulting from greenhouse warming effects on biodiversity in forests. in *Consequences of Global Warming for Biodiversity,* Proceedings of the World Wildlife Fund Conference, R.L. Peters and Thomas E. Lovejoy (Eds.), Yale University Press, New Haven, Conn.

DeLoach, C.J. 1971. The effect of habitat diversity on predation. *Proceedings Tall Timbers Conference on Ecological Animal Control by Habitat Management* 2:223–241.

Donald, C.M. 1968. The breeding of crop ideotypes. *Eurhytica* 17:385–403.

Dunne, T. and Luna B. Leopold. 1978. *Water in Environmental Planning,* W.H. Freeman, San Francisco, 818 pp.

Durrieu, G.M. Genard and F. Lescourret. 1984. Les micromammifères et la symbiose mycorhizienne dans une forêt de montagne. *Bull. Ecol.* 5:253–263.

Ehrlich, Paul R. 1985. Humankind's war against Homo Sapiens. *Defenders* 60:4–12.

Eiseley, L 1973. *The Man Who Saw Through Time,* Charles Scribner's Sons, New York, 125 pp.

Fetcher, N. and G.R. Shaver. 1990. Environmental sensitivity of ecotypes as a potential influence on primary productivity. *American Naturalist* 136:126–131.

Fogel, Robert and James M. Trappe. 1978. Fungus consumption (mycophagy) by small animals. *Northwest Science* 52:1–31.

Forsman, Eric D., E. Charles Meslow, and Howard M. Wight. 1984. Distribution and biology of the spotted owl in Oregon. *Wildlife Monograph* 87:1–64.

Franklin, Jerry F., Kermit Cromack, Jr., William Denison, Arthur McKee, Chris Maser, James Sedell, Fredrick Swanson, and Glenn Juday. 1981. Ecological Characteristics of

Old-Growth Douglas-Fir Forests, USDA Forest Service General Technical Report PNW-118, Pacific Northwest Forest and Research Experiment Station, Portland, Ore., 48 pp.

Franklin, Jerry F., Fredrick Hall, William Laudenslayer, Chris Maser, J. Nunan, J. Poppino, C.J. Ralph, and Thomas Spies. 1986. Interim Definitions for Old-Growth Douglas-Fir and Mixed-Conifer Forests in the Pacific Northwest and California, USDA Forest Service Research Note PNW-447, Pacific Northwest Forest and Research Experiment Station, Portland, Ore., 7 pp.

Franklin, Jerry. F. and Thomas A. Spies. 1984. Characteristics of old-growth Douglas-fir forests. in *New Forests for a Changing World,* Proceedings, Society of American Foresters National Conference, Bethesda, Md., pp. 328–334.

Franz, H. and W. Loub, 1959. Bodenbiologische Untersuchungen an Walddüngungsuersuchen. *Centralbl. Gesamte. Forstwes.* 76:129–162.

Fries, N. 1966. Chemical factors in the germination of spores of Basidiomycetes. in *The Fungus Spore,* M.F. Madelin (Ed.), Butterworths, London, pp. 189–199.

Fries, N. 1982. Effects of plant roots and growing mycelia on Basidiospore germination in mycorrhiza-forming fungi. in *Arctic and Alpine Mycology,* G.A. Laursen and J.F. Ammirati (Eds.), University of Washington Press, Seattle, pp. 493–508.

Fryer, J.H. and F. Thomas Ledig. 1972. Microevolution of the photosynthetic temperature optimum in relation to the elevational complex gradient. *Canadian Journal of Botany* 50:1231–1235.

Fyfe, W.S., B.I. Kronberg, O.H. Leonardos, and N. Olorunfemi. 1983. Global tectonics and agriculture: A geochemical perspective. *Agricultural Ecosystem Environment* 9: 383–399.

Galbraith, J.R. 1977. *Organizational Design,* Addison-Wesley, Reading Mass., 426 pp.

Gatto, M. and S. Rinaldi. 1987. Some models of catastrophic behavior in exploited forests. *Vegetatio* 69:213–222.

Gentry, A.H. and J. Lopez-Paradi. 1980. Deforestation and increased flooding of the upper Amazon. *Science* 210:1354–1356.

Golley, F.B. 1983. Nutrient cycling and nutrient conservation. in *Tropical Rainforest Ecosystems,* F.B. Golley (Ed.), Elsevier, Amsterdam, pp. 101–115.

Gronwall, O. and A. Pehrson 1984. Nutrient content in fungi as a primary food of the red-squirrel *Sciurus vulgaris* L. *Oecologia* 64:230–231.

Hall, E. Raymond. 1981. *The Mammals of North America,* Vol. 1, 2nd ed., John Wiley & Sons, New York, 600 pp.

Hansen, J., I. Fung, A. Lacis, D. Rind, S. Lebedeff, R. Ruedy, G. Russell, and P. Stone. 1988. Global climate changes as forecast by Goddard Institute for Space Studies three dimensional model. *Journal of Geophysical Research* 93:9341–9364.

Harley, J.L. and S.E. Smith. 1983. *Mycorrhizal Symbiosis,* Academic Press, New York, 483 pp.

Harmon, Mark E., William K. Ferrel, and Jerry F. Franklin. 1990. Effects on carbon storage of conversion for old-growth forests to young forests. *Science* 247:699–702.

Harmon, Mark E., Jerry F. Franklin, Fredrick J. Swanson, Phillip Sollins, Stanley V. Gregory, John D. Lattin, Norman D. Anderson, Steven P. Cline, N.B. Sumen, James R. Sedell, George W. Lienkaemper, Kermit Cromack, Jr., and Kenneth W. Cummins. 1986. Ecology of coarse woody debris in temperate ecosystems. *Advances in Ecological Research* 15:133–302.

Harris, Larry D. (Ed.). 1984. *The Fragmented Forest,* University of Chicago Press, Chicago, 211 pp.

Harris, Larry D. and Chris Maser. 1984. Animal community characteristics. in *The Fragmented Forest,* Larry D. Harris (Ed.), University of Chicago Press, Chicago, pp. 44–68.

Harris, Larry D., Chris Maser, and Arthur McKee. 1982. Patterns of old-growth harvest and implications for Cascade wildlife. *Transactions of the North American Wildlife and Natural Resources Conference* 47:374–392.

Harris, J.M. 1981. Effect of rapid growth on wood processing. in Proceedings, Division 5, 17th IUFRO World Congress, Japan, pp. 117–125.

Hill, S.B., L.J. Metz, and M.H. Farrier. 1975. Soil mesofauna and silvicultural practices. in *Forest Soils and Forest Land Management,* Proceedings Fourth North American Soils Conference, B. Bernier and C.H. Winget (Eds.), Les Presses de L'Université Laval, Québec, pp. 119–135.

Hopwood, A., F. Marshall, and D. Smith. 1988. A case study: Discipline, ethics and the forestry profession in British Columbia. *Forest Planning Canada* 4:15–23

Hoyt, E. 1990. Wild relatives. *Wilderness* Summer:45–54.

Hrebiniak, L.G. 1978. *Complex Organizations,* West Publishing, New York, 402 pp.

Hughes, J.D. 1989. Mencius' prescriptions for ancient Chinese environmental problems. *Environmental Review* 13:15–27.

Hummel, F.C. 1985. The use of forests as a source of biomass energy. in *Proceedings of the Third International Conference on Energy from Biomass,* W. Palz, J. Coombs, and D.W. Hall (Eds.), Elsevier Applied Science, London, pp. 90–98.

Hunter, Malcolm L. Jr. 1990. *Wildlife, Forests, and Forestry,* Prentice-Hall, Englewood Cliffs, N.J., 370 pp.

Iltis, H.H., J.F. Doebley, R.M. Guzman, and B. Pazy. 1979. *Zea diploperennis* (Gramineae), new teosinte from Mexico. *Science* 203:186–188.

Indonesia Forestry Community. 1990. Indonesia: Tropical forests forever. *Journal of Forestry* 88:26–31.

Jablanczy, A. 1988. Sustainable silviculture. *Forest Planning Canada* 4:7–9.

Jinjya Honcho. 1992. *Japan, Land of Festivals,* The United Association of Shinto Shrines, Tokyo, 16 pp.

Jones, C.O. 1970. *An Introduction to the Study of Public Policy,* Duxbury Press, Belmont, Calif., 70 pp.

Kaufman, Marc. 1988. Deforested Himalayas become staggering ecological disaster. *The Oregonian* (Portland), May 15.

Kershaw, J.A., C.D. Oliver, and T.M. Hinckley. 1993. Effect of harvest of old-growth Douglas-fir stands and subsequent management on carbon dioxide levels in the atmosphere. *Journal of Sustainable Forestry* 1:61–77.

Knight, H.A. 1987. The pine decline. *Journal of Forestry* 85:25–28.

Kotter, Martha M. and R.C. Farentinos. 1984. Formation of ponderosa pine ectomycorrhizae after inoculation with feces of tassel-eared squirrels. *Mycologia* 76:758–760.

Kriebel, H.B. 1957. Patterns of Genetic Variation in Sugar Maple, Ohio Agricultural Experiment Station Research Bulletin 791, Wooster, 56 pp.

Lashof, D.A. and D.R. Ahuja. 1990. Relative contribution of greenhouse gas emissions to global warming. *Nature* 344:529–531.

Ledig, F. Thomas and D.R. Korbobo. 1983. Adaptation of sugar maple along altitudinal gradients: Photosynthesis, respiration, and specific leaf weight. *American Journal of Botany* 70:256–265.

Leopold, Aldo 1966. *A Sand County Almanac, with Other Essays on Conservation from Round River,* Oxford University Press, New York, 269 pp.

Leopold, Luna B. 1974. *Water: A Primer,* W.H. Freeman, San Francisco.

Leopold, Luna B. 1990. Ethos, equity, and the water resource. *Environment* 2:16–42.

Li, C.Y. and Michael A. Castellano. 1985. Nitrogen-fixing bacteria isolated from within sporocarps of three ectomycorrhizal fungi. in *Proceedings 6th North American Conference on Mycorrhiza,* Randolph Molina (Ed.), Forest Research Laboratory, Oregon State University, Corvallis, p. 164.

Li, C.Y., Chris Maser, Zane Maser, and Bruce A. Caldwell. 1986. Role of three rodents in forest nitrogen fixation in western Oregon: Another aspect of mammal–mycorrhizal fungus–tree mutualism. *Great Basin Naturalist* 46:411–414.

Macfadyen, A. 1961. Metabolism of soil invertebrates in relation to soil fertility. *Annals of Applied Biology* 49:215–218.

Magnuson, John. J., C.J. Bowser, and A.L. Beckel. 1983. The invisible present: Long term ecological research on lakes. *L & S Magazine* Fall:3–6.

Malcjczuk, N., James M. Trappe, and Randolph Molina. 1987. Interrelationships among some ectomycorrhizal trees, hypogeous fungi, and small mammals: Western Australian and western North American parallels. *Australian Journal of Ecology* 12:53–55.

Manley, S.A.M. 1975. Genecology of Hybridization in Red Spruce (*Picea rubens* Sarg.) and Black Spruce (*Picea mariana* (Mill.) BSP.), Ph.D. dissertation, Yale University, New Haven, Conn., 154 pp.

Manley, S.A.M. and F. Thomas Ledig. 1979. Photosynthesis in black and red spruce and their hybrid derivatives: Ecological isolation and hybrid inviability. *Canadian Journal of Botany* 57:305–314.

Margalef, R. 1969. Diversity and stability: A practical proposal and a model of interdependence. in *Diversity and Stability in Ecological Systems,* Brookhaven Symposium of Biology 22, Brookhaven National Laboratory, Upton, N.Y., pp. 25–37

Marks, G.C. and T.T. Kozlowski. 1973. *Ectomycorrhizae—Their Ecology and Physiology,* Academic Press, New York, 444 pp.

Maser, Chris. 1967. Black bear damage to Douglas-fir in Oregon. *Murrelet* 48:34–38.

Maser, Chris. 1984. Human dignity: A diminishing resource. *Outdoors West* 7:8.

Maser, Chris. 1985. Rangelands, wildlife technology, and human desires. in Technologies to Benefit Agriculture and Wildlife, Office of Technology Assessment, U.S. Congress, U.S. Government Printing Office, Washington, D.C., pp. 83–92.

Maser, Chris. 1987. Framing objectives for managing wildlife in the riparian zone on eastside federal lands. in Managing Oregon's Riparian Zone for Timber, Fish and Wildlife, National Council of the Paper Industry for Air and Stream Improvement, Technical Bulletin No. 514, pp. 13–16.

Maser, Chris 1988. Ancient forests, priceless treasures. *Mushroom* 6:8–18.

Maser, Chris 1988. Restoration and the future of land management. *Journal Restoration and Management Notes* 6:28–29.

Maser, Chris. 1989. *Forest Primeval, The Natural History of an Ancient Forest,* Sierra Club Books, San Francisco, 282 pp.

Maser, Chris. 1990. On the "naturalness" of natural areas: A perspective for the future. *Natural Areas Journal* 10:129–133.

Maser, Chris. 1990. *The Forest Is a Living Trust,* Canadian Scholars' Press, Toronto, pp. 7–14.

Maser, Chris. 1990. The future is today: For ecologically sustainable forestry. *Trumpeter* 7:74–78.

Maser, Chris. 1991. Authenticity in the forestry profession. *Journal of Forestry* 89:22–24.

Maser, Chris. 1991. Economically sustained yield versus ecologically sustainable forests. in *Restoration of Old Growth Forests in the Interior Highlands of Arkansas and Okla-*

homa, Proceedings of the Conference, Sept. 19–20, 1990, D. Henderson and L.D. Hedrick (Eds.), Ouachita National Forest and Winrock International Institute for Agricultural Development, USDA Forest Service, Morrilton, Ariz., pp. 1–13.

Maser, Chris. 1992. *Global Imperative: Harmonizing Culture and Nature,* Stillpoint Publishers, Walpole, N.H., 267 pp.

Maser, Chris 1993. Adaptable landscapes are the key to sustainable forests. *Journal of Sustainable Forestry* 1:49–59.

Maser, Chris and Zane Maser. 1987. Notes on mycophagy in four species of mice in the genus *Peromyscus. Great Basin Naturalist* 47:308–313.

Maser, Chris and Zane Maser. 1988. Mycophagy of red-backed voles, *Clethrionomys californicus* and *C. gapperi. Great Basin Naturalist* 48(2):269–173.

Maser, Chris and Zane Maser. 1988. Interactions among squirrels, mycorrhizal fungi, and coniferous forests in Oregon. *Great Basin Naturalist* 48:358–369.

Maser, Chris, Zane Maser, Joseph W. Witt, and Gary Hunt. 1986. The northern flying squirrel: A mycophagist in southwestern Oregon. Canadian *Journal of Zoology* 64: 2086–2089.

Maser, Chris and James M. Trappe (Tech. Eds.). 1984. The Seen and Unseen World of the Fallen Tree, USDA Forest Service General Technical Report PNW-164, Pacific Northwest Forest and Range Experiment Station, Portland, Ore., 56 pp.

Maser, Chris and James M. Trappe. 1984. The fallen tree—A source of diversity. in *New Forests for a Changing World,* Proceedings of the Society of American Foresters National Conference, Bethesda, Md., pp. 335–339.

Maser, Chris, James M. Trappe, and C.Y. Li. 1984. Large woody debris and long-term forest productivity. in Proceeding Pacific Northwest Bioenergy Systems: Policies and Applications, Bonneville Power Administration, Portland, Ore., 6 pp.

Maser, Chris, James M. Trappe, and Ronald A. Nussbaum. 1978. Fungal–small mammal interrelationships with emphasis on Oregon coniferous forests. *Ecology* 59:799–809.

Maser, Chris, Robert F. Tarrant, James M. Trappe, and Jerry F. Franklin. 1988. From the Forest to the Sea, A Story of Fallen Trees, USDA Forest Service General Technical Report PNW-229, Pacific Northwest Research Station, Portland, Ore., 153 pp.

Maser, Zane, Chris Maser, and James M. Trappe. 1985. Food habits of the northern flying squirrel (*Glaucomys sabrinus*) in Oregon. *Canadian Journal of Zoology* 63:1084–1088.

Maser, Zane, Robert Mowrey, Chris Maser, and Wang Yun. 1985. Northern flying squirrel: The moonlight truffler. in *Proceedings 6th North American Conference on Mycorrhizae,* Randolph Molina (Ed.), Forest Research Laboratory, Corvallis, Ore., p. 269.

McCartney, S. 1986. Watering the west. Part 3: Growing demand, decreasing supply send costs soaring. *The Oregonian* (Portland), September 30.

McKeever, S. 1960. Food of the northern flying squirrel in northeastern California. *Journal of Mammalogy* 41:270–271.

Meeker, Joseph W. 1988. *Minding the Earth,* The Latham Foundation, Alameda, Calif., 110 pp.

Meslow, E. Charles, Chris Maser, and Jared Verner. 1981. Old-growth forests as wildlife habitat. *Transactions of the North American Wildlife and Natural Resources Conference* 46:329–335.

Miller, O.K. Jr. 1983. Ectomycorrhizae in the Agaricales and Gastromycetes. *Canadian Journal of Botany* 61:909–916.

Miller, Steven L., C.D. Koo, and Randy Molina. 1991. Characterization of red alder ectomycorrhizae: A preface to monitoring belowground ecological responses. *Canadian Journal of Botany* 69:516–531.

Miller, Steven L., C.D. Koo, and Randy Molina. 1992. Early colonization of red alder and Douglas fir by ectomycorrhizal fungi and *Frankia* in soils from the Oregon coast range. *Mycorrhiza* 2:53–61.

Molina, Randy and Michael Amaranthus. 1991. Rhizosphere biology: Ecological linkages between soil processes, plant growth, and community dynamics. in Proceedings—Management and Productivity of Western-Montane Forest Soils, Alan E. Harvey and Leon F. Neuenschwander (compilers), General Technical Report INT-280, USDA Forest Service, Boise, Idaho, pp. 51–58.

Molina, Randy, Hugues Massicotte, and James M. Trappe. 1992. Specificity phenomena in mycorrhizal symbioses: Community-ecological consequences and practical implications. in *Mycorrhizal Functioning,* Michael F. Allen (Ed.), Chapman & Hall, New York, pp. 357–423.

Monastersky, R. 1991. Hot year prompts greenhouse concern. *Science News* 139:36.

Morrison, Peter H. 1988. *Old Growth in the Pacific Northwest, A Status Report,* The Wilderness Society, Washington, D.C., 46 pp.

Morrison, Peter H. and Fredrick J. Swanson. 1990. Fire History and Pattern in a Cascade Range Landscape, USDA Forest Service General Technical Report PNW-GTR-254, Pacific Northwest Research Station, Portland, Ore., 77 pp.

Myers, N. and R. Tucker. 1987. Deforestation in Central America: Spanish legacy and North American consumers. *Environmental Review* 11:55–71.

Nault, L.R. and W.R. Findley. 1981. Primitive relative offers new traits to improve corn. *Ohio Report* 66 (Nov./Dec.):90–92.

Nelson, G. 1986. Preface. in *Conserving Biological Diversity in Our National Forests,* E.A. Norse, K.L. Rosenbaum, D.S. Wilcove, B.A. Wilcox, W.H. Romme, D.W. Johnson, and M.L. Stout, The Wilderness Society, Washington, D.C., p. i.

Norse, Eliott A., K.L. Rosenbaum, Dave S. Wilcove, B.A. Wilcox, W.H. Romme, D.W.

Johnson, and M.L. Stout. 1986. *Conserving Biological Diversity in Our National Forests,* The Wilderness Society, Washington, D.C., 116 pp.

Olson R.K. and A.S. Lefohn (Eds.). 1989. *Transactions Effects of Air Pollution on Western Forests,* Air and Waste Management Association, Pittsburgh, 577 pp.

Oort, A.J.P. 1974. Activation of spore germination in *Lactarius* species by volatile compounds of *Ceratocystis fagacearum.* Konikl. *Nebraska Academy Wetenschool Series C* 77:301–307.

Overton, W. Scott and L.M. Hunt. 1974. A view of current forest policy, with questions regarding the future state of forests and criteria of management. *Transactions of the North American Wildlife and Natural Resources Conference* 39:334–353.

Patterson, L.E. and S. Eisenberg. 1983. *The Counseling Process,* 3rd ed., Houghton Mifflin, Boston, 259 pp.

Perlin, John 1989. *A Forest Journey: The Role of Wood in the Development of Civilization,* W.W. Norton, New York, 445 pp.

Perry, David A. 1988. An overview of sustainable forestry. *Journal of Pesticide Reform* 8:8–12.

Perry, David A. 1988. Landscape pattern and forest pests. *Northwest Environmental Journal* 4:213–228.

Perry, David A. 1990. Interrelationship of terrestrial and aquatic ecosystems. in *Old Growth Forests,* Canadian Scholars' Press, Toronto, pp. 15–26.

Perry, David A. 1993. Biodiversity and wildlife are not synonymous. *Conservation Biology* 7:204–205.

Perry, David A. (in press). *Forest Ecosystems,* Johns Hopkins University Press, Baltimore.

Perry, David A., Michael P. Amaranthus, Jeffery G. Borchers, Susan L. Borchers, and R.E. Brainerd. 1989. Bootstrapping in ecosystems. *BioScience* 39:230–237.

Perry, David A. and Jeffrey G. Borchers. 1990. Climate change and ecosystem responses. *Northwest Environmental Journal* 6:293–313.

Perry, David A., Jeffery G. Borchers, Susan L. Borchers, and Michael P. Amaranthus. 1990. Species migrations and ecosystem stability during climate change: The belowground connection. *Conservation Biology* 4:266–274.

Perry, David A. and J. Maghembe. 1989. Ecosystem concepts and current trends in forest management: Time for reappraisal. *Forest Ecology and Management* 26:123–140.

Pessemier, E.A. 1982. *Product Management, Strategy and Organization,* 2nd ed., John Wiley & Sons, New York, 668 pp.

Petts, G.E. 1984. *Impounded Rivers, Perspectives for Ecological Management,* John Wiley & Sons, New York, 326 pp.

Phillips, D.R. and D.H. Van Lear. 1984. Biomass removal and nutrient drain as affected by total-tree harvest in southern pine and biomass removal and hardwood stands. *Journal of Forestry* 82:547–550.

Pimentel, David 1971. Population control in crop systems: Monocultures and plant spatial patterns. *Proceedings Tall Timbers Conference on Ecological Animal Control by Habitat Management* 2:209–220.

Pirozynski, K.A. and Malloch, D.W. 1975. The origin of land plants: A matter of mycotrophism. *BioSystems* 6:153–164.

Plochmann, Richard. 1968. *Forestry in the Federal Republic of Germany,* Hill Family Foundation Series, School of Forestry, Oregon State University, Corvallis, 52 pp.

Plochmann, Richard. 1989. *The Forests of Central Europe: A Changing View,* 1989 Starker Lectures, Forestry Research Laboratory, College of Forestry, Oregon State University, Corvallis.

Powers, R.F. et al. 1990. Sustaining site productivity in North American forests: Problems and prospects. in *Sustained Productivity of Forest Soils,* Proceedings 7th North American Forest Soils Conference, S.P. Gessel, D.S. Lacate, G.R. Weetman, and R.F. Powers (Eds.), Forestry Publication, University of British Columbia, Vancouver, pp. 49–79.

Public Relations Section for the Regular Removal of the Grand Shrine of Ise. 1992. *Jingu Shikinen Sengu,* The United Association of Shinto Shrines, Tokyo, 16 pp.

Rapport, David J. 1989. What constitutes ecosystem health? *Perspectives in Biology and Medicine* 33:120–132.

Rapport, David J., H.A. Regier, and T.C. Hutchinson. 1985. Ecosystem behavior under stress. *American Naturalist* 125:617–640.

Reisner, M. 1989. *Overtapped Oasis: Reform or Resolution for Western Water,* Viking Press, New York.

Robbins, William G. 1984. Timber town, market economics in Coos Bay, Oregon, 1850 to the present. *Pacific Northwest Quarterly* 75:146–155.

Robbins, William G. 1985. The social context of forestry: The Pacific Northwest in the twentieth century. *Western Historical Quarterly* 16:413–427.

Robbins, William G. 1988. *Timber Legacy: Work, Culture, and Community in Coos Bay, Oregon, 1850–1986,* University of Washington Press, Seattle.

Sachs, D. and Phillip Sollins. 1986. Potential effects of management practices on nitrogen nutrition and long–term productivity of western hemlock stands. *Forest Ecology and Management* 17:25–36.

Sanders, S.D. 1984. Foraging by Douglas Tree Squirrels (*Tamiasciurus douglasii*:Rodentia) for Conifer Seeds and Fungi, Ph.D. thesis, University of California, Davis, 95 pp.

Savonen, C. 1990. Ashes in the Amazon. *Journal of Forestry* 88:20–25.

Schindler, D.W., K.G Beaty, E.J. Fee, D.R. Cruikshank, et al. 1990. Effects of climatic warming on lakes of the central boreal forest. *Science* 250:967–970.

Schlesinger, M.E. and F.F. Mitchell. 1985. Model predictions of the equilibrium climatic response to increased carbon dioxide. in Projecting the Climatic Effects of Increasing Carbon Dioxide, DOE/ER-0237, U.S. Department of Energy, Washington, D.C, pp. 83–147

Schowalter, T.D. 1985. Adaptations of insects to disturbance. in *The Ecology of Natural Disturbance and Patch Dynamics,* S.T.A. Pickett and P.S. White (Eds.), Academic Press, New York, pp. 235–386.

Schowalter, T.D. 1988. Forest pest management: A synopsis. *Northwest Environmental Journal* 4:313–318.

Schowalter, T.D. 1989. Canopy arthropod community structure and herbivory in old-growth and regenerating forests in western Oregon. *Canadian Journal of Forest Research* 19:318–322.

Schowalter, T.D., W.W. Hargrove, and D.A. Crossley, Jr. 1986. Herbivory in forested ecosystems. *Annual Reviews in Entomology* 31:177–196.

Schowalter, T.D and J.E. Means. 1988. Pest response to simplification of forest landscapes. *Northwest Environmental Journal* 4:342–343.

Schowalter, T.D and J.E. Means. 1989. Pests link site productivity to the landscape. in *Maintaining the Long-Term Productivity of Pacific Northwest Forest Ecosystems,* D.A. Perry, R. Meurisse, B. Thomas, R. Miller, et al. (Eds.), Timber Press, Portland, Ore., pp. 248–250.

Schütt, Peter and E.B. Cowling. 1985. Waldsterben, a general decline of forests in central Europe: Symptoms, development and possible causes. *Plant Disease* 69:548–558.

Senft, J.F., B.A. Bendtsen, and W.L. Galligan. 1985. Weak wood, fast-grown trees make problem lumber. *Journal of Forestry* 83:477–484.

Shannon, Margret A. 1981. Sociology and public land management. *Western Wildlands* 7:3–8.

Sheffield, R.M., N.D. Cost, W.A. Bechtold, and J.P. McClure. 1985. Pine Growth Reductions in the Southeast, Resource Bulletin SE-83, USDA Forest Service, Southeastern Forest Experiment Station, Asheville, N.C., 112 pp.

Sheffield, R.M. and N.D. Cost. 1987. Behind the decline. *Journal of Forestry* 85:29–33.

Siccama, T.G., M. Bliss, and H.W. Vogelmann. 1982. Decline of red spruce in the Green Mountains of Vermont. *Bulletin of The Torrey Botanical Club* 109:163.

Sidle, R.C., A.J. Pearce, and C.L. O'Loughlin. 1985. *Hillslope Stability and Land Use,* Water Resources Monograph Series 11, American Geophysical Union, Washington, D.C., 140 pp.

Solo, R. 1974. Problems of modern technology. *Journal of Economic Issues* 8:859–876.

Spies, Thomas and Steven P. Cline. 1988. Coarse woody debris in unmanaged and managed coastal Oregon forests. in From the Forest to the Sea, A Story of Fallen Trees, Chris Maser, Robert F. Tarrant, James M. Trappe, and Jerry F. Franklin (Tech. Eds.), USDA Forest Service General Technical Report PNW-229, Pacific Northwest Research Station, Portland, Ore., pp 5–24.

Stoltmann, R. 1987. *Hiking Guide to the Big Trees of Southwestern British Columbia,* Western Canada Wilderness Committee, Vancouver, British Columbia, 144 pp.

Swetnam, Thomas. W. 1988. Forest fire primeval. *Natural Science* 3:236–241.

Swetnam, Thomas W. 1990. Fire history and climate in the southwestern United States. in Effects of Fire in Management of Southwestern Natural Resources, J.S. Krammers (Tech. Coord.), USDA Forest Service General Technical Report RM-191, Rocky Mountains Research Station, Fort Collins, Colo., pp. 6–17

Takekawa, J.Y. and E.O. Garton. 1984. How much is an evening grosbeak worth? *Journal of Forestry* 82:426–427.

Thomas, Jack Ward. 1984. Fee-hunting on the public lands?—An appraisal. *Transactions of the North American Wildlife and Natural Resources Conference* 49:455–468.

Thomas, Jack Ward, Eric D. Forsman, Joseph B. Lint, E. Charles Meslow, B.R. Noon, and Jared Verner. 1990. A Conservation Strategy for the Northern Spotted Owl, Report of the Interagency Scientific Committee to Address the Conservation of the Northern Spotted Owl, U.S. Government Printing Office, Washington, D.C., 427 pp.

Thomas, W.L. Jr., C.O. Sauer, M. Bates, and L. Mumford, (Eds.). 1956. *Man's Role in Changing the Face of the Earth,* University of Chicago Press, Chicago, 1193 pp.

Torgersen, Torolf. 1993. The forest immune system—Its role in insect pest regulation. *Natural Resource News* (Blue Mountain Natural Resource Institute), Special Edition, pp. 6–7.

Trappe, James M. 1981. Mycorrhizae and productivity of arid and semiarid rangelands. in *Advances in Food Producing Systems of Arid and Semiarid Lands,* Academic Press, New York, pp. 581–599.

Trappe, James M. and Robert D. Fogel. 1977. Ecosystematic functions of mycorrhizae. in *The Belowground Ecosystem: A Synthesis of Plant-Associated Processes,* J.K. Marshal (Ed.), Range Science Department Science Series 26, Colorado State University, Fort Collins, pp. 205–214.

Trappe, James M. and Chris Maser. 1976. Germination of spores of *Glomus macrocarpus* (Endogonaceae) after passage through a rodent digestive tract. *Mycologia* 68:433–436.

Trappe, James M. and Chris Maser. 1977. Ectomycorrhizal fungi: Interactions of mushrooms and truffles with beasts and trees. in *Mushrooms and Man: An Interdisciplinary Approach to Mycology,* Tony Walters (Ed.), Linn Benton Community College, Albany, Ore., pp. 165–178.

Trappe, James M. and Randolph Molina. 1986. Taxonomy and genetics of mycorrhizal fungi: Their interaction and relevance. in *Proceedings of 1st European Symposium on Mycorrhizae,* V. Gianinazzi-Pearson and S. Gianinazzi (Eds.), Institut National Researche Agronomique, France, pp. 133–146.

Tsukada, M. 1985. *Pseudotsuga menziesii* (Mirb.) Franco: Its pollen dispersal and late Quaternary history in the Pacific Northwest. *Japanese Journal of Ecology* 32:159–187.

Turner, M.G. and C.L. Ruscher. 1988. Changes in landscape patterns in Georgia, USA. *Landscape Ecology* 1:241–251.

Ure, Douglas and Chris Maser. 1982. Mycophagy of red-backed voles in Oregon and Washington. *Canadian Journal of Zoology* 60:3307–3315.

Waring, Richard H. and Jerry F. Franklin. 1979. Evergreen coniferous forests of the Pacific Northwest. *Science* 204:1380–1386.

Waring, Richard H. and W.H. Schlesinger. 1985. *Forest Ecosystems: Concepts and Management,* Academic Press, Orlando, Fla., 340 pp.

Warwick, D.P. 1975. *A Theory of Public Bureaucracy: Politics, Personality and Organization in the State Department,* Harvard University Press, Cambridge, Mass., 252 pp.

Weber, I.P. and S.D. Wiltshire. 1985. *The Nuclear Waste Primer, A Handbook for Citizens,* The League of Women Voters Education Fund, Nick Lyons Books, New York, 90 pp.

Weigl, Peter, Zane Maser, and Chris Maser, unpublished data on file with Dr. Weigl at Wake Forest University, Winston-Salem, N.C.

West, N.E. 1991. Benchmarks for rangeland management and environmental quality. in *Proceedings Noxious Range Weed Conference,* L.F. James et al. (Eds.), Westview Press, Boulder, Colo., pp. 1–13.

Wiens, J.A. 1975. Avian communities, energetics, and functions in coniferous forest habitats. in Proceedings Symposium on Management of Forest and Range Habitats, D.R. Smith (Ed.), USDA Forest Service General Technical Report WO-1, U.S. Government Printing Office, Washington, D.C., pp. 226–265.

Wilson, J.Q. (Ed). 1980. *The Politics of Regulation,* Basic Books, New York, 364 pp.

Worster, Donald. 1985. *River of Empire: Water, Aridity, and the Growth of the American West,* Pantheon Books, New York, 402 pp.

Zedaker, S.M., D.M. Hyink, and D.W. Smith. 1987. Growth declines in red spruce. *Journal of Forestry* 85:34–36.

Zengerle, M.W. and M.A. Allan. 1987. Status Reports on Selected Environmental Issues, Volume 2: Forest Decline—Environmental Causes, Electric Power Research Institute EA-5097S-SR, Volume 2, Research Project 2662, 5002, Palo Alto, Calif., pp. i–iii, 1–11, A-1–A-3; B-1–B-7.

Zhang, X.W., H. Guang, H.Y. Zhen, Y. Hong, X.Q. Zhou, and C.L. Zhou. 1980. Repeated plantation of *Cunninghamia lanceolata* and toxicisis of soil. in *Ecological Studies on the Artificial Cunninghamia lanceolata Forests,* Institute of Forests and Soil Science, Academy Sinica, People's Republic of China, pp. 151–158.